Interfacial Phenomena in Biotechnology and Materials Processing

Process Technology Proceedings

Process Technology Proceedings, 7

Interfacial Phenomena in Biotechnology and Materials Processing

Proceedings of the International Symposium on Interfacial Phenomena in Biotechnology and Materials Processing, August 3–7, 1987, Boston, Massachusetts, U.S.A.

Edited by

Yosry A. Attia

The Ohio State University, Department of Metallurgical Engineering, Mining Engineering Division, Columbus, Ohio, U.S.A.

Brij M. Moudgil

University of Florida, Department of Materials Science and Engineering, Gainesville, Florida, U.S.A.

and Associate Editor

S. Chander

The Pennsylvania State University, Mineral Processing Section, University Park, Pennsylvania, U.S.A.

Elsevier — Amsterdam — Oxford — New York —Tokyo   1988

ELSEVIER SCIENCE PUBLISHERS B.V.
Sara Burgerhartstraat 25
P.O. Box 211, 1000 AE Amsterdam, The Netherlands

Distributors for the United States and Canada:

ELSEVIER SCIENCE PUBLISHING COMPANY INC.
52, Vanderbilt Avenue
New York, NY 10017, U.S.A.

LIBRARY OF CONGRESS
Library of Congress Cataloging-in-Publication Data

International Symposium on Interfacial Phenomena in Biotechnology and
    Materials Processing (1987 : Boston, Mass.)
    Interfacial phenomena in biotechnology and materials processing :
proceedings of the International Symposium on Interfacial Phenomena
in Biotechnology and Materials Processing, August 3-7, 1987, Boston,
Massachusetts, U.S.A. / editors, Yosry A. Attia, Brij M. Moudgil ;
associate editor, S. Chander.
        p.   cm. -- (Process technology proceedings ; 7)
    Bibliography: p.
    ISBN 0-444-42980-8 : fl 300.00
    1. Surface chemistry--Congresses. 2. Biotechnology--Congresses.
3. Materials--Congresses.   I. Attia, Yosry A., 1945-   .
II. Moudgil, Brij M.   III. Chander, S., 1946-   .   IV. Title.
V. Series.
QD506.A1I58 1987
660'.6--dc19   88-16027
                                                        CIP

ISBN 0-444-42980-8 (Vol. 7)
ISBN 0-444-42382-6 (Series)

v

INTERNATIONAL SYMPOSIUM ON INTERFACIAL

PHENOMENA IN BIOTECHNOLOGY AND

MATERIALS PROCESSING

18th Annual Meeting of the Fine Particle Society
Boston, Massachusetts, August 3-7, 1987

EXECUTIVE COMMITTEE

Y.A. Attia
Ohio State University

B.M. Moudgil
University of Florida

S. Chander
Pennsylvania State
University

ORGANIZING COMMITTEE

Yosry A. Attia, Chairman
Ohio State University
Columbus, Ohio

K.P. Anatha
Union Carbide Corporation
Terrytown, New York

G. Barbery
Laval University
Sainte-Foy, Quebec, Canada

A. Bleier
Oak Ridge National Laboratory
Oak Ridge, Tennessee

S. Chander
Pennsylvania State University
University Park, Pennsylvania

D.W. Fuerstenau
University of California
Berkeley, California

R. Hogg
Pennsylvania State University
University Park, Pennsylvania

F.T. Hong
Wayne State University
Detroit, Michigan

W. Hu
University of Utah
Salt Lake City, Utah

R.G. Jenkins
Pennsylvania State University
University Park, Pennsylvania

M. Labib
David Sarnoff Research Center
SRI International
Princeton, New Jersey

A. Marabini
Institute per il trattamento
dei minerali
Rome, Italy

B. Moudgil
University of Florida
Gainesville, Florida

S. Rao
Rutgers University
Piscataway, New Jersey

D.O. Shah
University of Florida
Gainesville, Florida

R.H. Yoon
Virginia Polytechnic Institute
Blacksburg, Virginia

Cover design: Elsa Drake

PREFACE

The importance of interfacial phenomena in the processing of biological organic and inorganic materials has been increasingly recognized in recent years. Many processes such as separations, stabilization, aggregation, preparation, formation, transport, lubrication, adhesion, etc. depend primarily on interfacial behavior of the materials involved. The utilization of interfacial and colloidal properties of materials are numerous and include such applications as protein separation, biomembranes, contact lenses, bioengineering of surface-active materials, bacterial transport, bioleaching, superconducting materials, high-performance ceramics, advanced minerals separations, metals processing and composite materials. In materials processing, for example, the development of high-performance ceramics and superconductors requires component reliability with low cost. To meet these goals, refinement of existing forming processes or the development of new processes is required. An understanding of the interactions between particles constituting high-performance materials is recognized as a critical factor in the development of the required technology. In advanced materials processing via the chemical route, synthesis of inorganic, organic, polymer and surface chemistry is involved. It is recognized that controlling the interfacial chemistry during the early stages would lead to the production of materials with the desired properties. Interfacial phenomena are of major significance also in mineral processing. The efficiency of techniques such as selective flocculation for the beneficiation of finely disseminated low grade ores depends critically on the surface chemistry of the particles and solution chemistry of the polymers employed. Understanding the dispersion and aggregation behavior of fine particle suspensions with and without chemical additives is therefore a prerequisite to developing suitable solid-solid separation technology.

This book presents the edited proceedings of an international symposium at which researchers reported the latest discoveries and theoretical and experimental developments leading to the recent advances in the scientific principles and applications of interfacial phenomena in many fields of biotechnology and materials processing. All the papers included in this book, which were selected from the symposium proceedings of over 60 papers, have been peer-reviewed and revised accordingly. Major technical areas covered in this book are: Interfacial Phenomena in Biotechnology; Interfacial Phenomena in Advanced Materials Processing; Interfacial Phenomena in Minerals Processing; and Colloid Formation and Characterization. This book will be of particular interest to researchers, graduate students, and all persons involved in biotechnology, ceramics, superconducting materials, biomedical applications, minerals and energy conservation, as well as investors in industry and new technology.

The success of the symposium and the realization of this book was due to the great cooperation received from the authors, reviewers and the organizing committee, all of whom we would like to thank very much. The financial support of the Ohio State University, which made possible the editing of this book, is gratefully acknowledged. The book cover design was made possible by the University of Florida and we are indebted to them. We would like to thank Patty Permar, Toni DiGeranimo and Chris Schulz as well as Dr. Attia's graduate students, Catherine Dentan, Shanning Yu and Farshad Bavarian, for their help and assistance in editing the manuscripts in a timely manner.

February 5, 1988   Yosry A. Attia
Columbus, Ohio   Brij M. Moudgil
                                  Subhash Chander

TABLE OF CONTENTS

PART 1

INTERFACIAL PHENOMENA IN BIOTECHNOLOGY

Interfacial Phenomena in Biotechnology and Materials Processing,
edited by Y.A. Attia, B.M. Moudgil and S. Chander
Elsevier Science Publishers B.V., Amsterdam, 1988 — Printed in The Netherlands

BACTERIAL CELL ATTACHMENT AND SULFUR LEACHING IN MICROBIAL COAL DESULFURIZATION BY Sulfolobus Acidocaldarius

Charles C. Y. Chen, and Duane R. Skidmore
Department of Chemical Engineering, The Ohio State University,
Columbus, Ohio 43210

SUMMARY

Pre-leaching cell attachment and bacterial leaching were studied on finely divided Kentucky #9 coal and an Ohio coal mixture. Rates and extents of Sulfolobus acidocaldarius cell attachment on the coals were determined. Batch leaching of the coals by the microorganism were investigated by monitoring sulfate and iron ions released to the leaching solution. The Kentucky #9 coal adsorbed more cells and yielded higher sulfur and iron leaching rates. The results were interpreted in terms of mechanisms of microbial sulfur oxidation, surface areas of the coals and sulfur-containing grain sizes in the coals.

INTRODUCTION

Bacterial cell attachment is closely related to sulfur leaching in microbial coal desulfurization. The mechanism for pyritic sulfur oxidation was first proposed by Silverman (ref. 1) for Thiobacillus species as comprised of direct and indirect oxidative components. The direct mechanism requires direct contact between bacteria and pyrite since no extracellular enzymes are involved. In the indirect mechanism, the ferric iron ion chemically reacts with pyrite to give ferrous ions and elemental sulfur. The bacteria then oxidize ferrous ions to ferric ions and oxidize elemental sulfur to sulfate ions. The chemical and biological reactions are summarized as follows (ref. 1):

$$2\ FeS_2 + 7\ O_2 + 2\ H_2O \xrightarrow{\text{Bacteria}} 2\ FeSO_4 + 2\ H_2SO_4 \tag{1}$$

$$2\ FeSO_4 + 1/2\ O_2 + H_2SO_4 \xrightarrow{\text{Bacteria}} Fe_2(SO_4)_3 + H_2O \tag{2}$$

Overall reaction:
$$2\ FeS_2 + H_2O + 15/2\ O_2 \xrightarrow{\text{Bacteria}} Fe_2(SO_4)_3 + H_2SO_4 \tag{3}$$

Indirect mechanism:
$$FeS_2 + Fe_2(SO_4)_3 \longrightarrow 3\ FeSO_4 + 2\ S \tag{4}$$

$$2S + 3\ O_2 + 2\ H_2O \xrightarrow{\text{Bacteria}} 2\ H_2SO_4 \tag{5}$$

$$2 \text{ FeSO}_4 + 1/2 \text{ O}_2 + \text{H}_2\text{SO}_4 \xrightarrow{\text{Bacteria}} \text{Fe}_2(\text{SO}_4)_3 + \text{H}_2\text{O} \qquad (2)$$

In the absence of bacteria, the regeneration of ferric iron ions is the rate limiting step for pyrite oxidation. The bacteria increase the pyrite oxidation rate by oxidizing ferrous iron to ferric iron ions.

The application of the bacterial mechanisms derived for Thiobacillus ferrooxidans to Sulfolobus acidocaldarius is not well documented. One of the reasons is that the cell wall structure of Sulfolobus is different from that of Thiobacillus by the absence of a peptidoglycan layer. However, direct or indirect mechanisms, which are the result of bacterial activities, can be assumed to be valid for Sulfolobus species because:

1) Sulfolobus oxidizes Fe^{2+}, S^0 and pyrite at acidophilic conditions with production of Fe^{3+} and SO_4^{2-} (refs. 2-4),
2) Selective cell attachment of S. brierleyi and of S. acidocaldarius to pyrite have been observed (refs. 5-6)
3) Evidence of Fe and S at the cell attaching sites of S. acidocaldarius on coal was reported (ref. 7)

Irreversible adhesion of Thiobacillus ferrooxidans on substrate surfaces has been studied by many investigators (refs. 8-9). It is generally agreed that a wetting agent is responsible for cell adhesion. Murr and Berry studied cell attachment on mineral ores by scanning electron microscopy and concluded that adhesion is selective on sulfide surfaces (ref. 5). Adsorption of T. ferrooxidans on coal and other particles was studied by some investigators (refs. 10-11). Some researchers even found that surfactants enhanced contact between cells and the sulfide surface and therefore increased sulfur removal rates (ref. 12). The pyrite-selective adsorption characteristics were applied to improve physical separation of coal and pyrite as described in the literature (refs. 13-14). In those studies, the properties of pyrite particle surfaces were modified by cell adsorption. Then the coal was separated by oil agglomeration, selective flocculation or froth flotation.

The attachment of Sulfolobus to solid particles has been reported by some investigators (ref. 3, ref. 5, ref. 15). In their study of sulfur oxidation by S. acidocaldarius, Shivvers and Brock observed that the cells were attached to sulfur crystals until the late exponential stage and stationary stage were achieved (ref. 16). Weiss also studied the phenomenon of cell attachment for S. acidocaldarius and found that the cells attached to sulfur crystals by means of pili (ref. 15). He indicated that cell attachment was not necessary for sulfur oxidation. However, attachment provided a means for colonization in natural habitats. His study observed that cells obtained from bubbling pools did not attach to sulfur crystals while those from a flowing stream did. Usually the attached cell was separated from the solid surface by a short distance (ref. 15).

Murr and Berry applied scanning electron microscopy to the study of cell attachment to a mineral surface. They reported that a Sulfolobus-like bacterium preferentially adsorbed on the pyrite surface. No pili were observed to effect cell attachment (ref. 5, refs. 17-18). Similar conclusions were drawn for chalcopyrite, molybdenite and low grade sulfide ores in their study. Based on an attachment study, Murr and Berry proposed that attached cells promoted direct oxidation of pyrite, while the freely suspended cells oxidized the pyrite by an indirect mechanism (ref. 17).

The direct mechanism for pyrite oxidation suggests that a direct bacterial contact with the pyrite surface is necessary (ref. 1, ref. 19). Consequently, the rate of oxidation can be correlated with the number of cells attached to the pyrite surface.

In this study, pre-leaching cell attachment of S. acidocaldarius to two coals was investigated. Then, bacterial leaching of iron and sulfate from the coal were observed. The data obtained from cell attachment experiments and leaching experiments were compared to define their relationship.

MATERIALS AND METHODS

Culture Method

The medium for microbial growth contained a basal salts solution, and an energy source with or without supplements. The basal salt solution had the following composition (g/l): $(NH_4)_2SO_4$, 1.3; KH_2PO_4, 0.28; $MgSO_4 \cdot 7H_2O$, 0.25; $CaCl_2 \cdot 2H_2O$, 0.07; in 1 liter of tap water. In some cases, yeast extract was added as a supplement at 0.04% (w/v), and sublimed sulfur was added at 2.0 g/l as the energy source. Since the culture is a facultative autotroph, without the yeast extract supplement growth was slow and most of the cells attached to the sulfur particles (ref. 16). The pH of the medium was adjusted to 2.0 with 3.0 N H_2SO_4.

Four hundred ml. of the medium were placed in a 500 ml three necked, round bottom flask equipped with a condenser. The system was autoclaved at 121 $^{\circ}$C for 15 min and then inoculated with a culture of S. acidocaldarius. and kept in a water bath at 76 $^{\circ}$C. The flask was continuously aerated with air at 4.0 SCFH (2.0 l/min) to provide oxygen and carbon dioxide for microbial activities. The condenser served the purpose of condensing the moisture evaporated with the air leaving the flask so that the medium would not be dried out. The air was saturated in a saturator before entering the fermenter. The cooling water leaving the condenser flowed through the water jacket of the saturator to keep the temperature of the saturator at the temperature of the exit air (about 15 ^{O}C). With this design, the liquid level inside the flask remained nearly constant for a week. With temperature of the water bath controlled at 76 ^{O}C, the temperature of the medium in the flask was 72 ^{O}C. A cooling effect occurred from water evaporation, refluxed condensate and cooled inlet air. The culture was transferred every week. Mechanical agitation was provided by stirrers to enhance particle mixing and mass transfer of gases.

The original culture of S. acidocaldarius from Dr. Brierley at the New Mexico Institute of Mining and Technology and Dr. Kargi at Lehigh University was used throughout the investigation. The culture had been transferred in coal slurry since 1983.

The harvest of cells occurred 3-5 days after inoculation. As indicated in the literature, the cells attached to the sulfur surface in the late exponential phase and stationary phase (ref. 16). The maximum number of cells was recovered from the solution in 3-5 days. The culture was first centrifuged at 1000x g for 5 min to remove the sulfur particles. Then, the cell solution was centrifuged at 4000x g for 20 min to separate the cells. The cells were resuspended in basal salt solution without an energy source. These cells were used as inoculum for sulfur leaching from coal or for the cell attachment study, as will be described later.

More accurate cell counts were achieved by protein analysis calibrated with visual cell count under a phase contrast microscope equipped with a Petroff-Hausser counting chamber. The total protein was assayed by the modified Bradford Method suggested by Peterson (ref. 20). This method, commercialized and known as the Bio-Rad method, used coomassie brilliant blue G to stain the protein for quantitative analysis. The cells were first hydrolyzed by heating in 1.0 N NaOH at 95°C for 15 min to release the protein. Then the solution was subjected to dye solution for protein assay.

Coal Samples

The coals were Kentucky #9 coal from the Babcock and Wilcox Co. in Alliance, Ohio, and an Ohio coal mixture sampled from a pneumatic transport line in a power plant in Pickaway County, Ohio. The analyses of the coals are shown in Appendix A and Appendix B. The two coals were high sulfur coals with more than 2% pyritic sulfur and about 1.4% organic sulfur. The Kentucky #9 coal had a fine pyrite grain size. According to the information obtained from the Babcock and Wilcox Co., Alliance, Ohio, the pyrite grain size was

100% -10 micron. The Ohio coal mixture, however, was assumed to have a pyrite grain size of 60% -10 micron according to the statistics of pyrite grain size in Ohio coals (ref. 21). Particles for this study ranged narrowly in size: 270 to 325 mesh (45 to 53 microns). All the solid particles were washed with sterilized basal salt solution at 72 ^{o}C for 30 min. prior to adsorption tests in order to reduce possible analytical interferences by colored extract from the coal.

Surface areas of the coals were measured by nitrogen adsorption at 77 ^{o}C on a physical adsorption analyzer "AccuSorb 2100E" manufactured by Micromeritics. The surface areas, designated as A_s, are given in Table 1. Kentucky #9 coal had a surface area twice as great as that of Ohio coal mixture despite selection of similar particle sizes. The surface area obtained this way is a combination of the areas of external surface and the surface of pores greater than 12 Å in diameter (ref. 22).

Cell Attachment Experiments

Adsorption tests were carried out in a series of test tubes maintained in a water bath controlled at 72°C. In each test tube, 5 ml of the harvested cells were mixed with 1 g of coal particles, and the cell density in the solution was determined at 1, 5, 15, and 30 min.. The test tubes were periodically agitated to keep the particles in suspension.

To obtain the adsorption isotherm, different initial cell densities were employed and the cell density in the solution was determined over the 30 min. of cell-solid contact time. Amounts of adsorbed cells were calculated by the difference between the initial and final cell concentrations. The specific adsorption was calculated as number of adsorbed cells per unit volume of solid particles. The volume of each sample was computed from known weight and density. If spherical particle shape were assumed, unit volumes of different solid particles should have given the same external surface area.

Desorption of cells from solid particles after attachment was investigated. One gram of coal or pyrite particles was mixed with a cell solution of 3.6 x 10^9 cells/ml for cell attachment. Cell density in the solution was determined and the number of adsorbed cells defined by difference between the initial and final cell densities. Then the particles were separated and resuspended in 5 ml of basal salt solution. The mixture was thoroughly agitated and set aside for 15 min. Cell density in the solution was then measured.

Leaching Experiments

Leaching experiments were conducted in a three-necked flask as described previously. However, instead of elemental sulfur, the coals were added to the basal salt solution to make up 5% slurries. In each flask, 20 g of Kentucky #9 coal were mixed with 380 ml of basal salts solution. The slurry was inoculated with the cells harvested from the stock culture. Every 12-24 hours, 3 ml of slurry were taken from each flask and filtered through Whatman #1 filter paper. Concentrations of sulfate, ferrous, and ferric iron in the solution and total sulfur content in the coal were measured. Concentration of sulfate was determined by the turbidimetric methods specified by the ASTM (ref. 23). The analysis of ferrous ion concentration in solution was accomplished by using 1,10 phenanthroline as the indicator. Total iron ion concentration was determined by adding hydroxylamine hydrochloride to reduce ferric ions to ferrous ions and followed by the measurement of ferrous ion concentration. Ferric ion concentration was calculated as difference (total iron - ferrous = ferric) (ref. 24).

RESULTS AND DISCUSSION

Cell Attachment

The adsorption experiments in this study provided information about the "short term", or pre-leaching cell-particle interaction. Figure 1 shows the solution cell densities versus contact time of cells with different solid particles. Adsorption was a very fast process in that equilibria were reached in less than 5 min., as indicated in Figure 1.

A comparison with prior work is difficult since published literature is not available for S. acidocaldarius adsorption on coal. Furthermore, for T. ferrooxidans, none of the reports (refs. 10-11) mentioned the mixing condition (whether the mixture of coal and cell solution was agitated to provide good contact) so that direct comparisons of the adsorption rates with T. ferrooxidans could not be made. Differences in coal types, in microbial species, and in cell concentrations also prevented significant comparisons. Nonetheless, the results showed a very high rate of cell attachment and a profound difference in extent of adsorption for different particles.

Figure 2 shows a plot of the number of adsorbed cells versus equilibrium cell density in the solution for the two coals. As shown in Figure 2, both Kentucky #9 and Ohio coal mixture showed reversible cell attachment. However, the levels of specific adsorption obtained from desorption experiments were higher than those derived from adsorption. This result indicated that some adsorbed cells developed an irreversible

attachment through contact with the particles. Similar phenomena have been
reported in the literature, where some researchers showed that more
Sulfolobus cells attached to elemental sulfur or mineral sulfides after
longer incubation (refs. 17,18). Interpretation of such experimental results
is difficult without precise information on surface properties of the
microbial cell wall and the solid surfaces. These properties are total
external surface area of the solid particles, surface charge,
hydrophobicity, glycocalyx or surface appendages on the cell wall and others
(ref. 25).

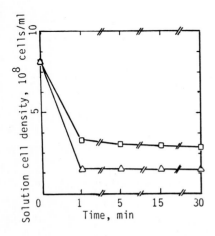

Figure 1. Cell adsorption on
coal particles, (△) Kentucky #9;
(□) Ohio coal mixture.

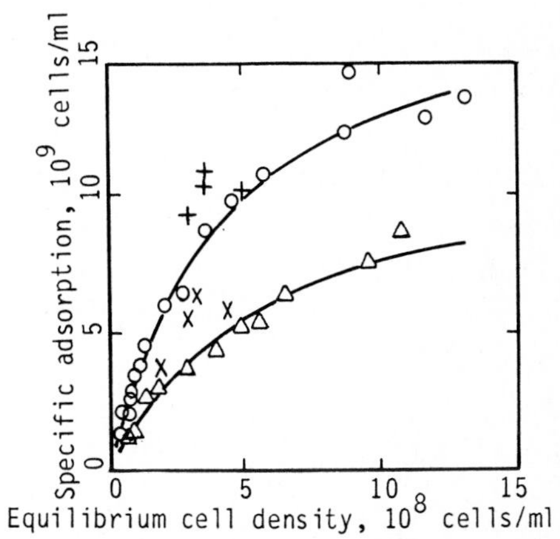

Figure 2. Adsorption isotherms.
(o) Kentucky #9; (△) Ohio coal
mixture; (+) Desorption data from
Kentucky #9; (x) Desorption data from
Ohio coal mixture.

To model the adsorption isotherms, the concept of Langmuir equation was
adapted (ref. 26):

$$\frac{1}{v_e} = \frac{1}{v_m} + \frac{1}{K\,v_m}\,\frac{1}{X_e} \tag{6}$$

where v_e is the specific adsorption at equilibrium, K is the equilibrium
constant defined as k_1/k_{-1}, and X_e is the equilibrium cell density in
solution. For a suitably modelled system, a plot of $1/v_e$ versus $1/X_e$ gives a
straight line with the system parameters v_m and K evaluated from the

intercept and slope of the line. The values of the parameters obtained by curve-fitting are given in Table 1. Greater K and v_m indicate that Kentucky #9 coal adsorbed cells at a higher rate and achieved a greater extent of cell attachment.

Bacterial Leaching

Figure 3 gives the iron release curves in a batch run of Ohio coal mixture and Figure 4 shows the concentration profiles in the batch leaching of Kentucky #9 coal. The initial cell densities of these two runs were about the same. The initial cell density in the batch leaching of Ohio coal mixture was 7.71×10^7 cells/ml, while that of Kentucky #9 coal was 6.81×10^7 cells/ml.

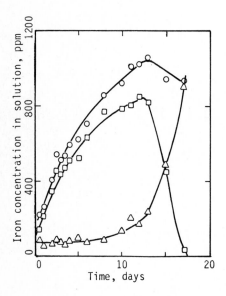 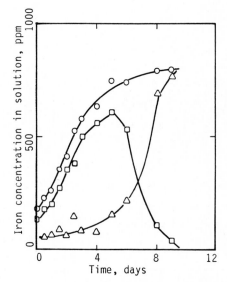

Figure 3. Iron release in batch leaching of a -270 +325 mesh Ohio coal mixture. (□) Ferrous iron, (△) Ferric iron, (o) Total iron.

Figure 4. Iron release in batch leaching of a -270 +325 mesh Kentucky #9 coal. (□) Ferrous iron, (△) Ferric iron, (o) Total iron.

The ferrous iron concentration started dropping at 5.0 days after inoculation for the run of Kentucky #9 coal, while that in the run of Ohio coal mixture did not drop until the 12th day. The maximum iron release rate from the Ohio mixture was 8.25 while that from Kentucky #9 coal was 9.33

mg/L hr. The ferrous ion concentration dropped only near the end of pyrite leaching. The results suggest that when pyrite was depleted, for ferrous ions, the rate of oxidation by the microbes exceeded the rate of generation from pyrite leaching.

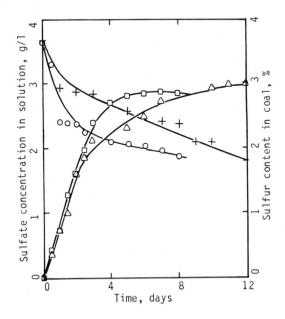

Figure 5. Sulfate release in batch leaching of -270 +325 mesh Kentucky #9 and Ohio coal mixture. Sulfate in solution: (□) Kentucky #9, (△) Ohio coal Mixture; Sulfur in coal: (o) Kentucky #9, (+) Ohio coal mixture.

The difference is even more evident in Figure 5. The rates of sulfate release were identical in the first 2.5 days. However, the rate in the run with Kentucky #9 coal (finer pyrite grain size) exceeded that in the run of Ohio coal mixture. The sulfur levels in the coals were similar. For better comparison, the data were fit into a first order rate equation According to the model proposed by Detz and Barvinchak (ref. 27), the rate equation is written as:

$$\frac{dS}{dt} = - k S \tag{7}$$

where S is the leachable pyrite concentration and k is the first order specific rate constant. Integration of this equation yields an exponential equation as follows:

$$S = S_0 \, e^{-kt} \tag{8}$$

where S_0 is the initial concentration of leachable sulfur. If sulfate is the only product of pyrite oxidation, the sulfate concentration can be described as:

$$[SO_4^=] = [SO_4^=]_0 (1 - e^{-kt}) \tag{9}$$

where $[SO_4^=]_0$ is the final sulfate concentration if all the leachable sulfur has been converted to sulfate. This equation fits most of the kinetic data in the literature where the initial cell density (inoculum size) was sufficiently high. The values of the rate constant k evaluated by curve-fitting are given in Table 1. Again, Kentucky #9 coal yielded a rate constant twice as large as that of Ohio coal mixture.

TABLE 1

Surface areas and values of parameters in cell adsorption and sulfur leaching models.

	As (m^2/cm^3)	K $(10^{-9}/cell)$	v_m $(10^{10} \, cells/cm^3)$	Max. $\frac{d[Fe]}{dt}$ $(mg/L \; hr)$	k (day^{-1})
Kentucky #9	5.48	2.162	1.884	9.33	0.787
Ohio Coal Mix.	2.92	1.875	1.161	8.25	0.301

Interpretation

The major factor that determines the sulfur removal rate could be the available pyrite surface area for oxidation. Finer pyrite grain size gave a higher oxidation rate as long as the coal was ground fine enough to expose the pyrite. The model proposed by Huber et al. (ref. 28) explains the effect of pyrite surface area available. In that model, the cells attached to the surface underwent exponential growth until the surface was saturated with cells; the so called "cell limiting regime". In this regime, the rate was determined by microbial growth, not the amount of available pyrite. Once the surface was saturated with microorganisms, no more cell increase was expected and the rate of sulfate release depended on the surface area of pyrite. The cells per unit area of pyrite remained constant in this regime and the rate of pyrite oxidation per unit surface area of pyrite remained constant. Kargi (ref. 29) reported that the number of attached cells increased while the number of free suspended cells remained constant during a batch leaching experiment. That result supports the concept of microbial growth on pyrite surface during pyrite leaching. In this study, more cells attached to Kentucky #9 coal and yielded higher sulfur and iron leaching rates. However, the data only indirectly support the ideas of selective cell attachment to pyrite and surface growth of the cells on pyrite, since the pyrite surface areas were not separately measured.

Mass balance on sulfur in the leaching experiment can be achieved with proper adjustments. As shown in Figure 5, the measured sulfur reduction in coal represents only about 90% of the sulfate increase in the leaching solution. One of the explanations is that the condenser on the flask did not completely retain the moisture carried by the air. Later, a careful study about the liquid level inside the flask showed that the solution lost 0 to 5 ml of water every day depending on the temperature and flow rate of the cooling water. When the cooling water temperature was 15 $^{\circ}$C, loss of moisture from the flask was about 5 ml a day. However when the temperature was lower than 10 $^{\circ}$C, loss of moisture was negligible. Loss of water from the solution simply concentrated the solution and gave a higher sulfate concentration. Table 2 gives the sulfur reduction calculated from both measurements: total sulfur in coal and sulfate in solution.

TABLE 2
Sulfur reduction in bacterial leaching of coals.

Coal type	Total S in raw coal	Total S in leached coal	% removed by S in coal	% removed by $SO_4^=$ in solution
KY #9	3.59	1.91	47	52
OH mixture	3.61	1.81	50	56

CONCLUSIONS

Pre-leaching cell attachment of S. acidocaldarius on Kentucky #9 coal and Ohio coal mixture indicated that the former coal adsorbed cells to a greater extent. The higher extent of cell attachment for Kentucky #9 coal can be interpreted as dependent upon a larger surface area and finer pyrite grain size. The cell attachment was reversible and the adsorption isotherms could be fit with the Langmuir model. A direct correlation of cell attachment to pyrite surface areas was not available since the pyrite

surface areas were not measured directly. However, for constant
concentration of pyritic sulfur, the Kentucky #9 coal could have a greater
surface area since the pyrite particles were finer.

Better adhesion correlated with better leaching since Kentucky #9 coal
showed higher rates for iron and sulfur release. The data obtained from cell
attachment and the leaching experiments support the concept of preferential
cell adsorption on pyrite surface for pyrite leaching. The iron release
curves showed that the organisms also oxidized ferrous ions to ferric iron
at high rates, which supports the idea of indirect mechanism. Consequently,
pyrite leaching can be significantly enhanced either by promoting cell
adhesion or by promoting bacterial ferrous ion oxidation. However, more
precise experiments are needed where measured pyrite surface areas and cell
attachment are monitored during leaching, to define the mechanism and
improve the understanding of microbial coal desulfurization.

REFERENCES

1 M. P. Silverman. Mechanism of bacterial Pyrite Oxidation, J. of
   Bacteriology, 94 (1967) 1046-1051.
2 C. L. Brierley and J. A. Brierley., A Chemoautotroph and Thermophilic
   Microorganism Isolated from an Acid Hot Spring, Can. J. Microbiol., 19
   (1973) 184-188.
3 T. D. Brock, S. Cook, S. Petersen and J. L. Mosser., Biogeochemistry and
   Bacteriology of Ferrous Iron Oxidation in Geothermal Habitats,
   Geochimica et Cosmochimica Acta., 40 (1976) 493-500.
4 C. L. Brierley. Thermophilic Microorganisms in Extraction of Metals from
   Ores, Development in Industrial Microbiology., 18 (1977) 273-284.
5 L. E. Murr and V. K. Berry, Direct Observations of Selective Attachment
   of Bacteria on Low-Grade Sulfide Ore and Other Mineral Surfaces,
   Hydrometallurgy, 2 (1976) 11-24.
6 C. Y. Chen and D. R. Skidmore, Attachment of Sulfolobus acidocaldarius
   cells on coal particles, Biotech. Progresses, (in press).
7 F. Kargi, Microbial Methods for Microbial Desulfurization of Coals,
   Trends in Biotechnology, 4(11) (1986) 293-296.
8 G. E. Jones and R. L. Startkey, Surface Active Substances Produced by
   Thiobacillus thioooxidans, J. of Bacteriology, 82 (1961) 788.
9 J. L. Takakuwa, T., T. Fugimori and H. Iwaski., J. of Gen. Microbiol.,
   25 (1979) 21.
10 A. S. Myerson and P. Kline, The Adsorption of Thiobacillus ferrooxidans
   on Solid Particles, Biotech. Bioeng., 25 1669-1676 (1983).
11 A. A. DiSpirito, P. R. Dugan and O. L. Tuovinen, Biotech. Bioeng. 25
   (1983) 1163-1168.
12 N. Wakao, M. Mishina, Y. Sakuai and H. Shiota, Bacterial Pyrite
   Oxidation II. The Effect of Various Organic Substances in Release of
   Iron from Pyrite by Thiobacillus ferrooxidans, J. of Appl. Microbiol.,
   29 (1983) 177-185.
13 A. G. Kempton, N. Moneib, R. G. L. McCready, and C. E. Capes, Removal of
   Pyrite from coal by Conditioning with Thiobacillus ferrooxidans Followed
   by Oil Agglomeration, Hydrometallurgy, 5 (1980) 117-125.
14 Y. A. Attia, Cleaning and Desulfurization of Coal Suspensions by
   Selective Flocculation, in Y. A. Attia (ed.), Processing and Utilization
   of High Sulfur Coal, Elsevier, Amsterdam, 1985, pp. 267-285.

15 R. L. Weiss, Attachment of Bacteria to Sulphur in Extremene
    Environments, J. of General Microbiology, 77 (1973) 501-507.
16 D. Shivvers and T. D. Brock, Oxidation of Elemental Sulfur by Sulfolobus
    acidocaldarius, J. of Bacteriology, 114 (1973) 706-710.
17 L. E. Murr and V. K. Berry, An Electron Microscope Study of Bacterial
    Attachment to Chalcopyrite: Mistructural Aspects of Leaching, in J. C.
    Yannopoulos and J. C. Agarwal, (eds.), Extractive Metallurgiy of Copper,
    The Metallurgical Soc. of AIME, New York, III, 1976, pp. 670-689.
18 L. E. Murr and V. K. Berry, Observation of Natural Thermophilic
    Microorganism in the Leaching of a Large Experimental, Copper-Bearing
    Waste Body, Metallurgical Transactions B., 10B, (1979) 523-531.
19 C. L. Brierley, J. A. Brierley, P. R. Norris and D. P. Kelly, Metal-
    tolerant Microorganisms of Hot, Acid Environments, Soc. Appl. Bacteriol.
    Tech. Ser. (1980) 39-50.
20 G. L. Peterson, Determination of Total Protein, in C. H. W. Hirs and S.
    N. Timasheff, (eds.), Methods in Enzymology-Enzyme Structure, Part I.,
    Academic Press, New York, 91 1983, pp. 95-119.
21 W. A. Kneller and G. P. Maxwell, Size, Shape and Distribution of
    Microscopic Pyrite in Selected Ohio Coals, in Y. A. Attia (ed.),
    Processing and Utilization of High Sulfur Coals, Elsvier, Amsterdam,
    1985.
22 P. L. Walker, Jr., H. Gan and S. P. Nandi, Nature of the Porosity in
    American Coal, Fuel, 51 (1972) 272.
23 ASTM Standard, Standard Test Method for Sulfate Ion in Water, 1986
    Annual Book of ASTM, D 516, 11.01, ASTM, Philadelphia, (1986) 700-706.
24 ASTM Standards, Standard Test Methods for Iron in Water, 1985 Annual
    Book of ASTM, ASTM, Philadelphia, 11.01, (1985) 507-511.
25 D. C. Savage and M. Fletcher (eds.), Bacterial Adhesion, Plenum Press,
    New York, 1985.
26 C. Y. Chen and D. R. Skidmore, Langmuir Adsorption Isotherm for
    Sulfolobus acidocaldarius on Coal Particles, Biotech. Letters, 9(3)
    (1987) 191-194.
27 C. Detz and G. Barvinchak, Microbial Desulfurization of Coal, Mining
    Congress J., 65(7) (1979) 75-82.
28 T. F. Huber and N. W. F. Kossen, Design and Scale-up of a Reactor forthe
    Microbial Desulfurization of Coal: A Kinetic Model for Bacterial Growth
    and pyrite oxidation, Proc. 3rd Europe Congress on Biotech., Frankfort,
    Sep. 10-14, III (1984) 151-159.
29 F. Kargi and J. M. Robinson, Biological Removal of Pyritic Sulfur from
    Coal by the Thermophilic Organism Sulfolobus acidocaldarius, Biotech.
    Bioeng., 27(1) (1985) 41-49.

APPENDIX A
Analyses of Kentucky #9 coal

Basis	As Received	Dry
Proximate Analysis(%)		
Moisture	3.78	--
Volatile Matter	38.14	39.64
Fixed Carbon	45.76	47.56
Ash	12.32	12.80
Ultimate Analysis(%)		
Moisture	3.78	
Carbon	66.95	69.58
Hydrogen	4.71	4.90
Nitrogen	1.47	1.53
Sulfur	4.42	4.59
Ash	12.32	12.80
Oxygen (Difference)	6.35	6.60
Total	100.00	100.00
Surfur forms, % as S		
Pyritic	2.88	2.99
Sulfate	0.05	0.05
Organic (Difference)	1.49	1.55
Total	4.42	4.59

APPENDIX B
Analysis of Ohio Coal Mixture

Proximate Analysis(%)	As Received
Moisture	8.12
Ash	10.13
Volatile matters	37.45
Fixed Carbon	40.95
Sulfur	3.68
Sulfur forms, as % S	
Pyritic	2.12
Sulfate	0.27
Organic (Difference)	1.29
Total	3.68

Interfacial Phenomena in Biotechnology and Materials Processing,
edited by Y.A. Attia, B.M. Moudgil and S. Chander
Elsevier Science Publishers B.V., Amsterdam, 1988 — Printed in The Netherlands

SURFACE TENSION EFFECTS ON BACTERIA TRANSPORT THROUGH POROUS MEDIA

Jau Ren Chen, Dawood Momeni, Jih Fen Kuo, and Teh Fu Yen.
School of Engineering, University of Southern California
Los Angeles, CA 90089.

SUMMARY

This study was undertaken in an attempt to assess the influence of surface tensions of bacteria, of solid substratum, and of suspending liquid on bacteria transport through porous media under simulated hydrodynamic conditions. The transport of Staphylococcus epidermidis cells through sandpack columns was investigated. The level of cell transport through the sandpack columns increased with increasing surface tension of the bacterial suspension applied. For a qualitative analysis, the thermodynamic model proposed by Neumann and coworkers (ref.1) in description of the static adsorption of bacteria on solid surface was taken into consideration. The results showed certain consistency between model predictions and experimental conclusion. The relative permeability data, derived from the pressure drop analysis across the sandpack column, for each case of injected bacterial suspension also indicated more bacteria deposition inside the porous media as solution surface tension was reduced.

INTRODUCTION

The extent of bacteria transport in porous solids has been of a major

concern in wide variety of biotechnological research areas. These comprise

such diversified studies as the transport of bacterial species through filter

beds and the application of bacterial species in in-situ biological treatment

of contaminated soils as well as of polluted groundwater. Elaborate

experiments have been undertaken by Yen and coworkers (refs. 2-6) in their

investigations of bacteria transport through porous media to identify the

controlling factors in bacteria-rock interactions, and to explore the

important parameters that govern the bacterial adhesion and penetration

processes. Based on a literature survey, from the studies of the particle

transport through porous media, three mechanisms have been identified to

control particle retension: (a) surface deposition, (b) straining or size

exclusion and (c) sedimentation or gravity settling. Since the sizes of

bacteria are relatively small when compared with pore sizes of sand column and

the density differences between bacteria and sand are not significant, the

governing process in the bacteria transport through sand column may be the

characteristics of interfacial tension among bacteria, sand and liquid media.

In recent investigations on the static testing of bacteria adhesion to various solid substrate (refs. 7-11), it was ascertained that the extent of adhesion was governed by the surface tensions of all three phases involved, i.e., the surface tensions of adhering particles, of the substratum, and of the liquid medium. The results proved that adhesion of bacteria to polymeric substrate from the aqueous phase follows a general pattern of behavior. Since the bacterial deposition on the solid surfaces would distinctively hinder their transport through porous media, it is therefore essential to inquire into these effects on the mechanism of bacteria transport in a hydrodynamic field. This paper specifically appraises the effects of surface tensions of bacteria, liquid, and solid phases on the bacteria transport through a porous solid and attempts to further extend thermodynamical description of the bacterial deposition on the porous solid surfaces under different media surface tensions and actual hydrodynamic conditions.

MATERIALS AND METHODS

An experimental apparatus was carefully designed in such a way that actual hydrodynamic conditions for bacteria transport in porous media could be simulated. The equipment, as it is sketched in Figure 1, mainly consisted of a

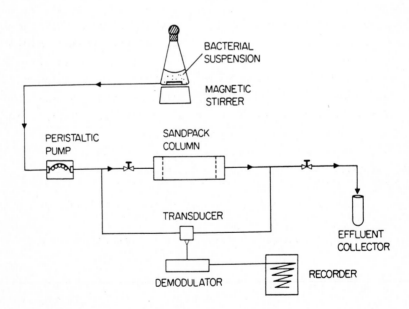

Fig. 1. Schematic diagram of the flow apparatus using a peristatic pump to supply a constant flow and a recorder to monitor pressure change.

source flask, a peristaltic pump for constant rate pumping of the bacterial
suspension from the source flask, a sandpack column, flow control valves, and
an effluent collector tube. The source flask was kept at a higher elevation to
minimize the bacterial settling in the inlet flow lines. Its content could
continuously be stirred with a magnetic stirrer. A pressure recording system
including a transducer, a demodulator, and a chart recorder was added in the
set-up for the purpose of continuous monitoring of pressure drop across the
sandpack column. The column was an acrylic cylinder of 9.5 centimeters in
length and 2.54 centimeters in diameter. Standard Ottawa sand, SX0070-3, 20-40
mesh size, obtained from MCB Manufacturing Chemicals, Inc., was used as the
packing material. Before packing the column, the sand was thoroughly washed
with distilled water to remove fines and then dried at $80^{\circ}C$ for 48 hrs. To
prepare the bacterial suspension, Staphylococcus epidermidis was grown in a
nutrient broth at $36^{\circ}C$ for five days. The bacteria were washed three times
by repeated centrifugation and resuspension in the appropriate solution. Four
bacteria suspensions were then prepared with different concentrations of
dimethyl sulfoxide (DMSO), ranging from 0% to 15% (Vol/Vol). Additions of DMSO
were to have different liquid surface tensions. For the four prepared
solutions, surface tensions varied from a high value of 73 to a low value of
63 erg/cm^2 (Table 1). The final bacteria concentrations were diluted to 8 to
8.5×10^6 bacteria per ml.

With known sand density of 2.65 g/cm^3, the sand grain volume in the
column could be readily calculated. The sand pore volume was then found, using
the difference between the column volume and the sand grain volume. For each
experiment a fresh sandpack column was prepared and its pore volume and
porosity (pore volume/bulk volume) were evaluated (Table 1). The column was
horizontally mounted in the set-up, and pressure recording devices were
connected across the column. Before pumping the bacterial suspension in each

Table 1
Properties of individual runs for bacteria transport experiments

Experiment (number)	Conc. of DMSO (Vol/Vol)	Surface tension (ergs/cm^2)	Porosity of sandpack (%)	Flow rate (ml/min)
1	0	73	36.8	1.30
2	5	70	38.6	1.31
3	10	67	37.9	1.25
4	15	63	38.3	1.24

run, an appropriate solution of 0.85 weight percent of salt was repeatedly passed through the flow system until no impurities could be detected by UV/Visible spectrophotometer in the effluent. During each experiment, the bacterial suspension injection rate was kept constant at 1.30 ml/min while the injected suspension liquid had a different surface tension value (Table 1).

Throughout the injection of bacterial suspension into the sandpack column, several important parameters were measured and evaluated. The effluent was periodically collected and tested at the same time. The cell concentration was determined based on the turbidity using a UV/Visible spectrophotometer, Beckman Model 25. The absorbance was measured at a wavelength of 240 nm because of its better sensitivity at this wavelength in our preliminary test. Based on an absorbance calibration curve, the number of bacteria in each sample was then estimated. As for the data interpretation, the volumes of liquid injected were converted to a dimensionless quantity, later identified as liquid pore volume, by simply dividing the actual liquid volume injected by the sandpack pore volume. The pressure drop across the sandpack column was continuously monitored and recorded as bacterial suspension was injected into the sandpack column. Due to the adsorption of bacteria on the sand solid surface the liquid permeability would vividly change. The permeability values at each interval of injection were calculated based on Darcy's equation,

$$K = \frac{Q \cdot L \cdot \mu}{A \cdot \Delta P} \tag{1}$$

where K is the permeability in Darcy, Q is the flow rate in cm^3/sec, ΔP is the pressure change in atmosphere, L is the column length in cm, μ is the liquid viscosity in cp, and A is the column cross sectional area in cm^2.

RESULTS AND DISCUSSION

The ratios of bacteria effluent concentration (C_e) to influent concentration (C_o) from transport experiments in sandpack columns under different fluid surface tensions are shown in Figure 2. It is evident from these results that at higher liquid surface tensions or lower DMSO concentrations the C_e/C_o ratio is higher indicating more cells penetrated through the sandpack column or less were adsorbed on the sand surfaces. An almost similar C_e/C_o ratio behavior is observed for two cases of zero and five percent DMSO concentration in solution. The ratio increases sharply for up to two pore volume injection of fluids, and increases steadily thereafter. On the other hand, for cases of ten and fifteen per cent DMSO concentration in solution, the C_e/C_o ratio behaves differently. It reaches a maximum after injection of 3.5 pore volume of fluid, and then drops significantly.

To better understand these results in effects of surface tensions on the bacteria transport through porous media a thermodynamics model, proposed by Absolom et al. (ref. 1), based on the studies of bacteria adhesion on polymeric surfaces in the static condition has been evaluated. This model implies that a property identified as thermodynamic potential, i.e., free energy, could be used as a major indication of the extent of bacteria adhesion to a solid surface. The bacteria adhesion would be enhanced by decreasing free energy and it would be reduced if the system free energy is increased. The free energy function (ΔF^{adh}), for systems in which the effect of electrical charges as well as biochemical interactions are neglected, is defined as:

$$\Delta F^{adh} = \gamma_{BS} - \gamma_{BL} - \gamma_{SL} \tag{2}$$

where ΔF^{adh} is the free energy of adhesion per unit surface area, γ_{BS} is the bacterium-substratum interfacial tension, γ_{BL} is the bacterium-liquid interfacial tension, and γ_{SL} is the substratum-liquid interfacial tension. Through the use of Young's equation, it becomes possible to obtain experimental data for various interfacial tensions involving solid surface. Young's equation is as follows,

$$\gamma_{SV} - \gamma_{SL} = \gamma_{LV} \cos\theta \tag{3}$$

where γ_{SV}, γ_{SL}, and γ_{LV} are, respectively, the interfacial tension between a solid substratum S and the vapor phase V, between S and the liquid L, and between L and V; θ represents the contact angle of the liquid on the solid. Only the liquid surface tension (γ_{LV}) and θ are readily determined experimentally. From the combination of Young's equation and an equation of state (ref. 12) the following formula is then derived (ref. 1):

$$\cos\theta = \frac{(0.015\,\gamma_{SV} - 2.00)\sqrt{\gamma_{SV}\gamma_{LV}} + \gamma_{LV}}{\gamma_{LV}(0.015\sqrt{\gamma_{SV}\gamma_{LV}} - 1)} \tag{4}$$

With aid of the above equation the surface tension of the substratum, γ_{SV}, may be determined from the easily measurable quantities γ_{LV} and θ. Overall, for a certain value of γ_{SV} it would be then possible to predict the extent of bacterial adhesion in a different liquid media by having a certain evaluation of the system free energy derived accordingly.

To apply this model for our experimental results, it is first necessary to evaluate the surface tension of the suspension liquids, (γ_{LV}), as well as the surface tension of bacteria (γ_{BV}) and that of the sand (γ_{SV}). The surface tension of S. epidermidis has been determined by Absolom et al. to be equal to 66.9 ergs/cm^2 (ref. 1). The surface tension of the sand is estimated to be 78 ergs/cm^2 which was basically the surface tension of the silica (ref. 13). The liquid surface tension values are summarized in Table 1. With the aid of a computer program (ref. 8), the interfacial tensions between bacteria and liquid, between bacteria and sand, and between sand and liquid, could be readily obtained. Through the input of these interfacial tensions a theoretical plot for the system under investigation was created and shown in Figure 3. From this figure it can be seen that at a substrate surface tension of 78 ergs/cm^2, i.e., surface tension of sand, the free energy of adhesion increases as DMSO concentration in liquid decreases. As it was stated earlier an increase in the free energy of adhesion would result in less adhesion of bacteria to the solid surface. This would therefore confirm our experimental results on the bacteria transport through the sandpack column as we observed. More bacteria are transported or less adsorbed at lower DMSO concentration with the corresponding higher liquid surface tension. In order to assess the influence of substratum surface tension on the extent of bacteria transport based on the data from Fig. 3, the free energy of adhesion for two different substrates of sand and polyethylene was evaluated and the results are sketched in Figure 4. It is quite interesting to note that the trend of the change in free adhesion energy is dependent upon the surface tension of substratum for a

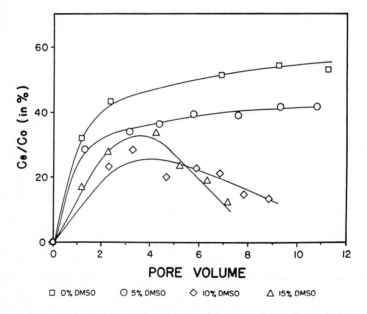

Fig. 2. Ratios of bacteria effluent concentration to influent concentration under different conditions of liquid surface tensions.

specific bacterium. As illustrated in the figure, if polyethylene which has a surface energy of 35 ergs/cm^2 (ref. 13) is used as substratum, the opposite trend of free adhesion energy change is demonstrated when compared to that of sand. Thus, an increase in liquid surface tension would cause a decrease in the free adhesion energy and consequently inhibit the transport of bacteria in this specific case.

To further elaborate our experimental results of the bacterial adhesion under hydrodynamic conditions, the changes in liquid relative permeability for different conditions of surface tensions were investigated and the results are shown in Fig. 5. These results are significant in assessment of permeability variation upon deposition of bacteria on solid surfaces inside the porous media. For the three bacteria suspensions tested, the case of one with 15% DMSO concentration shows the highest permeability reduction as injection volume is increased. These data well support our earlier prediction on high adsorption of bacteria as the liquid surface tension is reduced causing significant reduction in bacteria transportability through the porous media. For the case of five per cent DMSO solution, although the reduction in permeability is more significant at the beginning as compared with zero per cent DMSO solution but it is much less than the solution with high DMSO concentration.

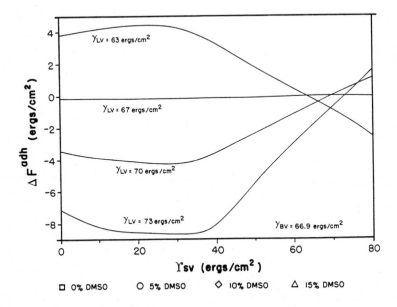

Fig. 3. Free energy of adhesion as a function of substratum surface tension for different liquid surface tensions.

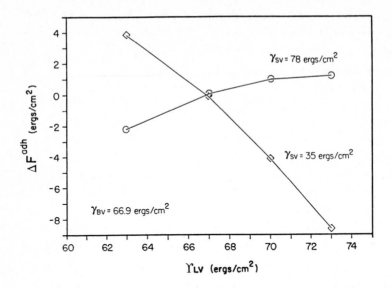

Fig. 4. Free energy of adhesion for two different substrates of sand and polyethylene under variable liquid surface tensions.

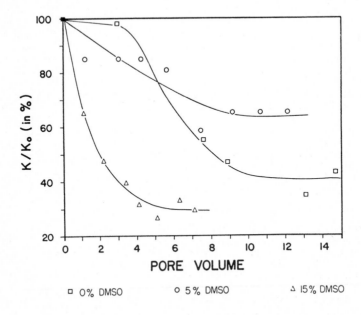

Fig. 5. The changes in liquid relative permeabilities for different condition of surface tensions.

In view of complexity and importance of bacteria transport processes, the results of this study significantly demonstrate distinct effects of solid, liquid, and bacteria surface tensions on the extent of cell movement through porous substrata. Without the appropriate understanding, controlling, and modifying these parameters, such unwanted phenomena as pore blockage might easily occur inside the porous structure of the solid material.

ACKNOWLEDGEMENT

Partial support from U.S. DOE cotract DEAS19-81BC10508 is appreciated. We are also grateful for the principal support of NSF grant CBTE-8614019.

REFERENCES

1   D. R. Absolom, F. V. Lambertic, Z. Policova, W. Zingg, C. J. van Oss and A. W. Neumann, Surface thermodynamics of bacterial adhesion, Appl. Env. Microb.,46 (1983) 90-97.
2   L. K. Jang, J. E. Findley and T. F. Yen, Preliminary investigation on the transport problems of microorganisms in porous media, in: J. E. Zajic, D. G. Cooper, T. R. Jack and N. Kosaric (Ed.), 28th IUPAC meeting, Vancouver, B.C., Canada, August 17-19, 1981, Microbial enhanced oil recovery, Penn Well Books, Tulsa, Okla., 1983, pp. 45-49,
3   L. K. Jang, M. M. Sharma, J. F. Findley, P. W. Chang and T. F. Yen, An investigation of the transport of bacteria through porous media, in Proc. Int. Symposium on Microbial Enhancement of Oil Recovery, U. S. Department of Energy, Bartlesville Energy Technology Center, Bartlesville, Okla. 1983, pp. 60-70. (Available through National Technical Information Service [CONF 8205140].).
4   L. K. Jang, P. W. Chang, J. E. Findley and T. F. Yen, Selection of bacteria with favorable transport properties through porous rock for the application of Microbial-Enhanced Oil Recovery, Appl. Environ. Microbiol., 46 (1983) 1066-1072.
5   L. K. Jang, and T. F. Yen, A theoretical model of convertive diffusion of motile and non-motile bacteria toward solid surface, in: E. C. Donaldson and J. E. Zajic (Ed.), Microbes and oil recovery, vol. 1, Bioresource Pub., Elpaso, Tx. 1984, pp. 226-246,
6   L. K. Jang, M. M. Sharma and T. F. Yen, The transport of bacteria in porous media and its significance in microbial enhanced oil recovery, in SPE 12770, Proceedings, SPE annual conference, Long Beach, CA., March 1984.
7   A. W. Neumann, D. R. Absolom, C. J. van Oss and W. Zingg, Surface thermodynamics of leukocyte and platelet adhesion to polymer surfaces, Cell. Biophys., 1 (1979) 79-92.
8   A. W. Neumann, O. S. Hum, D. W. Francis, W. Zingg and C. J. van Oss, Kinetic and thermodynamic aspects of platelet adhesion from suspension to various substrates, J. Biomed. Mater. Res., 14 (1980) 499-509.
9   D. R. Absolom, A. W. Neumann, W. Zingg and C. J. van Oss, Thermodynamic studies of cellular adhesion, Trans. Am. Soc. Artif. Intern. Org., 25 (1979) 152-156.
10 S. C. Dexter, Influence of substratum critical surface tension on bacterial adhesion, Colloid Interface Sci., 70 (1979) 346-354.

11 M. Fletcher, and G. I. Loeb, Influence of substratum characteristics on the attachment of a marine pseudomand to solid surfaces, Appl. Env. Microbiol., 37 (1979) 67-72.

12 A. W. Neumann, R. J. Good, C. J. Hope and M. Sejpal, An equation-of-state approach to determine surface tensions of low energy solids from contact angles, J. Colloid Interface Sci., 49 (1974) 291-304.

13 W. Stumm and J. J. Morgan, 2nd edn., Aquatic chemistry, John Wiley & Sons, New York, 1981. pp. 604.

Interfacial Phenomena in Biotechnology and Materials Processing,
edited by Y.A. Attia, B.M. Moudgil and S. Chander
Elsevier Science Publishers B.V., Amsterdam, 1988 — Printed in The Netherlands

27

THE ELECTRIC POTENTIAL PROFILE DUE TO DISCRETE CHARGES IN THE
INHOMOGENEOUS INTERFACIAL REGIONS

V.S. VAIDHYANATHAN
Department of Biophysical Sciences, State University of New York at Buffalo,
Main Street Campus, Buffalo New York 14214.

SUMMARY
    The computed exact electric potential profiles arising due to the
presence of a finite number of univalent charges on a lattice of a square
mesh, normal to the plane of the grid, are presented. The resultant electric
potential profiles, obtained by placement of such grids parallel to each
other, are shown to exhibit extrema values. These profiles are consistant
with the conclusions obtained, on the basis of inclusion of ion-ion interaction energy terms to the limiting expression for chemical potential of
charged species in aqueous interfacial region, viz., the signs of charge
density and the electric potential should be the same in major portion of
inhomogeneous interfacial regions. The justification for the existence of
oscillations in both charge density and electric potential profiles are
presented on the basis of the conflicting requirements of the mass conservation requirement and the requirement of the Gauss integral. It appears
that the simple Nernst relation between electric potential difference and
concentrations of ionic species, is not valid in interfacial regions.

INTRODUCTION

    There are several compelling reasons for interest in the theoretical

study of the inhomogneous interfacial regions. The ultimate aim is to under-

stand the mechanism of interactions, between an electrolyte and a surface

such as a biological membrane, and elucidation of the resultant distributions

of various ion concentrations and the electric potential profiles in inhomo-

geneous regions. This paper is concerned with certain fundamental questions

raised by our studies (refs. 1,2), regarding the relations existing between

concentration distributions of charged species and the electric potential

profile. The necessity for the existance of a dielectric profile and the

factors determining the finite nature of the extent of interfacial regions,

denoted by 'd', are examined. A critical examination of the classical theory

of ionic double layers (ref. 3) suggests that the assumption of constant

dielectric coefficient is never valid. In biophysical systems of interest,

the concentrations of ions in aqueous solutions are of the order of unimolar.

Hence, the results of strong electrolyte theory, whose validity is limited to

extremely dilute solutions are evidently not applicable without modifications.

In recent years, the techniques of statistical mechanics has given impetus to the study of inhomogeneous interfacial regions. A review of these advances are presented by Carnie and Torrie (ref. 4). One may classify these studies as the HNC/mean spherical approximation, the Modified Poisson-Boltzmann equation and the Born Green Yvon integral equation methods. In our opinion, a major deficiency of these approaches is the neglect of consideration of the variation with position of the dielectric coefficient in interfacial regions of finite molecular dimensions, where evidently the electric field is very large.

Though it has been long recognized that the traditional nonlinear Poisson-Boltzmann equation for charge density in diffuse portion of a double layer, suffers from lack of consistency the progress made to-date to address this question can be said to be only minimal. The so-called Debye-Approximation, viz., the replacement of energy term in Boltzmann expression for equilibrium distribution by the electric potential term, utilized in the derivation of Poisson-Boltzmann equation, is not valid for strongly inhomogeneous regions. The finite size of the molecules constituting the electrolyte fluid is known to modify the double layer structure. Another unsatisfactory aspect of the classical theory for electric double layer and the electrolyte theory, is that the solvent properties and its role is relegated as a uniform dielectric continuum. Our analysis appear to favor a quasi-lattice picture of assembly of charges (and molecules) near the interface, in interfacial regions.

The restricted primitive model calculations of D'Aguanno et al,(ref.5) has shown that for certain values of charge density on the surface, an oscillatory behavior in the density profile exists. This feature is not explainable on the basis of classical theory. Many of the thermodynamic properties of the concentrated electrolyte solutions are obviously related to the structural effects. The oscillating form of the radial distribution functions for the oppositely charged ions is characteristic of a quasi-lattice structure. The Monte Carlo calculations (refs. 6, 7) have found oscillations of the radial distribution functions. The quasi-lattice theories, valid for high concentrations have the same advantages and defects as the Debye-Huckel theory at comparable concentrations. Historically, the cube root dependence on concentrations of many properties of salt solutions, including the activity coefficients, preceded the square root dependence.

When one includes the ion-ion interaction energy term contributions to free energy and to the limiting expression for the chemical potential(refs.8,9) of ions in interfacial region, one obtains the conclusion that the signs of the electric potential and charge density must be similar. One concludes that in the interfacial region, the repulsion between similar charges play a more

dominant role in the determination of concentration profiles of charged species
than the electric potential. This aspect is also not evident from classical
Poisson-Boltzmann equation or the Nernst expression. Thus, the relation be-
ween concentration and electric potential is more involved than that expressed
by the simple Nernst equilibrium expression. The unreasonably high concentration of ions of electrolyte near the surface, for reasonable concentrations
in bulk aqueous solutions and values of electric potential predicted by the
classical theory is therefore avoided. If the signs of the charge density and
the electric potential at an arbitrary location in the interfacial region are
the same, then it is the requirement of Poisson equation that a dielectric
profile must exist. Our studies have shown that extremum values in charge
density and electric potential must exist in interfacial regions of finite
extent. One concludes that the value of the dielectric coefficient near the
interface should be very close to unity, if not unity. As a first order of
correction, the Poisson-Boltzmann equation needs to be modified with the
presence of terms denoting contributions from ion-dipole interactions and
a term denoting the local deviation from electroneutrality.

The Nernst expression and the Nernst-Planck equations of electrodiffusion
form the main basis of various theoretical investigations of the bioelectric
phenomena (ref. 10). Therefore, it is extremely important that the validity,
or the nonvalidity of stated divergent conclusions of our analysis and the
concepts of classical double layer theory is determined. This forms the main
object of this paper.

ADDITIONAL CONSIDERATIONS

The concept of an electrical double layer, at surfaces separating a liquid
and a solid phase, appears to have been originally suggested by Quincke in 1861.
(ref. 11). It was pictured then as two parallel layers of uniform charge
density, but of opposite signs, separated by a small distance. Since the electric potential vanishes at infinite distances, such a picture evidently, implies
an electric potential profile with two extrema points. Grahame (ref. 12),
Booth (ref. 11), Levine et al (ref. 13) have critically considered the problems
associated with the classical Gouy-Chapman theory. In almost all these reexami-
nations of the theory of ion distributions in interfacial regions, the position-
al variation of the dielectric coefficient is neglected. The electrical
permittivity of a medium differs from unity, due to polarization, involving the
appearance of induced charges in the dielectric, which are immobile. The
classical theory of dielectrics is restricted to the case, where the medium is
nonconducting (refs. 14, 15). Thus, the condition of electroneutrality both
in the macroscopic and microscopic senses are implied. Absence of mobile

charges, such that the medium is electrically nonconducting is required. It is
evident that these conditions are not satisfied in biological interfacial
regions. Far away from the interface, where homogeneity prevails, the dielectric coefficient of aqueous solution is of the order of 80.36 at 20°C. The
value of the dielectric coefficient of nonpolar liquids is of the order of 2.
The high frequency value of dielectric coefficient of water, quoted by Debye
(ref.16) is about 3. The dielectric coefficient of aqueous electrolyte varies
with the temperature, electric field and concentrations of ions (ref. 17).
The electric field near an ion is very large, and is of the order of 10^5 to
10^6 volts/cm. These fields are large enough to cause appreciable saturation
effect. At a distance of about 10^{-7} cm, from a central univalent ion, using
a value of 80 for the dielectric coefficient, Debye has computed the field
to be 1.8×10^5 volts/cm, in the vicinity of a monovalent ion. Thus, at a
distance of about 20-50 Angstroms, which is of molecular dimension, the
existance of a dielectric profile cannot be neglected. The computed value
of interfacial inhomogeneous region, d, is of the order of 2×10^{-7} cm,
when concentration of electrolyte in solution is unimolar. Debye has published
the calculated (approximate) dielectric profile near a monovalent ion in his
monograph. This S-shaped profile of dielectric coefficient, computed by
Debye, can be approximately expressed as,

$$\varepsilon(x) = \sum_{i=0} \varepsilon_i \, x^i; \quad \varepsilon'(o) = 0 = \varepsilon_1$$

$$\varepsilon(o) = 1 \text{ or } 3. \quad \varepsilon_2 = 2 \, \varepsilon_4 \, d^2 \, ; \quad \varepsilon_3 = -(8/3) \, \varepsilon_4 d$$

$$\varepsilon_4 \, d^4 = 3 \, \{\varepsilon(d) - \varepsilon(o)\} = 237; \quad d = 2 \times 10^{-7} \text{ cm.} \tag{1}$$

The classical theory of interfacial region starts with the premise that
if the surface contains, say, fixed negative charges, the electric potential
at the surface is negative. However, it ignored the contribution to the value
of electric potential at an arbitrary location in interfacial region, due to
charges present in aqueous electrolyte solution. In classical theory it is
assumed that the electric potential decays down to a null value, from a finite
value at the surface, in a monotonic manner at distances far from the surface.
If one assumes the presence of a specified number of charges, on a lattice
(square grid, with lattice parameter 'a' unit of length), one can compute the
electric potential profile caused by these charges, extending from the plane
to large distances, along a direction normal to the plane of the grid, exactly.
In this calculation, the permittivity of the medium can be assumed to equal
unity. One can now place additional grids, containing specified number of

charges of specified kind and specified locations away from the first grid. In this manner, one can simulate the ion distributions in the aqueous solution near a surface with fixed charges, as exhibited in Figure 1. The calculated values of electric potentials, normal to the surface arising from grids with four different number of total charges, viz., 168, 120, 80 and 64, are presented in Figure 2. In these computations it was assumed that the charges are negative and are univalent. The computed electric potential profiles for these grids, decay with increase of distance in a monotonic manner, as visualized in classical theory of double layer. The calculations with 64 charges were performed, assuming that the charges were placed at the lattice points of a lattice with lattice parameter '2 a', units of length apart. The calculations with grids containing 168, 120 and 80 charges were performed assuming that lattice points on these grids were 'a' unit of length. If a dielectric profile had been present, the decay of depicted electric potential profiles of Figure 2, would have been more steep. It may be noticed that these exact computed electric potential profiles are not exponential, emphasizing the nonlinear nature of the problem. If the lattice parameter 'a' is of the order 4×10^{-8} cm., the computed potential profiles span a distance of $5 - 10 \times 10^{-7}$ cm, which is of similar magnitude as distance spanned by membranes and interfacial regions of biological systems.

BRIEF CRITIQUE OF THE CLASSICAL THEORY.

The limiting expression for the chemical potential of an ion of kind σ, with charge Z_σ e, in solution at location x, μ_σ (x) is

$$\mu_\sigma(x) = \mu_\sigma^*(T,P) + kT \ln C_\sigma(x) + Z_\sigma e\, \emptyset(x) \tag{2}$$

where μ_σ *(T,P) is the composition, (hence position) independent part of the partial molar free energy of ions of kind σ. T is the temperature in Kelvin scale and P is the pressure. Z_σ and C_σ are respectively, the signed valence charge number and the concentration at location x, of ions of kind σ. k is the Boltzmann constant and e is the proton charge. It is well known that the logarithmic dependence of chemical potential on concentration, cf: eq.(2), is obtained from the derivation of ideal entropy of mixing of binary mixtures, where the mole fractions are replaced by concentrations, for solutes of dilute solutions, and the phenomenological addition of the electric potential term for ionic species. Thus, the validity of equation (2) for concentrated electrolyte solution is questionable. The use of activity coefficient term for concentrated solutions is only empirical. For equilibrium situation, the constancy of the value of chemical potential of a specified species everywhere, leads to the famous Nernst equilibrium expression, namely,

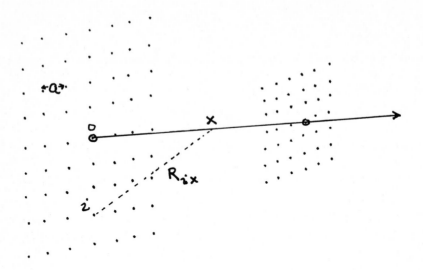

FIGURE 1. Grids of charge locations utilized in computations of electric potential profiles presented in figures 2 and 5. No charge is present at origin, in order to avoid singularity.

$$\emptyset(x) \quad = \quad N\,e \quad \sum_{i=1}^{N}\left\{R_{ix}^{-1}\right\}$$

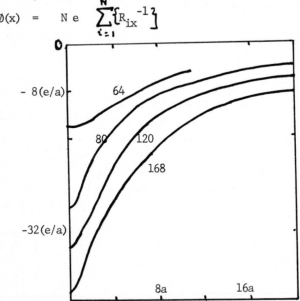

FIGURE 2. Plots of electric potential profiles, along direction normal to grids of charges, on a plane square lattice. The number of changes on the grids causing the profiles, are presented adjacent to plots.

$$\beta_{Nernst} = \left[1/Z_\sigma\right]\left\{\ln\ C_\sigma(x_1)/C_\sigma(x_2)\right\}$$

$$= (e/kT)\left\{\emptyset(x_2) - \emptyset(x_1)\right\} \tag{3}$$

Equation (3) relates in this manner, the concentrations of a specified ion at two specified locations, x_1 and x_2, with the difference in electric potential at these two locations. Thus, equation (3) requires the distributions of univalent cations like sodium and potassium to be similar. Many concepts in biology, such as the active transport, Hodgkin-Huxley equations, are based on the assumed validity of Nernst expression. The validity of equations (2) and (3) require evidently that near a surface with negative charges, the concentrations of positive ions decrease monotonically to its bulk value, $C^+(d)$, while the electric potential increase from some finite negative value at the surface to a null value at large distance away from the surface. The schematic plots of the concentration profiles, $C^+(x)$, $C^-(x)$, the electric potential profile $\emptyset(x)$ and the negative of the charge density profile $Y(x)$, as given by the classical theory are presented in figure 3. Therefore, it is the requirement of classical theory, that in the interfacial region, the ratios, $\{\emptyset(x)/Y(x)\}$ and $\{\emptyset'(x)/Y'(x)\}$ are positive definite for any value of x in the interfacial region. x is the position variable, defined normal to the plane of the surface. For brevity, the various orders of derivatives with respect to x, of various functions are denoted by appropriate number of primes.

It is interesting to note that the values of concentrations of charges and charge density distribution at regions very close to the interface are never seriously discussed in literature. A number of other criticisms can be advanced against the plausible validity of the classical theory of ionic double layer. 1. When the potential at the surface is about 100 millivolts, and the concentrations of salt in electrolyte is of the order of 0.1 molar, equation (3) predicts values of 5.46 molar for univalent and 298 molar for divalent ion concentrations at the interfacial surface. Evidently these values are impossible for ions to be accommadated. The average ion-ion separation for unimolar electrolyte can be computed to equal about 9×10^{-8} cm at normal temperatures. 2. The interionic interactions between ionic species in solution lead to some sort of ordering, resulting in reduced entropy and resultant structure, such as the ion atmosphere, in the configuration of ion distribution in concentrated solutions. Thus, one must include ion-ion interaction energy term contributions to free energy and the expression for chemical potential, even for dilute solution. Our attempts to include such contributions and modify equation (2) led to the conclusion that extrema in charge density and electric potential profiles in the interfacial region are expected. This result is schematically presented in the

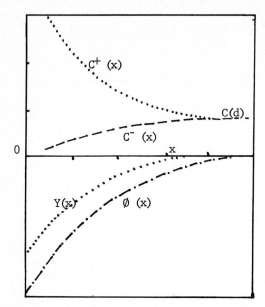

FIGURE 3. Schematic representation of profiles of $\emptyset(x)$, $Y(x)$, $C^+(x)$ and $C^-(x)$, near a negatively charged surface, given by classical theory.

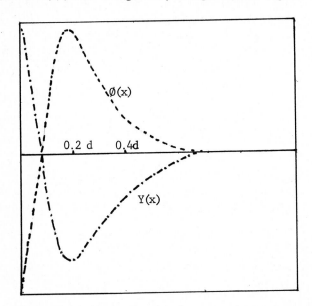

FIGURE 4. Schematic representations of the electric potential profile $\emptyset(x)$ and negative of the charge density profile $Y(x)$, given by the theory, when ion-ion interaction terms contribution to partial molar free energy are included. The scales in y-axis are arbitrary.

figure 4. 3. The assumption of constant dielectric coefficient, usually employed, in obtaining the solution of the Poisson-Boltzmann equation, and in various statistical mechanical approaches to theory of ionic double layer, can be easily shown to be nonvalid. If the dielectric coeffcient ε, is not a function of x, in interfacial region, one has from the Poisson equation,

$$\nabla \cdot \varepsilon(x) \nabla \emptyset(x) = Y(x) = -(4 \pi e) \sum_\sigma Z_\sigma C_\sigma (x) \qquad (4)$$

the conclusion that $Y(x)$ should be proportional to $\emptyset''(x)$. If one assumes, say the concentration profile, or $\beta(x)$, and computes precisely the values of $\emptyset(x)$, $\emptyset''(x)$, $Y(x)$ and $C_\sigma(x)$, using equations (2) and (3), one may verify that the computed values of $Y(x)$ are never proportional to $\emptyset''(x)$. 4. One has the exact result of the Gauss theorem,

$$\int_o^\lambda Y(x) \ dx = -\varepsilon(o)\emptyset'(o) \quad ; \text{ since } \emptyset'(d) = 0. \qquad (5_a)$$

$Y(d) = 0$. $\emptyset'(o)$ is the value of the electric potential gradient at the interface and $\varepsilon(o)$ is the value of the dielectric coefficient at the interface. One may verify also that for any assumed value of the electric potential profile, the computed values of the integral, (eq. 5) of the charge density, leading to calculated value of $\varepsilon(o)$, since $\emptyset'(o)$ is assumed and known, never equals the bulk homogeneous region value of the dielectric coefficient, $\varepsilon(d)$. The cited results of Debye, and our calculations show that $\varepsilon(o)$ has a value very close to unity, if not unity. 5. Finally, if no chemical reactions occur in interfacial region or at the surface, the presence of a charged surface immersed in an electrolyte solution causes perturbation in concentration distribution of ions. The mass conservation relations should still be satisfied. This may be expressed as

$$\int_o^\lambda C_\sigma(x) \ dx = C_\sigma(d) \ d \ ; \text{ since } \sum_\sigma Z_\sigma C_\sigma (d) = 0 \qquad (5 b)$$

where d is the extent of the interfacial region, and $C_\sigma(d)$ is the value of concentration of ions of kind σ in bulk homogeneous electrolyte. The imposition of this condition for interfacial region adjacent to a surface with fixed charges, will lead to the existence of extrema points in the concentration profiles of both anions and cations. This aspect, not considered in classical theory, is schematically shown in figure 6. The resultant change density profile will have two extrema points as a minimum requirement. If the electric potential profile is the resultant of the charge density profile, then extrema in electric potential profile can also occur.

BASIC EQUATIONS.

In this section, equations resulting from inclusion of ion-ion interaction terms and the results are presented. The neglect of ion-dipole interaction energy terms and inclusion of ion-ion interaction terms, enables one to express that the chemical potential of an ion of kind σ, of an electrolyte in the inhomogeneous interfacial region for the location x, as

$$\mu_\sigma(x) \;=\; \mu_\sigma{}^*(T,P) + kT \ln C_\sigma(x) + Z_\sigma \, e \, F(x)$$

$$F(x) \;=\; \left[\emptyset(x) - (H/4\pi)\, Y(x) \right]$$

$$Y(x) \;=\; -(4\pi e) \sum_\sigma Z_\sigma C_\sigma(x) \;=\; \varepsilon'(x)\emptyset'(x) + \varepsilon(x)\emptyset''(x) \tag{6}$$

where the Poisson's equation has been utilized. Equations (4) and (6) are independent of the state of the system. Equation (6) is obtained, by the addition of interaction energy terms, viz., $\left\{ \sum_\eta H_{\sigma\eta}\, C_\eta(x) + \sum_k H_{\sigma k} C_k(x) \right\}$ to equation (2), and the neglect of ion-dipole interaction terms, $H_{\sigma k}$. $H_{\sigma\eta}$ terms are expressed as $Z_\sigma Z_\eta e^2 H$, where H is the charge independent part of the molecular integral of pair potential energy of interaction and probability distribution functions. The utilization of Poisson's equation, results in equation (6). In the classical theory, the separation of anions and cations in interfacial region, resulting in a charge density profile, is effected by the electric potential $\emptyset(x)$. According to equation (6), this separation is mitigated by the presence of the charge density term, $Y(x)$, representing the local deviation from local electroneutrality. Thus, the tendency of the system to avoid charge separation, is proportional to deviation locally from electroneutrality. This factor acts as a balancing term in the determination of the concentration of ionic species, in addition to the electric potential term, in interfacial regions. This aspect also, is not evident from the classical theory.

The concentration profiles, electric potential profile and the dielectric profile in interfacial region, are thus determined by a set of coupled non-linear differential equations, viz.,

$$C_\sigma'(x) + (Z_\sigma e/kT) C_\sigma(x) F'(x) \;=\; \begin{cases} 0, & \text{for equilibrium state} \\ -(J_\sigma/D_\sigma), & \text{for steady state.} \end{cases} \tag{7a}$$

$$4\pi kT \quad A'(x) \;=\; Y(x) F'(x) \; ; \quad A(x) = \sum_\sigma C_\sigma(x); \quad A' = (dA/dx) \tag{7b}$$

$$K(x)\, \emptyset'(x) \;=\; Y'(x) \left\{ 1 + (H/4\pi) K(x) \right\} \tag{7c}$$

$$4\pi kT \quad K(x) A'(x) \;=\; Y(x) Y'(x) \tag{7d}$$

$$K(x) \;=\; \kappa^2(x)\, \varepsilon(x) \;=\; (4\pi e^2/kT) \sum_\sigma Z_\sigma{}^2 C_\sigma(x) \tag{7e}$$

J_σ and D_σ are respectively the stationary state flux and diffusion coefficient of species σ, in the interfacial region. The set of equations (7) are exact, subject only to the validity of equation (6). In place of equation (3), one now has the relation between electric potential and $\beta(x)$,

$$\beta(x) = (Z_\sigma)^{-1} \ln \left\{ C_\sigma(x)/C_\sigma(d) \right\} = (e/kT) \left\{ (H/4\pi)Y(x) - \emptyset(x) \right\} \quad (8)$$

It is reasonable to suspect that the value of the molecular integral, H, is given by equation (9), when a dielectric profile exists.

$$- (4\pi/H) = K(d) = \kappa^2(d)\, \varepsilon(d) \quad (9)$$

where $\kappa^2(d)$ is the value of the Debye-Huckel ion atmosphere parameter of the bulk electrolyte solution. $\varepsilon(d)$ is the value of dielectric coefficient of the bulk homogeneous electrolyte solution. One may verify that for any electrolyte, the value of $K(d)$ is always greater than the value of $K(x)$ for values of x, in the range, $0 < x < d$. The magnitude of the extent of the inhomogeneous region, d, is assumed to be finite and determined by the system in a self-consistant manner, by the system properties. Thus, it follows that, when equations, (6) and (7c) are valid, the validity of equation (9) requires that the ratio, $\emptyset'(x)/Y'(x)$ should be negative definite. The validity of equation (8), demands that the signs of the charge density and the electric potential should be the same at an arbitrary location in the interfacial region. Evidently, this conclusion is in disagreement with the conclusions of the classical theory. Though the validity of equation (9) may be in suspect, this conclusion is evidently in accordance with the double layer (condenser) picture of Quincke. If one presumes that the fixed negative charges on the surface, results in a negative value for the electric potential at the surface, the excess positive ions at some distance away from the surface in the interfacial region, will certainly contribute positive values to electric potentials. The actual sign and magnitude of the electric potential at these locations, depends on the relative magnitudes of the various contributions, and is essentially the algebraic sum of all contributions from all charges present in the system. The integralof equation (7d) results in the generalized form of the Maxwell's Osmotic Balance equation. When equation (9) is valid, one obtains for a symmetrical (1-1) ion system, the results,

$$\sinh \beta(x) - \beta(x) = (e/kT)\emptyset(x) \quad (10a)$$

$$16\,\pi^2 e^2\, A(x)\, A'(x) = Y(x)Y'(x) \quad (10b)$$

$$(8\pi\, kT/\varepsilon\,(d)) \left\{ A(x) - A(d) \right\} = \emptyset'(x)^2 - \emptyset'(d)^2 \quad (10c)$$

Equation (10b) is the same as the equation (7d) restricted to symmetrical ion systems. Equation (10c) is the familiar form of Maxwell's Osmotic Balance equation, obtained by quadrature, from (10b), with the stipulation that the dielectric coefficient is not a function of x. The validity of equation (10c) evidently contradicts the validity of constant field assumptions employed in biophysics. If the interfacial surface contains fixed negative charges, the concentrations of positive ions will dominate over the concentrations of negative ions, in the immediate vicinity of the interfacial surface. Thus, $Y(X)$ will be negative over the major portion of the interfacial region. The integral of eq. (5) will therefore be negative. Since the values of $\varepsilon(o)$ and $\emptyset'(o)$ are positive, an extremum value in charge density profile must exist. An extremum value in the electric potential profile may exist as a consequence.

The schematic forms of the electric potential and charge density profiles given by our analysis, are presented in figure 4. The magnitude of the extent of the inhomogeneous region has been approximately estimated to be about 20 Angstroms, when concentration of salts in the electrolyte is of the order of unimolar. For every tenfold decrease in concentration the magnitude of d increases by a factor of $10^{\frac{1}{2}}$. Our calculations indicate that the value of dielectric coefficient near the interface should be of the order of unity, a value consistant with Debye's calculation of electrical saturation, of equation (1).

PLAUSIBLE RESOLUTIONS OF THE CONFLICT.

Since almost all analysis of electrokinetic studies in biological systems are based on the assumed validity of Nernst equation and transport theories are based on the analysis of solutions of Nernst-Planck equations, it is extremely important to determine the validity of conclusions based on equations (6) and (7) and the results of classical Gouy-Chapman theory. The disagreements between the plots of figures (3) and (4) should be resolved. Unfortunately, all experimental determinations involve either the implicit assumption of the validity of equations (2) and (3), or based on the information obtained by performing experiments in regions of solution where electroneutrality is preserved. It is our opinion that an experimental determination of the validity or the nonvalidity of either of these two divergent conclusions cannot be accomplished at this time.

Two arguments can however, be advanced which appear to indicate the validity of the conclusion that the signs of charge density and electric potential should be the same at a chosen location in interfacial region. The first argument is as follows: If one has knowledge of the dielectric and electric potential profiles, then one can compute the sign and magnitude of $Y(x)$, using the Poisson equation (6). One can thereby compute $C_\sigma(x)$ and the sign of $\beta(x)$. The validity of equations (6) and (10a) require that $\beta(x)$ and $\emptyset(x)$ must have

same signs. On the other hand the validity of equations (2) and (3) require
that $\beta(x)$ and $\emptyset(x)$ must have mutually opposite signs. Assuming that the
profiles in question are continuous and analytic, and that these may be expanded
in Taylor series, one can express,

$$\epsilon(x) = \sum_{i=0} \epsilon_i x^i ; \quad (i!) \; \epsilon_i = (d^i \epsilon/dx^i)_{x=0}$$

$$\emptyset(x) = \sum_{i=0} \emptyset_i x^i \qquad (11)$$

One may truncate the Taylor series, retaining finite number of leading terms.
The retained Taylor coefficients may be evaluated utilizing the plausible condi-
tions, viz., $\epsilon'(d) = 0 = \epsilon''(d)$, $\emptyset(d) = \emptyset'(d) = \emptyset''(d) = 0$ and $Y'(d)=0$.
In this manner, one can calculate the values of $\epsilon(x)$, $\epsilon'(x)$, $\emptyset'(x)$ and $\emptyset''$.
Therefore, one can compute the values of $Y(x)$ and $Y'(x)$ for any desired
value of x. The results of such calculations are presented in Tables 1 and 2.
In these calculations, the leading (m+1) terms of dielectric profile and the
leading (n+1) terms of electric potential profiles are retained. The results
presented in these tables, indicate definitively that the ratios; $\{\emptyset'(x)/Y'(x)\}$
and $\{\emptyset(x)/Y(x)\}$ are negative definite. Since $\beta(x)$ has a sign opposite to
the sign of $Y(x)$, it follows that $\beta(x)$ and $\emptyset(x)$ must have same signs.

TABLE 1. Computed values of various order derivatives of the electric poten-
profile, using eq.(11), retaining the leading (n+1) terms.

	n = 4	n = 5	n = 6
$\emptyset(o)$	$\emptyset_4 d^4$	$- \; \emptyset_5 d^5$	$\emptyset_6 d^6$
$\emptyset'(o)$	$- 4 \emptyset_4 d^3$	$5 \emptyset_5 d^4$	$- 6 \emptyset_6 d^6$
$\emptyset''(o)$	$12 \emptyset_4 d^2$	$- 20 \emptyset_5 d^3$	$30 \emptyset_6 d^4$
$\emptyset'''(o)$	$- 24 \emptyset_4 d$	$60 \emptyset_5 d^2$	$- 120 \emptyset_6 d^3$

TABLE 2. Computed values of $Y(o)$ and $Y'(o)$, retaining the leading (n+1) terms
of the electric potential profile and the leading (m+1) terms of the
dielectric profile. The values of $Y(o)$ are in $\emptyset_n \Delta\epsilon \; d^{n-2}$.

m	n	Y(o)	Y'(o)	Y'(o)/Ø'(o)
2	4	- 6.8	93.6	- 23.4
2	5	8.0	- 84.0	- 16.8
2	6	- 9.0	60.0	- 10.0
3	4	-10.8	93.6	- 23.4
3	5	13.0	-144.0	- 28.8
3	6	-15.0	204.0	- 36.0

The second argument is based on the use of electric potential profiles calculated and presented in figure 2, on the basis of discrete charge distribution. The computed electric potential profile, when a grid containing 120 univalent positive charges on its lattice points, is placed parallel to the grid containing 168 univalent negative charges, at a distance of '8a' units of length apart, is presented in figure 5. This profile, denoted by plot I, exhibits an extremum value. In figure 5, the plot II, represents the resultant electric profile, obtained when a grid containing 64 univalent positive charges on its lattice, (with reduced charge density per unit area, since the lattice points on this grid are '2 a' units of length apart), placed at a distance of '12 a' from the grid containing 168 univalent negative charges. Plot II also exhibits an extremum value. The real situation of ion distribution in the interfacial region can now be simulated by placing additional grids with various number of charges at specified locations away from the grid containing 168 univalent negative charges. In plot III, the calculated electric potential profile, obtained by the placement of 120 charge grid at distance '8 a' and a grid of 80 univalent positive charge, placed at a distance '16 a' from the grid of 168 negative charge, at location $x = 0$. Again one obtains extrema values in the computed profile. From these kind of simulations, one concludes again that the sign of charge density and electric potential profiles should be the same in major portion of the interfacial regions.

However, the validity of this conclusion is based on the relative magnitude and locations of assumed charge distributions, since the value of electric potential at a specified point is an algebraic sum of contributions from all charges present in the system. The value of potential at $x = 0$, where the grid with 168 negative charges are present, is reduced by the presence of a grid with positive charges elsewhere in adjacent space. If the plate with 64 positive univalent charges had been placed closer to the grid with 168 negative charges, say at $x = 8a$, in place of 12a, the resultant value of electric potential will only slightly steeper than the electric potential profile of 168 charges depicted in figure 2. This will not exhibit an extremum at $x = 8a$. This situation may correspond to the conclusions of the classical theory, in which the contributions to electric potential at an arbitrary location in the interfacial region from charges of the aqueous electrolyte is ignored.

DISCUSSION.

In order that the charge density and the electric potential have similar signs, in locations of the interfacial region, it is necessary that a dielectric profile exists, as demanded by the Poisson equation (4). When a dielectric profile exists, the values of $Y(x)$, $\emptyset(x)$, $Y'(x)$ and $\emptyset'(x)$ listed in Tables 1

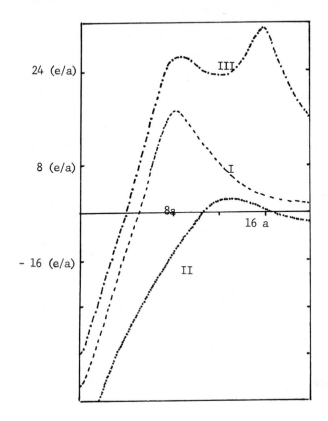

FIGURE 5. Computed values of the electric potential profiles, obtained when grids of different number of charges are placed parallel to each other, at specified locations. Plot I is obtained when two grids with 168 negative charges at x = 0, and another with 120 positive charges at x = 8a. Plot II is obtained when a grid with 168 negative charges is at x = 0, and a second grid with 64 positive charges is placed at x = 12a. Plot III is obtained by placement of 3 grids, 168 at x = 0, 120 at x = 8a and 80 at x = 16a.

and 2, lead to the conclusions that {$\emptyset(x)/Y(x)$} should be negative definite. This conclusion is independent of the validity or nonvalidity of equation (9). If {$\emptyset'(x)/Y'(x)$} is negative, then the existance of an extremum value in the charge density profile is required by the validity of the Gauss integral (5). Thus, it appears that the validity of equations (2) and (3) for the inhomogeneous interfacial region is extremely limited.

If the mass conservation conditions of equation (5b) are imposed, one obtains intuitively, the concentration profiles of anions and cations, presented schematically in figure 6, near a negatively charged surface. When these profiles are plausible, the charge density profile should exhibit two extrema values, as depicted in figure 6. These oscillations of charge density profile and electric potential profile are characteristic of a quasi-lattice picture of distribution of ions in the interfacial region. It is shown in the appendix, that when the leading six terms of a Taylor expansion of concentration profiles of ionic species are retained, and the roots of a resultant algebraic equation are analyzed, one obtains the results viz., charge density equals zero at location $x = (d/6)$, an extremum value in charge density occurs at location $x = (d/3)$ and that the inflection point of the charge density profile occurs at location $x = (d/2)$. Similarly, when the leading five terms of the concentration profile Taylor expansion are retained, one obtains the results, $Y(x) = 0$, at $x = 0.2d$, $Y'(d) = 0$ at $x = 0.4d$ and $Y''(x) = 0$, at $x = 0.6d$. If one does not impose the condition that $Y'''(d) = 0$, more than one extremum point in charge density profile can occur. The nonvanishing nature of higher order Taylor coefficients also permit the existence of more than one extremum point in the charge density and electric potential profiles.

Finally, it must be stressed that a fundamental conflict exists between the validity of the mass conservation equations and the Gauss integral of equations (5a) and (5b). If the mass conservation equations are strictly satisfied by all ionic species of the interfacial region, then the integral of charge density of equation (5a) should identically vanish. Thus, the system has to balance between satisfying equations (5a) and (5b). One possible consequence is that the value of the integral is small. In this case, either $\varepsilon(o)$ is very small, (Debye results) or the gradient of the electric potential $\emptyset'(o)$ becomes very small. It is premature to obtain definitive conclusions on these. The results of this paper, clearly indicates, the need for a serious reinvestigation of the ionic double layer theory.

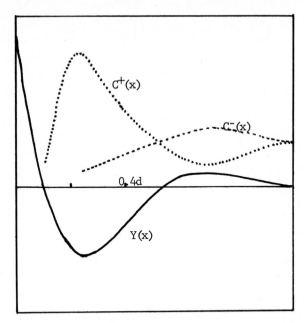

FIGURE 6. Schematic representation of the concentration profiles of anions and cations and charge density profiles $- Y(x)$ obtained when mass conservation relations (5b) are imposed.

APPENDIX.

When concentration profiles of ionic species in the interfacial region, are expressed in a Taylor series with the retention of the leading 6 terms, one has

$$C_\sigma(x) = \sum_\sigma C_{\sigma i} x^i$$

$$C_{\sigma i} = (i!)^{-1} dC_\sigma(x)/dx^i \big|_{x=o}$$

$$- Y(x)/4\pi e = \sum_i B_i x^i ; \quad B_i = \sum_\sigma Z_\sigma C_{\sigma i} \qquad (A.1)$$

Since,

$$Y(d) = 0, \sum_i B_i d^i = 0. \qquad (A.2)$$

The mass conservation relations (5b) lead to the results,

$$\sum_{i=o} \{(i+1)/(i+2)\} C_{\sigma i} d^i = 0. \qquad (A.3)$$

The conditions that $C_\sigma'(d) = 0 = Y'(d)$, yields

$$\sum_{i=o} i C_{\sigma i} d^{i-1} = 0. \qquad (A.4)$$

The imposition of the plausible conditions that $Y''(d) = Y'''(d) = 0$, enables one to express, with the use of algebra, $B_0 = -(1/6) B_5 d^5$; $B_1 = (5/3) B_5 d^4$; $B_2 = -5 B_5 d^3$; $B_3 = (20/3) B_5 d^2$ and $B_4 = -(25/6) B_5 d$. When $n = 5$, one obtaines the condition that $Y(x_1)$ will equal zero, when

$$6 \alpha^5 - 25 \alpha^4 + 40 \alpha^3 - 30 \alpha^2 + 10 \alpha - 1 = 0: \quad \alpha d = x_1$$

(A.5)

Similarly, if one states that $Y'(x_2) = 0$, $x_2 = \lambda d$, and that $Y''(x_3) = 0$, $x_3 = \zeta d$, one obtains that λ and ζ are the roots of the equations $(n = 5)$,

$$3 \lambda^4 - 10 \lambda^3 + 12 \lambda^2 - 6 \lambda + 1 = 0$$

$$2 \zeta^3 - 5 \zeta^2 + 4 \zeta - 1 = 0.$$

(A.6)

REFERENCES

1 V.S. Vaidhyanathan, A fundamental question about electrical potential profile in the interfacial region of biological systems, in: M. Blank (Ed.), Electrical Double Layers in Biology, Plenum Press, 1968, pp. 31-51.
2 V.S. Vaidhyanathan, The electric potential profile in inhomogeneous interfacial regions, Studia Biophysica, 110 (1985) 29-42.
3 E.J.W. Verwey and T.G. Overbeek, Theory of Stability of Lyophobic Colloids, Elsevier, New York, 1945.
4 S.L. Carnie and G.M. Torrie, The statistical mechanical theory of the electrical double layer, Advances in Chemical Physics, 56 (1984), 141-253.
5 B.D. Aguanno, P. Nielaba, T. Alts and F. Forstmann, A local HNC approximation for 2:1 restricted model electrolyte at a charged wall, Journal of Chemical Physics, 85 (1986) 3476-3481.
6 F. Vaslov, in: R.A. Horne (Ed.), Water and Aqueous Solutions, Wiley-Interscience, New York, Chapter 12.
7 H.L. Frisch and J.L. Lebovitz, The Equilibrium Theory of Classical Fluids, W.A. Benjamin Inc., New York, 1964.
8 V.S. Vaidhyanathan, Philosophy and phenomenology of ion transport and chemical reactions in membrane systems, in: G. Milazzo (Ed.), Topics in Bioelectrochemistry and Bioenergetics, Vol. 1, John Wiley & Sons, London, 1976, pp. 287-378.
9 V.S. Vaidhyanathan, Nernst-Planck analog equations and stationary state membrane electric potentials, Bulletin of Mathematical Biology, 41 (1979) 365-385.
10 N. Lakshminarayaniah, Transport Phenomena in Membranes, Academic Press, New York, 1969.
11 F. Booth, Recent work on the applications of the theory of ionic double layer to colloidal systems, in: J.A.V. Butler and J.T. Randall (Eds.), Progress in Biophysics and Biophysical Chemistry,3(1953), Pergamon Press, London, pp. 131-194.
12 D.C. Grahame, The electrical double layer and the theory of electrocapillarity, Chemical Reviews, 41 (1947) 441-501.
13 S. Levine and C.W. Outhwaite, Comparison of the theories of aqueous electric double layers at a charged plane interface, J. Chem. Soc. Faraday Trans., 74 (1978) 1670-1689.
14 H. Frohlich, Theory of Dielectrics, Oxford University Press, 1949.

15 G. Stell, G.N. Patey and J.S. Hoye, Dielectric constant of fluid models: stat. mech. theory and quant. implementation, Advances in Chemical Physics, 48 (1980) 183-328.
16 P. Debye, Polar Molecules (1912), Dover Publications, New York, 1929, Fig. 30, p. 118.
17 W. Schroer, The local field in the statistical mechanical theory of dielectric polarization, Advances in Chemical Physics, 63 (1985) 719-773.

Interfacial Phenomena in Biotechnology and Materials Processing,
edited by Y.A. Attia, B.M. Moudgil and S. Chander
Elsevier Science Publishers B.V., Amsterdam, 1988 — Printed in The Netherlands

NONLINEAR EFFECTS OF INTERFACIAL ELECTRICAL FLUCTUATIONS AND OSCILLATIONS ON MEMBRANE ENZYMES

R. DEAN ASTUMIAN[1] and T.Y. TSONG[2]

[1]Laboratory of Biochemistry, National Heart, Lung, and Blood Institute, National Institutes of Health, Bethesda, Maryland 20892 (U.S.A.)

[2]Department of Biological Chemistry, Johns Hopkins University Medical School, Baltimore, Maryland   20215 (U.S.A.)

SUMMARY
   Interfacial electric fluctuations and oscillations may play a significant role in governing the activity of membrane enzymes. This is discussed theoretically in the context of a simple model enzyme, where it is shown that many typical enzymes should be capable of absorbing energy from a dynamic electric field and using it to drive the reaction for which it is specific away from equilibrium. It is shown that this behavior is frequency specific and thus may serve as the basis of methods for determining kinetic properties of membrane enzymes. Additionally, the concepts may be important for understanding signal and free energy transduction accomplished at biological membranes.

INTRODUCTION

   Electrical fluctuations that occur spontaneously across biological membranes differ in a number of significant aspects from their homogeneous phase counterparts. While deviations of a thermodynamic parameter from its mean value in bulk solution are typically small ($\sim 1/V^2$) and dissipate on a time scale rapid as compared to most chemical events, excursions of the local membrane potential (caused by, e.g., opening and closing of ion channels) from the average value may be quite large and relax on a time scale determined by a protein conformational transition. Often these fluctuations display effectively coherent behavior, giving rise to macroscopically observable oscillations. For example, the voltage across β-pancreatic cells displays periodic burst activity, depending ont he amount of glucose in the perfusion media (ref. 1). The amplitude of these oscillations is 40 mV (i.e., a shift in the transmembrane electric field strength of $\approx$ 80 kV/cm) with a period on the order of 10 Hz. Similar phenomena have been observed in a wide range of other systems (refs. 2 and 3). It must be realized that these local fluctuations, and macroscopic oscillations, are manifestations of energy releasing processes, i.e., the movement of ions down their electrochemical gradients. This energy can be harnessed by many ordinary membrane proteins and enzymes for signal and energy transduction.

It might be thought that chemical reactions in the presence of external, or even energy driven, environmental fluctuations can be described in terms of simple deterministic kinetic equations, where the rate and equilibrium coefficients are those relevant for the average value of the fluctuating parameter. This is the case only if two conditions are met. First, the fluctuation amplitude must be small, such that non-linear effects due to the intrinsic exponential dependence of rate and equilibrium coefficients on thermodynamic parameters are not manifested. Second, the fluctuation dissipation rate (i.e., inverse correlation time) must be great to insure that cross correlations between the external "noise" (or oscillation) and the enforced variation of the concentrations of chemical reactants can be safely neglected. In general, neither of these conditions are met by membrane potential ($\Delta\psi$) fluctuations across biological membranes (refs. 4 and 5).

This is an important consideration in light of the fact that the conformational transitions of many transmembrane proteins have a significant electric susceptibility. It is particularly evident in the case of voltage gated channels, but will come into play for any conformational change involving either intramolecular charge transfer or rotation of dipole groups such as α-helices. This clearly includes, but is not limited to, all ion transport ATPases, ion carriers, and electron transfer proteins in the electron transfer chain (ETC) (refs. 5 and 6). In this paper, we will describe the effect of an oscillating or fluctuating electric field on a simple enzyme system and discuss how free energy can be transduced from the dynamic electric field and used by the enzyme to drive the reaction for which it is specific away from equilibrium, as was demonstrated previously by computational studies (refs. 7, 8, and 9).

It is anticipated that the concepts presented will be useful in understanding mechanisms of biological energy and signal transduction and how "noise" may be a source of order rather than disorder.

HOW THE MEMBRANE POTENTIAL INTERACTS WITH PROTEIN CONFORMATIONAL EQUILIBRIA

Enzymes of the membrane are more often than not sensitive to an electric field. The mechanisms by which these proteins accomplish their function involve conformational transitions. If intramolecular charge transfer or dipole moment changes occur during these transitions, the conformational equilibria will vary as a function of the electric field. Let us consider a simple conformational equilibrium

$$E(\mu,\alpha) \rightleftharpoons E^*(\mu^*,\alpha^*) \tag{1}$$

$\mu(\mu^*)$ and $\alpha(\alpha^*)$ are the permanent dipole moment and polarizability of the states $E(E^*)$, respectively. In general, the influence of an electric field on such an equilibrium can be written in terms of a generalized van't Hoff equation (refs. 5 and 7)

$$(d\ln K/d\varepsilon)_{T,p} = \Delta M/RT \tag{2}$$

If we consider that E in Eq. (1) is a membrane protein, where its rotational degrees of freedom are constrained (it can't rotate in response to the field, ΔM), the macroscopic molar polarization, may be written

$$\Delta M = (\mu^*-\mu) + (\alpha^*-\alpha)\cdot\varepsilon = \Delta\mu + \varepsilon\cdot\Delta\alpha \tag{3}$$

Integration of Eq. (2) yields

$$K_E \rightarrow K_0\cdot\exp\{[(\Delta\alpha\cdot\varepsilon)/2 + \Delta\mu]\varepsilon/RT\} \tag{4}$$

This equation implies that at low field strength, a positive membrane potential will favor the E* state, and negative $\Delta\psi$ will favor the E state (relative to the zero field condition) since the magnitude of $\Delta\psi$ is usually much large than that of $\Delta\alpha$. At very high field strengths, however, the quadratic term due to the polarizability will take over, and the E* state will be favored irrespective of the sign of the field. Although this quadratic dependence on the field may be of importance in explaining the optimal field strengths for various phenomena observed experimentally (ref. 7), for the remainder of this paper we shall consider that $(\Delta\mu\cdot\varepsilon)/2 \ll \Delta\mu$. Then Eq. (6) may be simply rewritten

$$K_E = K_0\cdot\exp(\Delta\mu\cdot\varepsilon/RT) = K_0\cdot\phi \tag{5}$$

and thermodynamically consistent rate constants are

$$k_{f,\varepsilon} = k_{fo}\cdot\phi^\delta \tag{6}$$

$$k_{r,\varepsilon} = k_{ro}\cdot\phi^{(\delta-1)} \tag{7}$$

where δ represents the apportionment of the electric field dependence amongst forward and reverse processes.

MODEL AND CALCULATIONS

General approach to nonlinear behavior due to fluctuation

The kinetic equation governing reaction eq. (1) is

$$dE/dt = -(k_{fo}\phi^\delta + k_{ro}\phi^{\delta-1})E + k_{ro}\phi^{\delta-1} \tag{8}$$

where E represents the fraction of total protein existing in the form E and use has been made of the fact that $E + E^* = 1$. If the electric field strength is constant and zero, the (equilibrium) steady state ($dE/dt = 0$) is given by

$$K_{eq} = k_{ro}/(k_{fo} + k_{ro}) \tag{9}$$

If, however, ε is taken to be a fluctuating quantity symmetric about zero, $\langle\varepsilon\rangle = 0$, we write instead of equation (8)

$$\langle dE/dt\rangle = \langle(k_{fo}\phi^\delta + k_{ro}\phi^{\delta-1})\cdot E\rangle + \langle k_{ro}\phi^{\delta-1}\rangle \tag{10}$$

Clearly, since the rate constants depend exponentially on the varying field, in general, $\langle k_{ro}\phi^{\delta-1}\rangle \neq k_{ro}$ if $\varepsilon\Delta\mu \gtrsim kT$, and the steady state average $\langle E\rangle_{ss}$ need not equal E_{eq} calculated for constant zero field. If, however, we take $\varepsilon\Delta\mu \ll kT$ (the "linear regime"), ϕ^δ and $\phi^{\delta-1}$ can be written

$$\phi^\delta = 1 + \delta\lambda; \quad \phi^{\delta-1} = 1 + (\delta-1)_\lambda \tag{11}$$

where $\lambda(t) = (\Delta\mu/kT)\cdot\varepsilon(t)$ and $\langle\lambda(t)\rangle = 0$. We may now write

$$dE/dt = -\{k_{fo}[1 + \delta\lambda] + k_{ro}[1 + (1 - \delta)\cdot\lambda]\}\cdot E + k_{ro}[1 + (1 - \delta)\lambda] \tag{12}$$

The steady state relation is

$$\langle dE/dt\rangle = 0 = A(k_{fo} + k_{ro})\langle E\rangle - [k_{fo}\delta + k_{ro}(\delta-1)]\langle\lambda E\rangle + k_{ro} \tag{13}$$

Only if $\delta = k_{ro}/(k_{fo} + k_{ro})$ will the average $\langle E\rangle_{ss}$ in the fluctuating field be the same as E_{eq} at zero constant field at arbitrary frequency. At very high frequency the fluctuation dissipation will be much fster than the response of the system, and $\langle\lambda\cdot E\rangle = \langle\lambda\rangle\langle E\rangle$. Since $\langle\lambda\rangle = 0$, the average state of the system will be its nonfluctuating equilibrium state. If such is not the case, even small fluctuations will drive the system away from equilibrium to a steady state given by

$$\langle E\rangle_{ss} = \{k_{ro} - [k_{fo}\delta + k_{ro}(\delta-1)]\langle\lambda E\rangle\}/(k_{fo} + k_{ro}) \tag{14}$$

We note that if $k_{ro} = k_{fo}$ and $\delta = \frac{1}{2}$, equation (11) may be written

$$dE/dt = -(k_{fo} + k_{ro})E + k_{ro} + [k_{ro}(1-\delta)\lambda]$$

which is formally the same as a Langevin equation (ref. 10). This is not unanticipated since indeed, e.g., for diffusion, the probability of motion of a particle in all directions is a priori equal (i.e., $k_{fo} = k_{ro}$) and the influence of random collisions from any direction will result in equal average displacements ($\delta = \frac{1}{2}$).

The ability of random noise to shift a system away from equilibrium except in a special case is at first paradoxical, since fluctuations arise even at equilibrium. We must remember, however, that in modelling external noise, it is assumed that the noise influences the system but not vice versa. This cannot be true of equilibrium noise (refs. 4 and 10) and implies that the fluctuations are caused by an energy dissipating process. Without going into the precise nature of this process, we will adopt this point of view and continue on to describe the effects of energy driven dynamic oscillations and fluctuations on a model enzyme system.

Model enzyme

A protein conformational change may be a part of an enzyme catalytic cycle. For example, we may write

$$(16)$$

This describes the catalytic conversion of A to B, where α_f^t and β_f^t are second order rate coefficients. For the purpose of this paper, we take the output reaction A $\longrightarrow$ B to be independent of the electric field strength, and the equilibrium constant to be unity. The differential equation describing the rate of change of E is

$$dE/dt = -(\alpha_f^t \cdot A + \alpha_r + \beta_f^t B + \beta_r)E + (\alpha_r + \beta_r) \qquad (17)$$

And the principle of detailed balance implies $\alpha_f^t \beta_r = \beta_f^t \alpha_r$. Eq. (16) is a nonlinear differential equation since two a priori variable quantities, $A \cdot E$ (or $B \cdot E$) are multiplied together. Under many circumstances of experimental and physiological relevance, the concentrations A and B may be considered to be constant. In this case, A and B may be subsumed into the pseudo first order rate coefficients α_f and β_f, respectively. In this case, we can write equation (16) in the form

$$dE/dt + PE = Q \qquad (18)$$

where $P = (\alpha_f + \alpha_r + \beta_f + \beta_r)$ (which may be recognized as the inverse relaxation time of the system) and $Q = (\alpha_r + \beta_r)$. This is a first order differential equation, but the coefficients will not be constant in a fluctuating electric field. Even so, the formal solution to this equation is

$$E(t) = E_o \exp(-\int Pdt) + \exp(-\int Pdt)\int Q\exp(\int Pdt) \cdot dt \qquad (19)$$

When the rate constants are modified based on the fluctuating field according to

$$\alpha_f = \alpha_{fo}\phi^\delta; \quad \alpha_r = A\alpha_{ro}^!\phi^{\delta-1}; \quad \beta_f = B\beta_{fo}^!\phi^\gamma; \quad \beta_r = \beta_{ro}\phi^{\gamma-1} \qquad (20)$$

where δ and γ are the apportionment constants for the upper (α) and lower (β) branches of reaction eq. (16), respectively. The instantaneous rates of binding of A and B to the enzyme, J_α and J_β, respectively, may be calculated in terms of $E(t)$

$$J_\alpha(t) = (\alpha_f + \alpha_r)E(t) - \alpha_r \qquad (21)$$

$$J_\beta(t) = (\beta_f + \beta_r)E(t) - \beta_r \qquad (22)$$

and the average rate of conversion of A to B (catalytic flux), $\langle J_{AB}\rangle$, is

$$\langle J_{AB}\rangle = \langle J_\alpha - J_\beta\rangle/2 \qquad (23)$$

Since the free energy of the output reaction, ΔG_{AB}, is uninfluenced by a fluctuating field, the free energy dissipation due to the process A $\longrightarrow$ B may be written as

$$\langle \Phi_{AB}\rangle = \Delta G_{AB} \cdot \langle J_{AB}\rangle \qquad (24)$$

We have shown (refs. 8-11) that dynamic fields may cause net flux of A → B, as furthermore, that Φ_{AB} may be negative, corresponding to energy transduced from the field. Naturally, the total free energy dissipation must be positive. The resolution of this paradox is that the electric fluctuation enforced protein conformational transition results in positive free energy absorption, Φ_{enz}, as can be realized by consideration of the fluctuation-dissipation theorem of non-equilibrium thermodynamics (see e.g. DeGroot and Mazur, ref. 12). This is given by

$$\langle \Phi_{enz}\rangle = \langle (\varepsilon\Delta M/RT) \cdot (dE/dt)\rangle \qquad (25)$$

The quantities in equations (20), (21), and (22) have been evaluated analytically in the case that ε could be described as a square wave or as Markovian dichotomous noise (ref. 9). Figure (1a) shows that clockwise (A → B) flux is induced by an oscillating electric field when **A** = .017 and B = 1. If E is a transmembrane protein, the fluctuating field used in calculating Figure (2a) was 142 mV where the ΔM was 240 Debye. Figure (1b) illustrates the frequency response characteristics of the components of the free energy dissipation function.

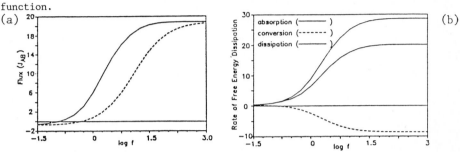

Fig. 1. Results of calculations for behavior of model enzyme in a dynamic electric field. Parameter values used were $\alpha_{fo}^t = \alpha_{ro} = \beta_{fo}^t = \beta_{ro} = 1$, ϕ = 256, $\delta = 1$, $\gamma = 0$, and A = .017 with B = 1. (a) The flux induced by a regularly oscillating (---) or stochastically fluctuating (----) electric field as a function of log frequency. (b) Components of the dissipation function representing contribution from the protein conformational transitions (absorption), output reaction A → B (conversion), and net free energy dissipation.

In Figure (1a) the solid line represents flux induced by a regularly oscillating square wave electric field, while the dashed line indicates results obtained with Markovian dichotomous noise electric perturbation. Figure (1b) demonstrates the breakdown of the free energy dissipation function (see, e.g., ref. 11) into three components. "Absorption" represents the rate at which free energy is absorbed (and released) by the enzyme as it moves through its conformational transitions impelled by the electric field. Some of this is transduced into doing thermodynamically uphill work or the chemical reaction A → B ("conversion") while the rest is degraded into heat ("dissipation"), which represents the net rate of free energy dissipation.

Fig. 2a, b, and c presents schematically the mechanism as to how the asymmetric response of the enzyme kinetic characteristics to a symmetric oscillating perturbation allows for energy to be transduced. In the positive phase of the field more A is bound than B in the time soon after the step perturbation to +ε while right after the perturbation to -ε more B than A is released, as explained in the figure caption and below. The asymmetry we have used in calculations ($\delta = 1$, $\gamma = 0$), which requires all of the electrical sensitivity to reside in the forward process in the α-branch and in the reverse process for the β-branch, may seem somewhat contrived. However, a more realistic four-state mechanism in which A binds much more tightly to the enzyme than B, reduces to precisely this form with the steady state assumption for two of the unstable intermediate states even if the potential dependence is taken to be a priori symmetric (ref. 9).

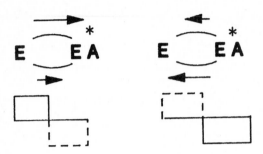

Fig. 2.   When a perturbation is applied to an enzyme system such as shown in Eq. 16, the system must adjust to the new condition, and there will be a redistribution between states E and E*A. This redistribution can take place via two paths, one involving binding and release of A (the α branch) and the other binding and release of B (the β branch). Because of the kinetic asymmetry at any moment in time, the α-flux, J_α, may be different than the β-flux, J_β. (a) This figure illustrates schematically the α and β fluxes a function of time. Notice that the two curves cross during each period such that $\oint J_+ dt = 0$, if a constant perturbation were allowed to persist long enough for equilibrium to be attained. The reversal of the field, however, "catches" the system at a point where net clockwise transport has occurred for both signs (phases) of the perturbation. (b) The flux, J_+, as a function of time, where $J_+ = (J_\alpha - J_\beta)$. (c) Demonstration of the net clockwise flux as a function of time.

DISCUSSION

Fluctuations and oscillations, particularly of the electric field strength, are an inherent part of the environment of a membrane protein. These may play an important role in both signal and energy transduction. Indeed, Serpersu and Tsong (refs. 13 and 14) have shown that the NaK ATPase of the erythrocyte membrane can pump Rb^{+2} up its electrochemical gradient without the hydrolysis of ATP. These results point to a very important role for the inherent conformational flexibility of proteins, for it is this flexibility which gives to an enzyme the ability to absorb free energy from <u>energy</u> <u>driven</u> fluctuations in its environment. If the enzyme has an appropriate asymmetricity with respect to its kinetic electrical response, energy can be transduced from the dynamic field to do work on an output reaction which might have no intrinsic thermodynamic dependence on the field whatever. Although we have discussed this phenomenon in the context of electrical fluctuations, the concepts presented can equally pertain to any dynamic thermodynamic parameter, including temperature, pressure, as well as the chemical potentials of the ligands A and B (ref. 9). Furthermore, the results shown are not particular to the simple model discussed. More complicated models have been computationally studied (refs. 4, 7, and 8) and display qualitatively similar and even richer behavior.

The ideas discussed here have been extended to treat energy transduction between two reactions, where the coupling has been explicitly modeled to originate through coulombic between two enzymes localized spatially close to one another (refs. 4 and 15).

ACKNOWLEDGEMENT

We are grateful for many interesting and fruitful discussions with Drs. P. Boon Chock and Hans. V. Westerhoff. The work of TYT is support in part by NIH, NSF, and ONR grants.

REFERENCES

1 P.M. Dean and E.K. Mathews, Glucose Induced Electrical Activity in Pancreatic Islet Cells, J. Physiol., 210 (1970) 255-264.
2 P.E. Rapp, An Atlas of Cellular Oscillations, J. Exp. Biology, 81 (1979) 281-306.
3 P.E. Rapp, Why Are So Many Biological Systems Periodic, Prog. Neurobiol., 29 (1987) 261-273.
4 R.D. Astumian, P.B. Chock, T.Y. Tsong, Y.-d. Chen, and H.V. Westerhoff, Can Free Energy Be Transduced From Electric Noise?, Proc. Natl. Acad. Sci. U.S.A., 84 (1987b) 434-438.
5 T.Y. Tsong and R.D. Astumian, Electroconformational Coupling and Membrane Protein Function, Prog. Biophys. Molec. Biol. (1987) in press.
6 T.Y. Tsong and R.D. Astumian, Electroconformational Coupling: How Membrane-bound ATPase Transduces Energy From Dynamic Electric Fields, Ann. Rev. Physiol, 50 (1988) in press.
7 T.Y. Tsong and R.D. Astumian, Absorption and Conversion of Electric Field Energy by Membrane-Bound ATPases, Bioelectrochem. Bioenerg., 15 (1986) 457-476.
8 H.V. Westerhoff, T.Y. Tsong, P.B. Chock, Y.-d. Chen, and R.D. Astumian, How Enzymes Can Capture and Transmit Free Energy From An Oscillating Electric Field, Proc. Natl. Acad. Sci. U.S.A., 83 (1986) 4734-4738.
9 R.D. Astumian, P.B. Chock, T.Y. Tsong, and H.V. Westerhoff, Effect of Energy Driven Fluctuations on Enzyme Dynamics, J. Chem. Phys. (1987a) sub-

mitted.

10 H. Risken, The Fokker-Planck Equation, Springer, Berlin, 1984.

11 H.V. Westerhoff and K. van Dam, Thermodynamics and Control of Biological Free Energy Transduction, Elsevier, Amsterdam, 1987.

12 S.R. DeGroot and P. Mazur, Non-Equilibrium Thermodynamics, North Holland, Amsterdam, 1969.

13 E.H. Serpersu and T.Y. Tsong, Stimulation of a Oubain Sensitive Rb^+ Uptake in Human Erythrocytes With an External Electric Field, J. Membrane Biol., 74 (1983) 191-201.

14 E.H. Serpersu and T.Y. Tsong, Activation of Electrogenic Rb^+ Transport of (NaK)-ATPase by an Electric Field, J. Biol. Chem., 259 (1984) 7155-7162.

15 R.D. Astumian, P.B. Chock, H.V. Westerhoff, and T.Y. Tsong, Energy Transduction by Electroconformational Coupling, in: P.B. Chock, C.Y. Huang, C.L. Tsou, and J.H. Wang (eds.), Enzyme Dynamics and Regulation, Springer, Berlin, 1987, pp. 247-260.

Interfacial Phenomena in Biotechnology and Materials Processing,
edited by Y.A. Attia, B.M. Moudgil and S. Chander
Elsevier Science Publishers B.V., Amsterdam, 1988 — Printed in The Netherlands

USE OF A HYDROPHOBIC MOLECULAR SIEVE FOR THE SEPARATION OF ALCOHOL FROM DILUTE
AQUEOUS SOLUTIONS

C. D. CHRISWELL and R. MARKUSZEWSKI

Ames Laboratory, Iowa State University, Ames, Iowa 50011

SUMMARY

　　Ethanol, butanol, and other low-molecular weight organic compounds which
are suitable for fuel use can be efficiently separated from dilute aqueous
solutions, such as fermentation beers, by selective adsorption on a
hydrophobic molecular sieve known as silicalite. This unusual adsorbent is a
commercially available synthetic, crystalline polymorph of silica with the
unique property of being hydrophobic and having pores which are about six
Angstroms in diameter. Thus, it has an affinity for small organic molecules.
Adsorption of ethanol is rapid, selective, and efficient over a wide range of
concentration, flow rate, temperature, and composition of the aqueous-
alcoholic solution. The adsorbed organic components can be recovered from the
molecular sieve by heating at moderate temperatures, application of a vacuum,
or stripping (displacement) with a gas such as carbon dioxide. The adsorption
of ethanol by a molecular sieve is the basis for potential alternatives to
distillation for the isolation of alcohol fuels from fermentation beers. It
has been found that ethanol with a purity of 98% is retained within the pores
of silicalite. During recovery, this high-purity ethanol is contaminated with
fermentation beer present within the interstices of the molecular sieve.
Continuing research is aimed at prevention of this contamination so that
high-purity ethanol can be produced in a single absorption-desorption cycle.

INTRODUCTION

Background

　　Ethanol is the only synthetic fuel produced in significant quantities in

the United States. As depicted in Figure 1, ethanol is produced typically by

the fermentation of corn-derived sugars. The overall process is

straight-forward, consisting of the following steps: preparation of the corn

by well-established wet milling procedures, hydrolysis of corn starches to

fermentable sugars, yeast fermentation of sugars to produce a beer typically

containing 12-14% ethanol, distillation of the beer to yield a 95% ethanol

solution, and finally upgrading of the distillate to absolute ethanol for use

in blending with gasoline. In addition to ethanol, the process yields corn

oil, carbon dioxide and distillers dry grain (DDG) as salable by-products.

Sales of DDG, for example, account for up to 30% of the sales revenues of

58

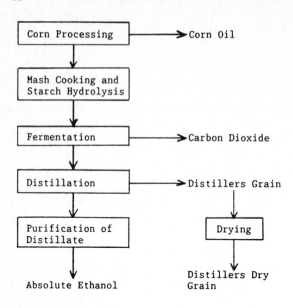

Fig. 1. Typical Ethanol Production Flow Diagram.

ethanol plants, but equipment for drying distillers grain account for about
20% of the capital costs, and drying distillers grain is the largest single
source of energy consumption in typical plants.

Separation of Ethanol

Distillation is currently the only widely used method of separation of
ethanol from fermentation beers. The technology is well-developed, effective,
and energy-efficient when applied to the beers containing 12-14% ethanol
resulting from yeast fermentations of sugars. However, ethanol can also be
produced by the bacterial fermentation of low-cost cellulosic crop residues,
forestry wastes and municipal refuse. But bacterial fermentations yield
solutions with only 1-2% ethanol. Distillation cannot be economically applied
to separation of ethanol from these dilute solutions. The equipment used must
be of a massive scale to accommodate the very dilute solutions, and thus it is
more expensive. The energy required for distillation of 1% ethanol solutions
is more than an order of magnitude greater than that required for distillation

of 12-14% solutions. In addition, for bacterial fermentations, the use of continuous fermenters in which ethanol is removed at a rate it is formed is highly desirable, but distillation is not readily coupled with continuously operating fermenters.

Alternative Ethanol Production Processes

One alternative to current processes used for the production of ethanol would consist of the continuous fermentation of cellulose by bacteria, removal of ethanol from the fermentation broth at the rate it is formed so that levels toxic to bacteria do not exist, and use of an energy-efficient adsorption process for the separation of ethanol continuously. A flow diagram of such an alternative process is depicted in Figure 2. In this proposed alternative process, it is anticipated that two adsorption units would be operating continuously, one removing ethanol from fermentation broths and the other being regenerated. It is also anticipated that desorption of ethanol would be accomplished by purging with a heated gas such as air or carbon dioxide from the fermenter.

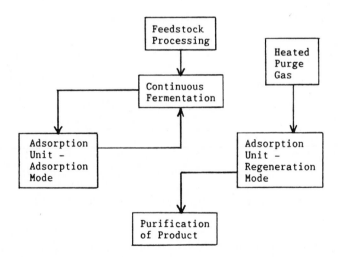

Fig. 2. Generic Ethanol Adsorption Process Flow Diagram.

In addition to application to dilute solutions arising from bacterial fermentations, adsorption processes offer one other potential advantage. Small-scale adsorbers typically offer the same operating efficiencies as do large-scale ones. In small-scale ethanol plants, distillers grain could be fed on site without drying, leading to reduced capital and operating costs even for plants using corn as a feedstock and producing beers with 12-14% ethanol.

Adsorbents for Ethanol Accumulation

Adsorption processes for separation of ethanol from aqueous solutions are not presently used due to the efficiency of established distillation processes. In the future, the use of lower cost cellulose feedstocks may well dictate the replacement of distillation with efficient adsorption processes. All studies to date indicate that adsorption processes based upon adsorption of ethanol on silicalite molecular sieve could form the basis for an efficient process.

Adsorption processes actually predate distillation as an ethanol purification process. It is reported that ancient Macedonians produced potent alcoholic beverages by passing wine through sponges impregnated with olive oil. The alcohol was retained on the sponge, made hydrophobic by the olive oil, and the water passed through. The retained ethanol was then recovered by simply squeezing the sponge.

The key to use of an adsorption process for separation of ethanol from fermentation beers is, of course, the availability of an effective adsorbent. In past work at the Ames Laboratory, several potential adsorbents were evaluated (see Table 1). In these preliminary evaluations, one of the important criteria was the capacity of the adsorbent for ethanol accumulation. The ability to retain large amounts of ethanol allows for the use of relatively small adsorption beds, which are obviously less expensive to construct than are large beds. In addition, small beds should produce less back-pressure than large ones and thus have lower operating costs for pumping.

Another important criterion evaluated in preliminary investigations was the selectivity of the adsorbents for ethanol over other constituents of fermentation beers such as water, inorganic salts, sugars, starches, cellulose, yeast cells, and bacteria. A perfect adsorbent would accumulate only compounds having value as liquid fuels. The cost of the adsorbent was another factor considered in preliminary investigations.

TABLE 1

Potential Adsorbents for Use in Ethanol Separation

Adsorbent	Capacity	Selectivity	Cost
Activated Carbon	Low	Low	Low
Carbonaceous Resin (Amberlite XE-340)	Fair	Fair	Moderate
Macroreticular Resin (Amberlite XAD-4)	Fair	Fair	Moderate
Hydrophobic Molecular Sieve (Silicalite)	High	Good	High

Of the adsorbents evaluated, the molecular sieve silicalite (ref. 1) has by far the most desirable properties for ethanol accumulation.

Properties of Silicalite Molecular Sieve

Some of the most important characteristics of silicalite molecular sieve are given in Table 2 and its applications are listed in Table 3. About 33% of the total crystal structure of silicalite is comprised of pores (ref. 1). This large pore volume gives rise to a high adsorption capacity. The pores in silicalite, depicted in Figure 3, are 5.2 and 6.0 Angstroms in diameter (ref. 1). Ethanol has a solution diameter of about 4 Angstroms and thus can enter the pore structure and be retained. However, larger molecules such as cellulose, sugars, and starches are too large to enter the pore structure and cannot be retained by silicalite (ref. 2). Silicalite is a polymorph of silicon dioxide containing essentially no aluminum or exchangeable cation sites. The lack of charge fields from polar or ionic sites leads to the hydrophobic nature of silicalite (ref. 3). As a silicon dioxide polymorph, silicalite exhibits stabilities similar to that of quartz. It is stable at temperatures up to 1100°C and is unaffected by common corrosive materials (ref. 4).

TABLE 2

Properties of Silicalite Molecular Sieve

Pore Volume - 33% of crystal volume, i.e., 0.19 ml/gram

Pore Diameters - 5.2 Å for circular and 6.0 Å for elliptical pores

No Aluminum or Exchangeable Cations - Giving Rise to Hydrophobicity

Temperature Stability - Stable to 1100°C, reverts to α-quartz at 1300°C

Chemical Stability - Dissolved by hydrofluoric acid, but unaffected by other corrosives studied

---

TABLE 3

Previous Applications of Silicalite Molecular Sieve

Application	References
Removal of chloroform from drinking water	(5)
Determination of low-molecular-weight contaminants in drinking water	(6)
Recovery of sulfur dioxide from stack gases	(7,8)
Determination of sulfur dioxide in stack gases	(9)
Gas chromatographic column-packing material	(10-12)
Separation of n-alkanes from petroleum streams	(13)
Separation of alcohols from aqueous solutions	(3,14-17)

Previous Applications of Silicalite

    Because of its unique structure and hydrophobic nature, silicalite has been previously used for diverse applications (see Table 3). Potential

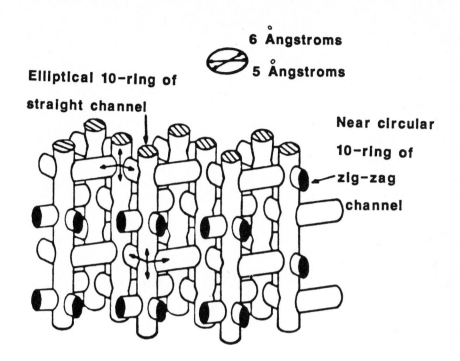

6 Ångstroms
5 Ångstroms

Elliptical 10-ring of
straight channel

Near circular
10-ring of
zig-zag
channel

Fig. 3. Pour Structure of Silicalite.

applications have been reported for the removal of chloroform from drinking

water (ref. 5) and for the determination of low-molecular-weight organic

contaminants in drinking water (ref. 6). Both rely upon the hydrophobic

nature of silicalite, and both benefit from the selectivity of silicalite for

accumulation of small molecules in the presence of the more prevalent high

molecular weight materials such as humic acids in water. Procedures for the

recovery (ref. 7, 8) and for the determination (ref. 9) of sulfur dioxide from

combustion gases rely upon the hydrophobicity of silicalite because of the

presence of about 10% water vapor in typical stack gases. In addition, these

procedures have an advantage due to the excellent stability of silicalite in

corrosive atmospheres. The use of silicalite as a gas chromatographic column-

packing material (ref. 10-12) has been shown to offer separations somewhat

better than those obtainable with the widely used molecular sieve 5A; but in contrast with molecular sieve 5A, silicalite is not deactivated by water or

carbon dioxide in samples. As with conventional molecular sieves, silicalite

can be used for the separation of n-alkanes from petroleum streams (ref. 13).

It is likely that the lack of polar functionalities in silicalite would reduce

the amount of cracking of hydrocarbons during separation. The application of

silicalite for the adsorptive separation of alcohol is the topic of this

report (ref. 3, 14-17).

EXPERIMENTAL

Adsorbent Preparation

Silicalite was obtained from Union Carbide Corporation (Tarrytown, New

York) in the form of fine powders, 20 x 60 mesh granules, and 1/8 in. and 1/16

in. pellets. Before use as adsorbent for ethanol accumulation, the virgin

silicalite was calcined at 800°C to remove any manufacturing impurities. The

calcined silicalite was then slurried with water and decanted to remove any

fines present. The slurried silicalite was packed into adsorption columns of

various sizes and geometries.

Adsorption of Ethanol

A flow diagram depicting adsorption studies performed is presented in

Figure 4. Ethanol solutions were prepared in concentrations ranging from 0.5

to 20% ethanol. In some studies, sodium chloride at concentrations up to 10%

was added to the ethanol-water solutions. In addition, actual fermentation

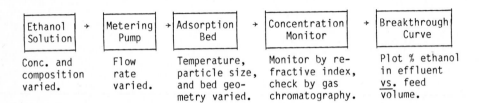

Figure 4. Ethanol Adsorption Studies

beers containing 12% ethanol and large volumes of dispersed solids were obtained from Archer Daniels Midland (ADM) in Decatur, Illinois.

The aqueous-alcoholic solutions were pumped through adsorption beds using a metering pump. Flow rates were varied to determine effects. Adsorption beds were maintained at temperatures ranging from 20 up to 80°C to determine the effects of temperature on ethanol adsorption.

The ethanol concentrations in column effluents were monitored by measuring the refractive index and were checked periodically by gas chromatography. Plots of the concentration of ethanol in effluents <u>vs</u>. the volume of ethanol pumped through the bed were prepared. One of these plots, which are referred to as breakthrough curves, is shown in Figure 5. The breakthrough curves obtained under various conditions were analyzed to

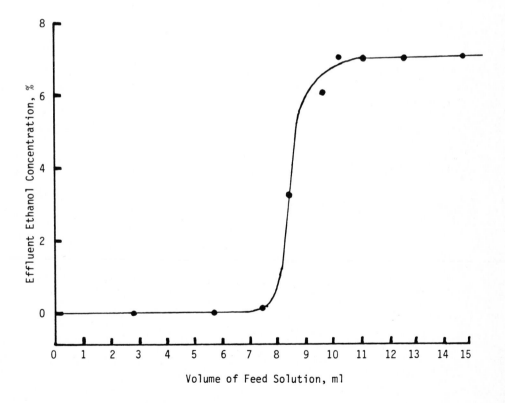

Fig. 5. Ethanol Breakthrough Curve.

determine the capacity of silicalite for ethanol accumulation, the minimum bed depth required for ethanol accumulation, and the rate of ethanol accumulation.

Desorption of Ethanol

The recovery of ethanol from adsorption beds was tested by application of a vacuum, by purging with a gas, by heating the beds with microwave energy, by heating the beds with a hot purge gas flowing through the beds, and by application of microwaves to the beds while purging with a gas.

RESULTS AND DISCUSSION

Adsorption Capacity

Of all parameters studied, only the concentration of ethanol in the influent had significant effects on the amount of ethanol that a given weight of silicalite could accumulate. As shown in Figure 6, at ethanol concentrations greater than about 15%, silicalite would accumulate 0.19 ml of ethanol per gram of adsorbent. This corresponds to the pore volume of silicalite, and thus all pores available were essentially saturated with ethanol. At lower ethanol concentrations, silicalite had a lower capacity. However, even at concentrations as low as 1% of ethanol in the influents, appreciable amounts of ethanol were still adsorbed by silicalite. These data indicate a significant utility of silicalite for the separation of ethanol from dilute solutions, such as are produced by bacterial fermentation of cellulose or are present in wastes destined for disposal.

Flow rates, temperatures, bed geometries, and particle sizes had no significant effects on the capacity of silicalite for ethanol accumulation (see Table 4). Addition of up to 10% sodium chloride to the ethanol-water solutions did increase the capacity for ethanol accumulation slightly, but the amount is not expected to result in a significant effect. The absence of deleterious effects with the addition of salts indicates the feasibility of coupling silicalite adsorption with continuously operating fermenters in which salt concentrations would eventually build up. Of course, salt build-ups could also affect the viability of bacteria or yeasts responsible for fermentation, but the evaluation of these effects needs to be carried out in future work.

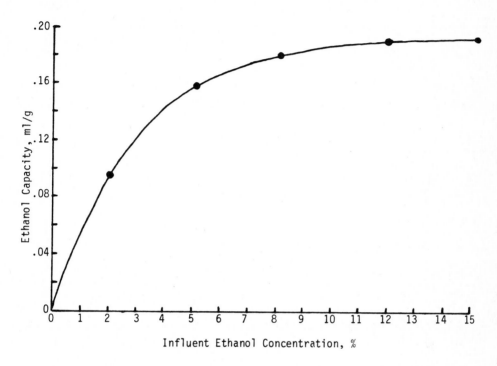

Fig. 6. Adsorption capacity (ml ethanol per g silicalite) vs. influent ethanol concentration.

TABLE 4

Effect of Experimental Parameters on Ethanol Capacity

Variable	Effects
Flow Rate	No effects
Temperature	No effect between 25°C and 80°C
Salt Content	Insignificant increase in absorption capacity with 10% NaCl
Real Beer Constituents	No effects

Adsorption of ethanol from real fementation beers provided by ADM resulted in breakthrough curves which were indistinguishable from those obtained from synthetic ethanol-water mixtures with the same ethanol content. Although it was anticipated that actual fermentation beers could cause problems due to their high solids contents, the yeast bodies and other solids in these ADM samples passed through the adsorption beds with no apparent retention.

Other Adsorption Parameters

The minimum bed depth required for efficient adsorption is an important parameter because large minimum bed depths would result in the formation of high back-pressures and increased pumping costs. As shown in Table 5, extremely low minimum bed depth requirements were obtained even when using the largest particle sizes of silicalite available. Because of the low minimum bed depth requirements, it is likely that far deeper beds could be used in ethanol adsorption processes simply to increase the total bed capacity. In studies performed to date, back pressures created by silicalite beds have been too low to measure with any degree of accuracy. This does not mean that there is no back pressure but only that in small-scale studies the connecting tubes contribute almost all of observable back-pressure. The rate of adsorption dictates how fast a beer solution can be pumped through a bed of silicalite. In all studies performed, the rate of adsorption as calculated from the slope of the breakthrough curve at its inflection point was too large to measure accurately. This effect is real and not an artifact of the scale of studies. It is typical of all molecular sieve adsorbents to have extremely high adsorption rates, and therefore this finding is not surprising.

TABLE 5

Adsorption Parameters

Minimum Bed Depth - 0.5 cm for 20 x 60 mesh particles

                   - 4.0 cm for 1/8 in. pellets

Back Pressure   - Too low to measure accurately on a small scale

Adsorption Rate   - Too high to measure accurately

Adsorption of Other Potential Fermentation Products

Ethanol is not the only potential fuel that can be produced by fermentation in dilute aqueous solutions. Distribution coefficients were obtained for various compounds typical of those that might be present as impurities in ethanol fermentation beers or might be produced by other fermentation processes. The distribution coefficient, D_g water, is simply the ratio of the concentration of an analyte adsorbed on silicalite to its concentration in the bulk of solution, when equilibrium is attained. It serves as a numerical indicator of the relative affinity of silicalite for various analytes. The higher the value for D_g water, the greater the affinity that silicalite has for a given analyte. As shown in Table 6, most analytes studied had higher distribution coefficients than did ethanol and thus would be retained. The D_g water for butanol, a potential fuel which is produced along with acetone in the Weizmann synthesis was not measured in this study, but Milestone and Bibby (ref. 14) previously reported excellent retentions of butanol on silicalite.

Summary of Adsorption Studies Findings

Significant findings about the adsorption of ethanol are summarized below:

- Ethanol adsorption is efficient over widely varying experimental conditions.

- Ethanol adsorption is likely applicable to bacterial fermentation broths containing only 1 to 2% ethanol.

- Ethanol adsorption could likely be performed in small-scale facilities with integrated feedlots using wet distillers grain.

- Other potential fuels produced by fermentation can be adsorbed by silicalite.

These findings as a whole indicate that silicalite is an excellent adsorbent for ethanol with potential applicability to commercial separations of ethanol from fermentation processes.

TABLE 6

Distribution Coefficients (D_g water) at 25°C for Selected Compounds

Compound	D_g water
Ethanol	65
Propanol	250
Butanol	high
Acetaldehyde	100
Crotonaldehyde	1300
Furfural	1100
Propanal	1400
Ethyl acetate	5000
Acetone	270
Methyl isobutyl ketone	2800
Acetic acid	72
Aceonitrile	150

Recovery of Ethanol from Silicalite

Various techniques have been tested for the recovery of adsorbed ethanol from silicalite beds. Basic findings are summarized in Table 7. The two most desirable techniques are heating the bed while passing a purge gas through the bed or heating the bed with microwaves while purging with a gas (ref. 18). As shown in Figure 7, during the initial stages of desorption, relatively high concentrations of water are eluted. This water was retained in the interstices of the bed. Even during the later stages of ethanol desorption, however, significant amounts of water are eluted. It has proven difficult to

TABLE 7

Desorption of Ethanol from Silicalite

Desorption Process	Comments
Simple heating	Ethanol desorbs slowly
Gas purging	Ethanol desorbed incompletely
Heating and purging	Ethanol desorbed rapidly and completely
Microwave heating	Ethanol desorbs slowly
Microwave heating and purging	Ethanol desorbed rapidly and completely
Vacuum stripping	Vaporization of ethanol cools the bed, causes interstitial water to freeze, and fractures silicalite particles.

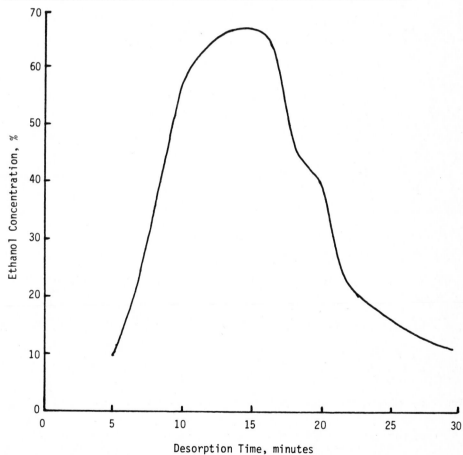

Fig. 7. Ethanol concentration in effluent during desorption with microwaves.

produce ethanol with purities of greater than about 75% by using this adsorption process. Although 75% ethanol solutions can be readily upgraded to absolute ethanol, it is unsatisfying to produce these lower concentrations because it has been shown that the ethanol retained in the pores has a purity of greater than 98% (ref. 16). In effect, very high purity ethanol is retained within the pores of silicalite and very low purity ethanol (mostly water) is retained in the interstices between silicalite particles; at the present time, further research is needed to recover the pure ethanol from the pores without contamination by the water in the interstices.

The major unresolved issue concerning ethanol purification by adsorption on silicalite molecular sieve is the above mentioned contamination of adsorbed alcohol by interstitial beer during desorption. If efficient means were developed for removing the interstitial beer before desorption of the pure ethanol from the pores of silicalite, the need for final purification of the product would be largely eliminated or simplified, and the overall cost of production of ethanol fuels would decrease significantly.

CONCLUSIONS

Silicalite is a hydrophobic molecular sieve with many applications in separation processes. Because it has small enough pores, it is also useful for the selective adsorption of alcohol from dilute aqueous solutions such as fermentation beers. The accumulation of ethanol on silicalite is a function of the influent ethanol concentration; the flow rate, temperature, bed geometry, or particle size of silicalite have no significant effect. The capacity of silicalite for ethanol corresponds to the pore volume of this molecular sieve.

The behavior of silicalite with actual fermentation beers is the same as that with synthetic aqueous-alcoholic solutions, proving its applicability to the separation of alcohol produced by fermentation processes. Special utility is indicated in the separation of alcohols from solutions of low concentrations, such as those obtained from bacterial fermentation, resulting during the production of butanol, or occurring in wastes in need of disposal.

There are several methods for desorbing ethanol from the silicalite after accumulation. However, the concentration of the desorbed ethanol is lowered (to about 75%) by the interstitial beer which is simultaneously removed during the desorption step. Further work is necessary to improve the desorption step to recover the pure ethanol accumulated on the silicalite without contamination from the interstitial beer.

ACKNOWLEDGEMENTS

Ames Laboratory is operated for the U. S. Department of Energy by Iowa State University under Contract No. W-7405-ENG-82.

This work was supported in part by the Iowa Corn Growers Association and in part by the Assistant Secretary for Conservation and Renewable Energy through the Solar Energy Research Institute.

REFERENCES

1 E. M. Flanigen, J. M. Bennett, R. W. Grose, J. P. Cohen, R. L. Patton, R. M. Kirchner, and J. V. Smith, Nature, 1978, 271, 512.
2 C. S. Oulman and C. D. Chriswell, U. S. Patent No. 4,277,635, July 7, 1981.
3 E. M. Flanigen, Pure Appl. Chem. 1980, 52(9), 2191.
4 G. M. W. Shultz-Sibbel, D. T. Gjerde, C. D. Chriswell and J. S. Fritz, Talanta, 1982, 29, 447.
5 C. D. Chriswell, C. S. Oulman, T. Moore and P. Miller, unpublished work, Iowa State University, 1979, presented to AWWA Research Foundation, 1979.
6 C. D. Chriswell, D. T. Gjerde, G. M. W. Shultz-Sibbel, J. S. Fritz, and I. Ogawa, "An Evaluation of the Adsorption Properties of Silicalite," 1983, NTIS, Springfield, VI (PB 83-148 502).
7 C. D. Chriswell and S. V. Gollakota, Am. Chem. Soc. Div. Fuel Chem. Proc., 1987, 32(1), 505.
8 S. V. Gollakota and C. D. Chriswell, Ind. Eng. Chem. Res., 1988, in press.
9 C. D. Chriswell and D. T. Gjerde, Anal. Chem., 1987, 54, 1911.
10 C. D. Chriswell, Y.-Y. Chen, C. W. McGowan and R. Markuszewski, Chromatogr., 1987, 405, 213.
11 P. J. Thomas, H. Izod and J. A. Duisman, U. S. Patent No. 4,375,568, February 4, 1981.
12 D. J. Campbell, B. M. Lowe, A. G. Rowling and R. Williams, Analytica Chimica Acta, 1985, 172, 347.
13 R. W. Neuzil and S. Kulprathipanja, U. S. Patent No. 4,455,444, June 19, 1984.
14 N. B. Milestone and D. M. Bibby, J. Chem. Tech. Biotech., 1981, 31, 1.

74

15 W. W. Pitt and D. D. Lee, Proc. 15th Intersociety Energy Conversion Engineering Conf., Aug. 18-22, 1980, Am. Inst. of Aero. and Astro., Seattle, WA.

16 S. M. Klein, "Adsorption of Ethanol and Water Vapor by Silicalite, a Hydrophobic Molecular Sieve," 1982, M. S. Thesis, Iowa State University, Ames, Iowa.

17 S. M. Klein and W. H. Abraham, AIChE Symposium Series, No. 230, Vol. 79, 1983, 53.

18 H. R. Burkholder, G. E. Fanslow and D. D. Bluhm, Ind. Eng. Chem. Fund., 1986, 25, 414.

Interfacial Phenomena in Biotechnology and Materials Processing,
edited by Y.A. Attia, B.M. Moudgil and S. Chander
Elsevier Science Publishers B.V., Amsterdam, 1988 — Printed in The Netherlands

A MICROBIAL BIOSURFACTANT: GENETIC ENGINEERING AND APPLICATIONS

W. R. FINNERTY and M. E. SINGER

Department of Microbiology, University of Georgia, Athens, Georgia 30602 U.S.A.

SUMMARY

A soil microorganism, identified as <u>Rhodococcus</u>, produces an extra-cellular product with exceptional surface-active properties. This product is a trehalose-containing glycolipopeptide (GLP) consisting of glycerol, trehalose, glucose, gluconic acid and peptide plus saturated and mycolic fatty acids. The GLP has been purified and characterized with respect to its surface properties. Interfacial tension values of 0.00006 mN/m at a CMC of 1 mg/ml are observed with either 0.5% butanol or 0.5% pentanol as coagents. A shuttle vector has been constructed which can be used in the transformation of <u>Rhodococcus</u> and <u>Escherichia coli</u>. Cloning of chromosomal genes into this shuttle vector, transformation of biosurfactant-negative mutants and screening for biosurfactant-positive transformants allows for the isolation of genes encoding biosurfactant synthesis. This biosurfactant forms stable oil-in-water emulsions of heavy crude oils, significantly improving the rheological properties of heavy oils. This product has application to the transportation and pipelining of heavy oils and improved enhanced oil recovery technologies.

INTRODUCTION

Biosurfactants are surface-active compounds derived from biological sources which, like synthetic surfactants, exhibit characteristic physical and chemical properties. The production of surface-active agents by microorganisms has been of general long-standing recognition with a systematic characterization of such products slow to emerge. Studies from various laboratories document the production of surface-active compounds by diverse microbial species. Sources, types and applications of biosurfactants have been reviewed in broad generalities (refs. 1, 2) and in relation to bacterial alkaline metabolism (ref. 3). This paper reports on the chemical and physical characterization of an extracellular biosurfactant produced by a new isolate of <u>Rhodococcus</u>, strain improvement through genetic means and application of this biosurfactant to viscosity reduction of heavy crude oil and enhanced oil recovery.

METHODS

Culture Conditions

Rhodococcus sp H-13A was grown in either 0.8% nutrient broth - 0.5% yeast extract (NBYE) medium or in a mineral salts medium containing (in g/liter): K_2HPO_4, 10 gms; NaH_2PO_4, 5 gm; $(NH_4)_2SO_4$, 2 gm; $MgSO_4 \cdot 7H_2O$, 0.2 gms; $CaCl_2 \cdot 2H_2O$, 0.001 gm; $FeSO_4 \cdot 7H_2O$, 0.001 gm, pH 7.0 supplemented with 2.5% n-hexadecane. All cultures were grown at 28°C on a rotary shaker operating at 300 rev/min.

Analytical Methods

The culture medium of bacteria grown on hexadecane or NBYE was centrifuged and filtered to remove bacterial cells prior to measurements of interfacial tension. Interfacial tensions (IFT) were determined by either the drop-weight method or a spinning-drop tensiometer. Viscosity of water-saturated oil following biosurfactant treatment was determined in a Brookfield Rheolog/thermocel viscometer. Fatty acid methyl esters derived from purified biosurfactant were analyzed by gas chromatography using a Tracor 560 gas chromatograph equipped with flame ionization detectors and 8 in. x 4 mm glass columns containing 5% DEGS-PS or SP-2100.

Extraction and Chromatography

Spent culture media was extracted with ethyl acetate/methanol (2:1, v/v). The organic-soluble material was designated total crude lipid which was fractionated by silicic acid column chromatography. Neutral lipids were eluted with chloroform, glycolipids with acetone and phospholipids with chloroform/methanol (2:1 v/v). Glycolipids were quantified by the anthrone method (ref. 4), phospholipids by the phosphate method (ref. 5) and neutral lipids were visualized by spraying with 50% H_2SO_4 and charring at 120°C for 20 min.

Polar lipids were separated by thin-layer chromatography (TLC) on 0.4 mm layers of silica gel G in a solvent system of chloroform-methanol-5 N NH_4OH (65:30:5 v/v/v). Glycolipid was detected by the orcinol spray reagent and phospholipids were detected by the phosphate spray reagent (ref. 6). Neutral lipids were separated on silica gel G in a solvent system of petroleum ether-diethyl ether-glacial acetic acid (80:20:1 v/v/v) and visualized as indicated.

DNA Chemistry

Plasmid DNA was isolated by the alkaline lysis method (ref. 7) and was purified by centrifugation in $CsCl_2$-ethidium bromide density gradients (ref. 8). The boiling method of Holmes and Quigley (ref. 9) was used for rapid, small-

scale isolation of plasmid DNA. Plasmid transformation of Esherichia coli DH1
was by the method of Hanahan (ref. 10). Restriction endonuclease digestions
were performed as per the manufacturer's directions. DNA fragments were
separated by horizontal gel electrophoresis using gels prepared with 0.7%
(wt/vol) agarose using 0.04 M Tris-acetate, 0.002 M EDTA electrophoresis buffer,
pH 8.0 at 100 volts. Gels were strained with ethidium bromide (0.5 g/ml) and
DNA was visualized with ultraviolet light. DNA size was determined by comparison with Hind III - digested linear phage lambda DNA fragments and with a
1 kilobase linear ladder DNA standard. Procedures for protoplast formation,
regeneration and transformation of Rhodococcus sp H-13A have been developed
(ref. 11).

RESULTS

Chemical Properties of Biosurfactant

The extracellular biosurfactant produced by Rhodococcus sp H-13A
fractionated in the acetone cut from the silicic acid column, indicating a
glycolipid-type molecule. The glycolipid consisted of 1 major component
($\sim$ 90%) and 5-6 minor components. The major glycolipid was purified to 99%
homogeneity by preparative TLC for chemical and physical analyses. The
proposed structure of the major glycolipid as represented in Figure 1
contains glycerol, trehalose, glucose, gluconic acid and peptide with
O-acylated ester-linked fatty acids. The fatty acid composition consists
of C-35 to C-40 mycolic acids plus n-saturated fatty acids varying in chain
length from C8 through C18. The position of fatty acyl substitution on the
carbohydrate residues has not been determined. The peptide appears linked
to gluconic acid by a peptide linkage. Glycine and histidine are the major
amino acid residues followed by alanine, leucine and phenylalanine. The
glycolipid is slightly acidic in water and exhibits 3 ionization constants
at pH 5.5, pH 8.0 and pH 9.5.

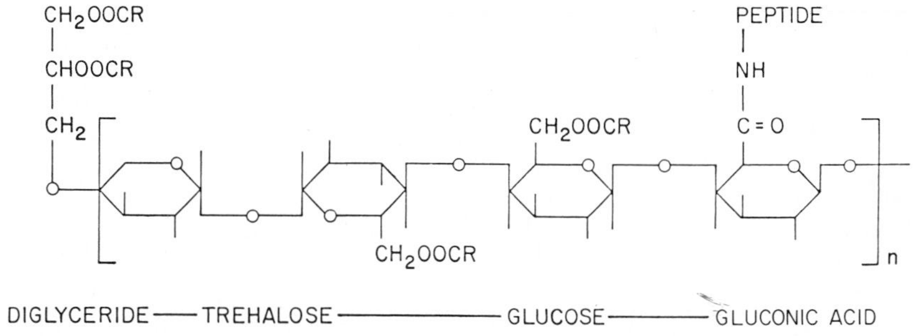

Fig. 1. Proposed Structure of H-13A Biosurfactant

Physical Properties of Biosurfactant

The critical micelle concentration (CMC) of crude biosurfactant was 1.5
mg/ml; the CMC for the purified biosurfactant was 1.0 mg/ml. The minimum
interfacial tension, as measured against hexadecane at 25°C, was 0.25 mN/m
for the crude biosurfactant and 1.0 mN/m for the purified biosurfactant,
indicating that both major and minor components promote maximum surface
activity.

The IFT of crude biosurfactant was measured as a function of alkane
chain length to determine the equivalent alkane carbon number (EACN). The
minimum IFT exhibited by crude biosurfactant was 0.02 mN/m against decane
(Table 1). Addition of 0.5% (v/v) pentanol as a co-surfactant yielded a
minimum IFT of 0.00006 mN/m against undecane. Addition of a co-surfactant
resulted in a shift from decane to undecane as the EACN as well as in a
significant IFT reduction with isopropanol and butanol equally effective
as co-surfactants. A 1:1 mixture of hexadecane and hexane (EACN:11) yielded
a minimum IFT of 0.00005 mN/m.

TABLE 1

Biosurfactant[a] interfacial tension values versus alkane carbon number

Alkane	mN/m[b]	
	minus pentanol	plus 0.5% pentanol
Hexane (C6)	0.570	0.16
Octane (C8)	0.30	0.05
Nonane (C9)	0.150	0.02
Decane (C10)	0.020	0.001
Undecane (C11)	0.030	0.00006
Dodecane (C12)	0.090	0.00014
Tridecane (C13)	0.070	0.00028
Tetradecane (C14)	0.160	0.06
Hexadecane (C16)	0.250	0.07
Octadecane (C18)	0.300	0.09
Hexadecane (C16)+Hexane (C6) (1:1 v/v)	0.029	0.00005

a) Biosurfactant concentration: 1.8 mg/ml; 1.7% salt.
b) Measurements conducted at equilibrium with a spinning drop tensiometer at
40°C.

Effects of Physical and Chemical Agents on Minimum Interfacial Tension

The effects of various physical and/or chemical agents on the minimum
interfacial tension of the biosurfactant demonstrated less than 10%
increases in IFT values (Table 2), indicating that the biosurfactant was not
significantly influenced in its surface-active properties by the presence of
salt, divalent cations or extreme pH values.

TABLE 2

Effect of physical/chemical parameters on minimum interfacial tension of
biosurfactant[a]

Condition	mN/m
Biosurfactant + 0.5% pentanol	0.00006
plus 7.0 % NaCl	0.00070
plus 1 mM $MgSO_4$	0.00095
plus 1 mM $CaCl_2$	0.00050
plus 1 mM $FeSO_4$	0.00060
plus 3 mM cation solution[b]	0.00025
pH 4.0	0.00016
pH 13.0	0.00004

a) Minimum IFT determined with a spinning drop tensiometer against undecane
   at 40°C.

b) Solution contained 1 mM $FeSO_4$, 1 mM $CaCl_2$, 1 mM $MgSO_4$.

Genetic Engineering in Rhodococcus

Strain improvement through molecular genetics requires a suitable
vector for cloning and expressing the genes encoding biosurfactant
synthesis. The general strategy employed for construction of this cloning
vector is shown in Figure 2. An Escherichia coli replicon, pIJ30, was used
as the cloning vector for shuttle plasmid construction. pIJ30 is a pBR322
derivative containing a 1.9 Kb BamHI insert encoding a thiostrepton-
resistance gene (tsr) derived from Streptomyces azureus. pIJ30 has a E. coli
origin of replication (ori) plus the pBR322 ampicillin-resistance gene (Ap^R).
The single Hind III site in pIJ30 was selected for cloning DNA fragments of
the Rhodococcus sp H-13A indigenous plasmid, pMVS300. pMVS300 was digested
with Hind III yielding restriction fragments of 3.7 Kb and 11.0 Kb. These DNA
fragments were ligated into the Hind III-digested, alkaline phosphatase-
treated vector, pIJ30 (Figure 2).

E. coli DH1 was transformed with the ligation mixture and Ap^R-resistance
transformants were selected. Ap^R transformants containing recombinant
plasmids were detected by colony hybridization with ^{32}P-labeled pMVS300
plasmid DNA as probe.

Two recombinant 10.0 Kb plasmids were detected, both containing a 3.7 Kb
HindIII fragment of pMVS300 cloned in 2 orientations, as mapped relative to
an internal BamHI site. The recombinant plasmids were designated pMVS301
and pMVS302.

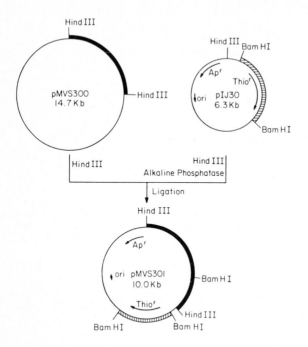

Fig. 2. General Cloning Strategy for Construction of pMVS301

Plasmid Transformation in Rhodococcus sp H-13A

A protoplast transformation system, designed for the uptake of plasmid DNA by Rhodococcus was developed using the E. coli - Rhodococcus shuttle plasmid, pMVS301. The method involved: i) generation of protoplasts, ii) polyethylene glycol-assisted transformation of protoplasts with plasmid DNA; and iii) regeneration of protoplasts and selection for transformants.

Thiostrepton^R (Thio^R)/Ap^R transformants of Rhodococcus sp AS-50 (a strain of H-13A lacking pMVS300) were detected using pMVS301 as donor DNA. pMVS301 was shown to replicate in Rhodococcus as an independent plasmid replicon. Thio^R/Ap^R transformants contained a 10.0 Kb plasmid identical in electrophoretic mobility to pMVS301. This 10.0 Kb plasmid was isolated from Rhodococcus 2B1 and transferred to E. coli DH1, selecting for Ap^R transformants. Restriction analysis of the 10.0 Kb plasmid isolated from E. coli and Rhodococcus transformants showed pMVS301 to be identical in each case, without rearrangements or deletions in either organism. Thio^R/Ap^R

transformants of Rhodococcus AS-50 were not detected in control experiments using pIJ30 as donor DNA. Also, spontaneous $Thio^R/Ap^R$ mutants were not detected in control experiments. Accordingly, the 3.7 Kb Hind III fragment of pMVS300 contains an Rhodococcus origin of replication allowing for independent replication of the hybrid plasmid, pMVS301, in Rhodococcus.

The bifunctional plasmid, pMVS301, is capable of interspecies transfer and replication in E. coli and Rhodococcus without rearrangement of the plasmid DNA. The shuttle plasmid contains 2 antibiotic resistance markers, ampicillin-resistance which is expressed in E. coli and Rhodococcus, and thiostrepton-resistance, which is expressed in Rhodococcus.

Transformation of Rhodococcus-derived pMVS301 into protoplasts of Rhodococcus strain AS-50 yielded 10,000-fold higher transformation frequencies (1 X 10^5 transformants/µg DNA) than when E. coli - derived pMVS301 was used as donor DNA (2.5 X 10^1 transformants/µg DNA). The presence of a restriction/modification system in Rhodococcus sp. H13-A is indicated.

The number of transformants obtained were linear up to 75 ng DNA/ml, using 2 X 10^6 protoplasts per ml. pMVS301 is 100% stable in E. coli without ampicillin selection. In Rhodococcus, the plasmid shows a 7% loss over 25 generations in the absence of antibiotic selection, but is stabilized in the presence of low levels of ampicillin.

Restriction Map of pMVS301

There are 12 unique restriction sites in the vector (Table 3). The Pst I and Sca I sites appear to be useful for cloning in Rhodococcus by insertional activation of the Ap^R gene. The 1.9 Kb Bam HI fragment containing the thiostrepton-resistance gene has 3 unique sites, all of which lie outside the tsr (thiostrepton resistance gene) coding region. Whether cloning into these sites results in insertional inactivation of the $thio^R$ gene is unknown. The Sph I site represents another potential cloning site in pMVS301. The essential regions in the Hind III fragment containing the Rhodococcus origin of replication remains undetermined. Figure 3 illustrates the restriction endonuclease sites defined in pMVS301.

Applications

Growth of Rhodococcus sp H-13A on heavy oils (Venezuelan Monagas or Cerro Negro crudes), with API gravity values between 8-12, results in biosurfactant production and subsequent formation of stable oil-in-water emulsions causing a reduction in the relative oil viscosity (Table 4). Monagas and Cerro Negro heavy crude oil was reduced 98% and 99%, respectively, in their relative viscosities. Visual characteristics associated with bacterial-treated oil

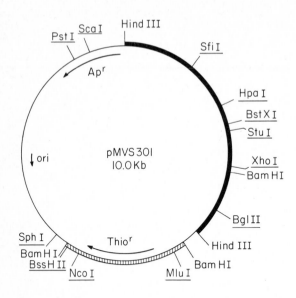

Fig. 3. Partial Restriction Map of pMVS301

were: i) a uniform distribution of the crude oil occurred through the aqueous medium; ii) the volume of the oil recovered was 2-3 times greater than the original oil volume; iii) the crude oil recovered was significantly less adherent to glass surfaces, tending to separate clearly from surfaces, and, iv) the recovered oil exhibited improved rheological properties, with flow characteristics significantly different from the original crude oil.

The increased volume of oil was determined to result from the formation of a stable oil-in-water emulsion containing 30% water as the continuous phase. The oil phase had a conductivity of 2.43×10^{-3} mho/cm.

Bacterial growth and extracellular biosurfactant production occurs solely at the expense of the paraffins present in the crude oil. Accordingly, such biological treatment serves to down-grade oil quality. This fact precludes the direct application of such biological agents for oil treatment. Attention was then directed to the spent culture broth possessing extracellular biosurfactant. The formation of stable oil-in-water emulsions with a corresponding reduction of the relative oil viscosity by greater than 90% occurs through volume adjustment of the culture broth by diafiltration and ultrafiltration (Table 4). This procedure allows for the adjustment of biosurfactant concentration to any desired amount, thereby increasing the effectiveness of the biosurfactant preparation. These results demonstrate that the cell-free spent culture medium is capable of reducing heavy oil viscosity by greater than 90% at appropriate biosurfactant concentrations.

TABLE 4

Viscosity changes in biosurfactant-treated oil[a]

Sample	Temperature	Viscosity(cps)	% Decrease
Monagas crude (wet control)[b]	40°C	6510	-
	60°C	1070	-
Cerro Negro (wet control)	40°C	>25,000	-
	60°C	6,500	-
Bacterial-treated Monagas crude	40°C	145	98
	60°C	76	93
Bacterial-treated Cerro Negro crude	40°C	275	99
	60°C	100	99
Spent culture medium-treated Monagas crude (300 mg biosurfactant/liter)	40°C	3252	50
	60°C	678	37
Spent culture medium-treated Monagas crude (600 mg biosurfactant/liter)	40°C	185	97
	60°C	39	96

a) All experiments were conducted with 50 gms of crude oil in 1000 ml of aqueous medium.
b) Samples designated wet represent the recovered oil as a stable emulsion.

Studies were conducted on the application of spent culture medium as an effective biosurfactant solution for displacement of oil from a rock matrix. The biosurfactant-containing spent culture medium was supplemented with a synthetic brine at 365.3 ppm total dissolved solids. Core flood studies employed a radial core configuration consisting of a 6" diameter by 2" thick Berea sandstone core. The core pore volume was 175 ml with a porosity of 0.2 and an absolute permeability of 749 md. The procedure used for core flooding was core saturation by evacuation in produced water, determining the permeability to produced water by injection, saturating the core with crude oil by injection until the pressure stabilized, determine the permeability to oil at residual water saturation, water flood the core until pressure stabilized, determine permeability to water at residual oil saturation and inject the biosurfactant solution at the rate of 1 ft/day. Figure 4 summarizes the results of these preliminary experiments. Tertiary oil recovery was approximately 15%. Operational parameters identified were loss of pH control within the core matrix and adsorption/fractionation of the biosurfactant, as evidenced by retardation of the biosurfactant slug by 1.5 - 2.0 pore volumes, indicating an approximate 2-fold dilution of the biosurfactant. However, greater than 95% of the biosurfactant injected was recovered from the core, indicating insignificant losses due to adsorption to the rock matrix.

84

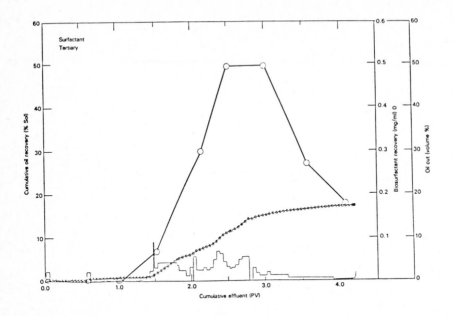

Fig. 4. Cumulative oil recovery and oil cut versus cumulative effluent from biosurfactant core flood. Symbols: -0-0-0- biosurfactant concentration; -o-o-o- cumulative oil recovery.

DISCUSSION

This project was initiated to evaluate possible role(s) for microbial processes in the modification of heavy crude oils with objectives relating to such applications as enhanced oil recovery, transportation of heavy oil and tar sand or oil shale displacement technologies. A soil microorganism has been isolated and characterized as Rhodococcus which has the physiological properties of producing an extracellular biosurfactant when grown at the expense of n-alkanes. This biosurfactant is a glycolipopeptide exhibiting surface-active properties of 0.02 mN/m at a CMC of 1.0 mg/ml and 0.00006 mN/m in the presence of the cosurfactant, pentanol. The biosurfactant exhibits excellent surface activity in the presence of brine, divalent cations, is active between pH 4-13, is stable to 150°C, forms stable oil-in-water emulsions and exhibits phase behavior with formation of a middle phase. Fatty acyl groups are covalently linked to specific hydroxyl groups of the oligosaccharide-peptide backbone. Deacylation of fatty acids destroys the surface activity of the molecule.

Biosurfactant application studies have demonstrated a greater than 90%
viscosity reduction of heavy oils, oil displacement from sand packs and tar
sands and a potential for oil displacement from oil-bearing cores.
Preliminary experiments with H-13A biosurfactant indicates a potential for
its application as a chemical enhanced oil recovery agent either as an
alternative to or in formulation with synthetic surfactants. Surfactants
are used to reduce the interfacial tension between oil and water in situ to
effect oil displacement (ref. 12). Certain biosurfactants, such as H-13A
glycolipopeptide, exhibit the appropriate surface properties required for oil
displacement.

The surface properties of several surfactants (biological and synthetic)
are listed in Table 5 for comparative purposes. Sodiumdodecylsulfate
(anionic), cetyltrimethylammonium bromide (cationic) and Tween 20 (nonionic)
all exhibit minimum surface tensions of 30 mN/m and interfacial tensions in
the range of 6-8 mN/m with CMC values ranging 0.6 to 1.5 gms/liter.
Theoretically, interfacial tensions in the range of 10^{-3} to 10^{-4} mM/m are
required for oil displacement from rock matrices in EOR processes (refs. 13,14).
Currently, petroleum sulfonates have been the surfactants of choice for EOR.
Two petroleum sulfonates, pyronate and TRS 10-80 (Witco Chemical Co.) were
studied. TRS 10-80 exhibited excellent surface-active properties with a
minimum IFT of 10^{-3} mN/m without coagent. The IFT was reduced to 10^{-4} in
the presence of 1.25% NaCl and 1% pentanol or 1% isopropanol. Biosurfactant
H-13A was at least as effective in reducing surface or interfacial tension
as the synthetic surfactants. Surfactin (a lipopeptide) exhibits the low
surface tension value of 27 mN/m, sophorolipids measure interfacial tension
values in the range of 1-2 mN/m and trehalose dimycolates fall in the range
of 16-18 mN/m. With the exception of the H-13A biosurfactant, these
biosurfactants have much lower CMC values than the synthetic surfactants.
These comparisons demonstrate that biosurfactants exhibit surface properties
comparable or superior to those of synthetic surfactants. Other
biosurfactants tested as oil recovery or displacement agents are few.
Jenneman et al. (ref. 15) have reported on a biosurfactant-producing Bacillus
lichenformis which mobilized crude oil with oil bank formation in sand packs.
Biosurfactants have been reported as effective in the release of bitumen from
tar sands (ref. 16).

TABLE 5

Comparison of surface properties of synthetic and biosurfactants

SURFACE-ACTIVE AGENT	Surface Tension mN/m	IFT mN/m	CMC mg/liter
Biosurfactants			
Rhamnolipid	40	–	50.0
Sophorolipid	36	1-2	82.0
Trehalose dimycolate	30	18	0.7
Surfactin	27	1-2	25.0
H-13A Glycolipopeptide	30	0.01 (0.00001)[a]	1000.0
Synthetic			
Sodium dodecyl sulfate	30	7.7	1500.0
Cetyltrimethylammonium bromide	30	5.8	1300.0
Tween 20	30	5.8	600.0
Petroleum Sulfonates			
Pyronate	–	0.1[a]	–
TRS 10-80	–	0.1 (0.0001)[a]	–

a) Numbers in parentheses ( ) represent the IFT in presence of 0.5% pentanol.

Biosurfactants appear to offer some advantages over synthetic surfactants for use in EOR or other production or transportation technologies. Biosurfactants offer a range of structural and physical characteristics that make them comparable if not superior to synthetic surfactants in effectiveness and efficiency. Biosurfactants are non-petroleum derived products and are biodegradable. The chemical structure of biosurfactants can be modified by either genetic, biological or chemical manipulations allowing one to tailor-make biosurfactants for specific applications. The cost effectiveness of surfactant-flooding operations for EOR using either synthetic or biosurfactants has been questioned. Cost-effectiveness depends on the cost of surfactants, coagents and other chemicals, operational costs, and also on the success of the operation in oil production. The cost-effectiveness of biosurfactants for use in EOR will depend on our ability to produce high yield, low-cost biosurfactants with demonstrated efficacy in oil recovery. Biosurfactant production can be optimized to increase product yields through strain improvement either by classical genetics or by recombinant DNA technology. The use of low cost, nonhydrocarbon growth substrates will also decrease biosurfactant production costs, as demonstrated for sophorolipid production (ref. 17). The construction of a bifunctional E. coli - Rhodococcus shuttle vector and development of an efficient transformation system in Rhodococcus represents the initial steps in gene cloning and amplification of biosurfactant genes for improved production yields. The cloning of biosurfactant structural and regulatory genes are currently in progress.

ACKNOWLEDGEMENT

This work was supported by U.S. Department of Energy Grants AS19-81BC10507, AS09-80ER-10683 and FG09-86ER-13588. The participation of SURTEK, Inc., Golden, CO is gratefully acknowledged in the evaluation and testing of H-13A biosurfactant for production of tertiary oil. This work was conducted without cost to SURTEK, Inc. and performed under the direction of Jim Ball and Malcolm Pitts, SURTEK, Inc.

REFERENCES

1 W. R. Finnerty and M. E. Singer. Biotechnology 1 (1983) 47-54.
2 D. G. Cooper and J. E. Zajic. Adv. Appl. Microbiol. 26 (1980) 229-253.
3 M. E. Singer and W. R. Finnerty, Microbial alkane metabolism, in: R. Atlas (Ed.), Petroleum Microbiology, Macmillan Publ. Co., New York (1984), pp. 1-50.
4 G. Ashwell, Methods Enzymol. 3 (1957) 73-105.
5 J. C. Dittmer and M. A. Wells. Methods Enzymol. 14 (1969) 486-487.
6 J. C. Dittmer and R. L. Lester. J. Lip. Res. 5 (1964) 126-127.
7 H. C. Birnboim and J. Doly. Nucl. Acids. Res. 7 (1979) 1513-1523.
8 T. Maniatis, E. F. Fritsch and J. Sambrook. Molecular Cloning. A Laboratory Manual Cold Springs Harbor Laboratory, New York, 1982.
9 D. S. Holmes and M. Quigley. Anal. Biochem. 114 (1981) 193-197.
10 D. J. Hanahan, Mol. Biol. 166 (1983) 557-580.
11 M. E. Singer-Vogt and W. R. Finnerty. J. Bacteriol. in press.
12 R. L. Reed and R. M. Healy, Some physicochemical aspects of microemulsion flooding: a review, in: D. O. Shah and R. S. Schechter (Eds.), Improved Oil Recovery by Surfactant and Polymer Flooding, Academic Press, New York, 1977, pp. 383-437.
13 J. C. Melrose and C. F. Brandner. J. Can. Petrol. Technol. 13 (1974) 54-62.
14 J. C. Morgan, R. S. Schecter and W. H. Wade, Recent advances in the study of low interfacial tensions, in: D. O. Shah and R. S. Schechter (Eds.), Improved Oil Recovery by Surfactant and Polymer Flooding, Academic Press, New York, 1977, pp. 101-118.
15 G. E. Jenneman, M. J. McInerney, R. M. Knapp, J. B. Clark, J. M. Feero, D. E. Revus and D. E. Menzie. Dev. Ind. Microbiol. 24 (1983) 485-492.
16 J. E. Zajic and J. Akit, Biosurfactants in bitumen separation from tar sands, in: J. E. Zajic, D. G. Cooper, T. R. Jack and N. Kosaric (Eds.), Microbial Enhanced Oil Recovery, Pennwell Publ. Co., Tulsa, OKLA., 1983, pp. 50-54.
17 D. G. Cooper and D. A. Paddock. Appl. Environ. Microbiol. 47 (1984) 173-176.

Interfacial Phenomena in Biotechnology and Materials Processing,
edited by Y.A. Attia, B.M. Moudgil and S. Chander
Elsevier Science Publishers B.V., Amsterdam, 1988 — Printed in The Netherlands

INTERFACIAL PHENOMENA IN PIGMENT-CONTAINING BIOMEMBRANES

FELIX T. HONG

Department of Physiology, Wayne State University, Detroit, MI 48201

SUMMARY

Pigment-containing biomembranes respond to pulsed light stimulation by rapid charge separation inside the membrane and at the membrane-solution interfaces. We have studied two such systems: electron transfer reactions in a magnesium porphryin membrane coupled to an aqueous redox gradient, and proton transfer reactions in a bacteriorhodopsin membrane. In both cases, interfacial charge transfers much faster than transmembrane charge movememt. Therefore, interfacial processes can be treated as conventional heterogeneous bimolecular reactions: one of the two reactants is the membrane-bound pigment, and the other is the aqueous charge donor or acceptor. Relaxation time is typically in the microsecond range. Each interfacial reaction depends on the chemical composition of the adjacent aqueous phase but not on that of the opposite aqueous phase. These reactions are also modulated by surface charges of the membrane which are either associated with the polar head groups of the phospholipids or generated by the light-induced charge transfer reaction at the interfaces. The surface potential modulates the interfacial concentrations of the aqueous reactants. The transmembrane electric field acts directly upon the membrane-bound pigment and could alter the second order rate constant. A better understanding of these phenomena is crucial to the biotechnological development of biological solar energy conversion devices, biosensors, and molecular electronic devices.

INTRODUCTION

The biomembrane is a universal building block of living organisms and is the site of many important physiological functions. It forms the outer cell envelope which serves as the boundary between the content of a cell (cytoplasm and intracellular organelles) and the external aqueous environment. It also lines the intracellular organelles so that the interior of a cell is further compartmentalized. Being an imperfect insulator, a biomembrane allows for restricted and regulated passages of ions or charges. Almost all ion and charge movements across the membrane are mediated by membrane-bound protein components (integral proteins).

Two types of transmembrane ion movements can be distinguished: (1) a passive diffusion of ions driven by a pre-existing electrochemical potential gradient through an aqueous channel which usually shows selective permeability to certain ions, and (2) an active transport or translocation of ions by an integral protein which acts as an ion pump and which requires the expenditure of energy. The active transport proteins invariably maintain a fixed and asymmetric orientation with respect to the membrane so that the direction of active ion movement is unidirectional. An active ion transport or translocation thus consists of at least three separate processes: uptake (binding) of ions at one membrane-solution interface, translocation of ions across the membrane, and release of ions at the other interface. Most investigations in this area ignore the two interfacial processes and focus upon the translocation step. This paper attempts to analyze the interfacial processes in one subclass of active transport proteins that depend on absorbed photons as the source of energy. The prime examples are the thylakoid membranes of chloroplasts in higher plants and cyanobacteria, and the photosynthetic membranes of chromatophores in purple photosynthetic bacteria.

The advent of pulsed lasers and high speed electronic instrumentation makes it possible to investigate the kinetics of the interfacial charge transfer process by means of a relaxation method. Pigment-containing biomembranes respond to pulsed light stimulation by rapid charge separation. Because of the asymmetric orientation of the integral protein pigment, this charge separation results in the generation of an electric field in a direction perpendicular to the membrane. In the present paper, the electrical manifestation of this rapid charge separation is emphasized. Our analysis is based on two experimental systems: (1) a magnesium porphyrin-containing artificial lipid bilayer membrane coupled to a redox gradient, and (2) a membrane reconstituted from the purified purple membrane of Halobacterium halobium. Many of the conclusions so reached may be generalized and applied to more complex systems such as the photosynthetic membrane in purple photosynthetic bacteria and in higher plants. The crystallization and complete X-ray crystallographic elucidation of the reaction center of Rhodopseudomonas viridis has spurred an intense and renewed interest in photosynthesis (refs. 1 and 2). Further elucidation of the mechanistic operation of this already well characterized structure will enhance our understanding of active transport processes. Insights into the interfacial phenomena in biomembranes are highly relevant to research in photosynthesis and vision and to development of biological solar energy conversion devices, biosensors, and biomolecular electronic devices.

STRUCTURE OF A BIOMEMBRANE

As shown in Fig. 1, a biomembrane consists of two molecular layers of phospholipid as the basic structural framework. The naturally occurring phospholipids are molecules with a hydrophilic head group (charged or zwitterionic) and two long hydrocarbon chains. One of the stable configurations of a phospholipid is a bilayer structure with the hydrophilic head groups exposed to the aqueous environment and with the hydrophobic hydrocarbon chains buried inside the membrane. Such a configuration was hinted at by the classical experiment of Gorter

and Grendel (ref. 3), who demonstrated that phospholipids extracted from red
blood cells formed a monolayer with an area twice the total surface area of the
red cells. Subsequently, the bilayer configuration was morphologically demonstrated as two dark staining lines in electron micrographs (unit membrane concept of Robertson) (ref. 4). That the bilayer configuration is a stable one was
demonstrated by the formation of bilayer lipid membranes <u>in vitro</u> pioneered by
Mueller, Rudin, Tien and Wescott (ref. 5): application of a phospholipid mixture
with an organic solvent (such as decane) to an inert plastic support with a hole
leads to the spontaneous formation of a membrane with a thickness of about 60 -
100 Å, the combined length of two molecules of phospholipid.

A plain lipid bilayer membrane with no additional components exhibits a conductivity which is approximately ohmic and is of several orders of magnitude
lower than that of a naturally occurring biomembrane (e.g., Chap. 1 in ref. 6).
It is well recognized that many functions and properties of a biomembrane are
attributable to the integral proteins that are embedded in the phospholipid
bilayer. These integral proteins are macromolecules with both hydrophobic and
hydrophilic domains. This amphipathic property allows for interactions between
the integral proteins and the phospholipid bilayer so as to form a stable structure. The widely accepted fluid mosaic model of Singer and Nicolson (ref. 7)
stipulates that integral proteins float in the fluid environment of the phospholipid bilayer.

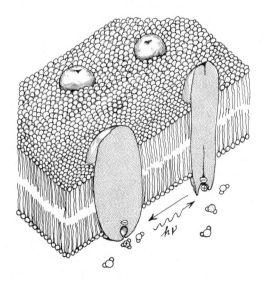

Fig. 1. The fluid mosaic model of a biological membrane. The phospholipid is
depicted as a sphere with two long tails, and the integral protein spans the
entire thickness of the lipid bilayer membrane. A light-induced conformational
change of the integral protein as well as a light-induced interfacial proton
uptake are also shown.

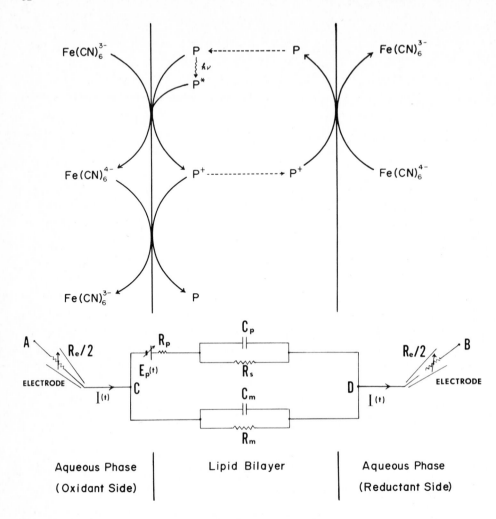

Fig. 2. A simple model membrane system of coupled interfacial electron transfer reactions with membrane-bound magnesium porphyrin and a redox gradient of ferricyanide ions and ferrocyanide ions, and its equivalent circuit representation. The ground state pigment (P), the excited state pigment (P*) and the photogenerated pigment cation (P+) are confined to the membrane phase but are free to diffuse inside the membrane phase. The light-induced interfacial pigment oxidation and the reverse reaction is shown at the left interface where the ferricyanide concentration is higher than the ferrocyanide concentration. The combined action of the forward and reverse reactions accomplishes no net electron transfer, and is represented by a displacement photocurrent. The access impedance, R_e, is an important parameter in an electrical measurement. The interfacial photoreaction is driven by the photoemf, E_p, with an interfacial resistance, R_p. The photocurrent is partly ac-coupled through the chemical capacitance, C_p, and partly dc-coupled through the transmembrane resistance, R_s, which is related to the efficiency of transmembrane diffusion of P and P+. The ordinary membrane resistance and capacitance are represented by R_m and C_m, respectively. (Reproduced and modified from the Proceedings of the National Academy of Sciences, ref. 10)

THE EXPERIMENTAL SYSTEMS

The majority of light-activated active transport systems utilize magnesium porphyrin derivatives as the light-sensitive component (chromophore) of the reaction center where active transport is being performed. These membrane-based reaction centers contain a number of redox components so arranged that electrons are transported outward across the photosynthetic membrane of the subcellular organelles - chloroplasts in higher plants and cyanobacteria, and chromatophores in purple photosynthetic bacteria (ref. 8). Essentially, the light-induced charge translocations in a photosynthetic membrane can be treated as a series of coupled charge transfer reactions. Despite significant differences between the purple bacteria and the higher plants, one common outcome is the generation of a proton electrochemical gradient, which is utilized to drive the synthesis of ATP. ATP is an energy-rich phosphate compound frequently utilized as an immediate energy source for many important biochemical processes, including the generation of ion gradients (e.g., ref. 9).

In our first experimental system, we used a simple model of coupled interfacial electron transfer reactions without the numerous intramembrane components of the electron transport chain (Fig. 2; ref. 10). Essentially, the artificial lipid bilayer contains only a single lipid-soluble component, magnesium octaethyl porphyrin or magnesium mesoporphyrin esters of long chain alcohols. This membrane is then coupled to a redox gradient of potassium ferricyanide and potassium ferrocyanide. Unlike the naturally occurring photosynthetic mem-

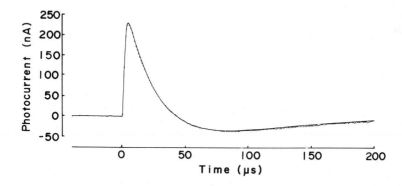

Fig. 3. A typical photocurrent elicited from a magnesium porphyrin membrane by a microsecond laser pulse. It was measured with an access impedance of 5.1 kΩ. A computed photoresponse based on the equivalent circuit model is superimposed on the measured photocurrent. The photocurrent has a near zero time-integral, indicating that the photocurrent is predominantly a displacement photocurrent (ac-coupling). The true photochemical relaxation time constant, as obtained by deconvolution of the measured signal, is 44 µsec, which is the inverse of the pseudo first order rate constant of the reverse electron transfer reaction. (Reproduced from the Proceedings of the National Academy of Sciences, ref. 10)

branes, there is no inherent asymmetry in the membrane. Rather, the asymmetry
resides in the two aqueous phases: one side contains predominantly the electron
acceptor potassium ferricyanide and the other side contains predominantly potassium ferrocyanide. Stimulation by light promotes an electron flow from the
reductant-rich side to the oxidant-rich side. This effect can be detected as a
photocurrent or photovoltage (Fig. 3). The coupling of the two interfacial
electron transfer reactions is apparently accomplished by transmembrane diffusion of the photogenerated pigment cation and the neutral ground state pigment. The coupling mechanism via mobile carriers is probably inefficient in the
charge transport across the membrane but has been suggested to be the lateral
coupling mechanism between the photosystem II and the cytochrome b_6-f complex of
the photosynthetic apparatus in higher plants (ref. 11). However, the simplicity of this model system facilitates a rigorous mathematical analysis of the
interfacial processes, based on first physical principles. We have relied on
this system to develop the relevant concept and technique for analyzing interfacial processes in biomembranes (refs. 10 and 12).

Our second experimental system is a reconstituted bacteriorhodopsin membrane.
Bacteriorhodopsin is the sole "reaction center" of the purple membrane of <u>Halo-
bacterium halobium</u> (refs. 13 and 14). It is a single chain polypeptide with 248
amino acid residues and with a covalently bound retinal (vitamin A aldehyde) as
the chromophore, and is more closely related to the visual pigment rhodopsin
than to the chlorophyll-protein complex in photosynthetic reaction centers.
However, the purple membrane functions primarily as a light-driven proton pump,
i.e., a photosynthetic apparatus. The generation of a proton electrochemical
gradient without the intervention of electron transfer is utilized to drive ATP
synthesis in the red membrane of <u>H. halobium</u>. The inherently asymmetric orientation of bacteriorhodopsin in the purple membrane ensures one-way proton pumping from the intracellular space to the extracellular space. From an experimental point of view, the absence of the multiple components of the electron
transport chain is a major simplification.

DISPLACEMENT PHOTOCURRENTS

In a photosynthetic membrane, the interfacial processes of proton uptake and
release are crucial for proper function. It is much less apparent that interfacial phenomena are functionally important in the visual membrane, which also
contains light-sensitive integral proteins (refs. 15 and 16). Nevertheless, the
first light-induced electric signal that is related to interfacial processes in
membranes was discovered in the visual photoreceptor membrane of monkey retina
(ref. 17). This photosignal, known as the early receptor potential (ERP),
belongs to a class of bioelectric signals that differed considerably from other
bioelectric signals known at that time (reviewed in ref. 18). First, the ERP

showed no detectable latency following the light pulse (submicrosecond rise-time). In contrast, the late receptor potential reflects the change of sodium ion permeability in the photoreceptor membrane and has a latency of 1.7 msec. Second, the ERP is generated by rapid charge displacements rather than by altered ionic fluxes secondary to changes of ionic permeabilities in the membrane. Third, the ERP defies a rigorous analysis by means of conventional electrophysiological techniques, as will be described later.

ERP-like photoelectric signals have subsequently been demonstrated in numerous pigment-containing tissues, including chloroplasts and a reconstituted bacteriorhodopsin membrane (refs. 19-21). The ERP-like signal in a reconstituted bacteriorhodopsin membrane bears a more than superficial resemblance to the ERP (refs. 22 and 23). This signal consists of a faster B1 component which has an ultrafast rise time and is temperature-insensitive like the R1 component of the ERP, and a slower B2 component of opposite polarity which is reversibly inhibited by low temperature ($0^{\circ}C$) like the R2 component of the ERP. Having only a single integral protein component, the purple membrane thus epitomizes the essence of a photosynthetic membrane and is most likely to provide the clues for the minimum requirements for such a membrane to function. Being chemically similar to the visual pigment rhodopsin but functionally similar to a photo-synthetic reaction center, bacteriorhodopsin also serves to bridge the conceptual gap between vision and photosynthesis.

INTERFACIAL PROPERTIES OF A PIGMENT-CONTAINING BIOMEMBRANE

The membrane-solution interface is by and large similar to the interface between two immiscible liquids. Equilibrium treatment of the partitioning of the charged species between the two phases is often a valid approach. In general, partitioning of small ions into the membrane phase is not favored, mainly because of the low dielectric constant of the membrane phase. An ionophore accomplishes its action by forming a complex with a small ion and thus significantly reduces the Born charging energy in the membrane phase and enhances the partition of the complexed ions into the membrane phase (ref. 24). The interfacial processes accompanying the action of ionophores in membranes have been reviewed in a number of treatises (refs. 25-27).

The cornerstone for the mathematical analysis of interfacial phenomena in biomembranes is the diffuse double layer theory of Gouy and Chapman (e.g., ref. 28). This theory has scored greater success in biomembrane research than in the analysis of electrodic processes, for which the theory was originally proposed. Interested readers are urged to consult the authoritative review by McLaughlin (ref. 26). It is worth noting that there is a sharp difference in space charge densities across the interface. A diffuse double layer in the aqueous phase is excellent description of the space charge region there, nevertheless, it is often possible to ignore the space charge in the membrane phase and approximate the membrane phase with a constant electric field (constant field approximation). However, the analysis of interfacial events in biomembranes is not a simple extension of that of an electrode-solution interface. The somewhat hetero-geneous structure of the membrane phase leads to complications not usually encountered in electrode-solution interfaces or in junctions between two dif-ferent semiconductors. The most remarkable feature in a biomembrane is the extremely close juxtaposition of the two membrane-solution interfaces (approxi-mately 60 Å apart). Thus, the biological membrane has a relatively large capa-citance of 1 $\mu F/cm^2$. Furthermore, the appearance of a modest transmembrane potential difference results in a large electric field inside the membrane. The ultrathinness of biomembranes also has a significant impact on the interpre-tation of displacement photocurrents, as will be shown later.

Since the photosynthetic pigments are membrane-bound, the uptake and release of protons are heterogeneous bimolecular reactions: one of the reactants is the membrane-bound pigment, and the other is the aqueous proton donor or acceptor. The two interfacial bimolecular reactions are coupled by the translocation of charges across the membrane; the products of one interfacial reaction serve as reactants in the interfacial reaction at the opposite interface. In the purple membrane, the proton is the only charge that is involved in the entire process of active transport. In the photosynthetic membranes of purple bacteria, cyano-bacteria and higher plants, the primary event is an electron transfer reaction; subsequent coupling via the electron transport chain to the quinone system and/or the ferredoxin-NADP oxidoreductase and the water-splitting enzyme leads to proton uptake at the outer membrane surface and proton release at the inner surface with the formation of a transmembrane proton electrochemical gradient (e.g., refs. 8 and 29).

Both of our experimental systems demonstrated that the rate-determining step of charge transfer is transmembrane coupling, i.e., the two interfacial reac-tions are decoupled on the time scale of interfacial relaxation (ref. 30). This feature renders steady state experiments somewhat irrelevant for studying the interfacial phenomena. However, it greatly simplifies the kinetic analysis of the interfacial reactions if the system is studied by means of a relaxation method using pulsed laser light excitation: each interfacial charge transfer reaction can be treated as independent of the other at the opposite interface. Furthermore, each interfacial reaction depends on the chemical composition of the adjacent aqueous phase but not on that of the opposite aqueous phase (con-cept of local reaction conditions). This concept is especially important in reference to macromolecules such as bacteriorhodopsin. Although the proton pumping function in the purple membrane is performed by a single protein compo-nent, separate descriptions of the experimental conditions for the two interfacial reactions are necessary. A comparison of the spectroscopic data of bacteriorhodopsin obtained in bulk aqueous suspensions and the electrical data measured in intact membranes may be misleading. A number of anomalous phenomena observed spectroscopically have simple explanations rooted in this peculiar effect. Confusion can probably be avoided if one treats the two interfacial reactions as if they involved two separate membrane-bound protein components (refs. 30 and 31).

CONCEPT OF CHEMICAL CAPACITANCE

Although the two interfacial reactions in a membrane can be considered chemically decoupled during the process of chemical relaxation (typically in the microsecond range), the electrical signals generated at both interfaces are nevertheless electrically coupled (ac-coupled). This feature is a consequence of the ultrathinness of the membrane and has an important consequence with regard to interpretation of displacement photosignals as observed by means of electrical recording.

The study of electrical responses from pigment-containing biomembranes was pioneered by Tien (ref. 32). The techniques have been an extension of classical electrophysiological methodology established more than a quarter century ago in the analysis of nerve excitation. However, for reasons to be mentioned later, the direct application of classical electrophysiological methodology has met with rather limited success and has led to some confusion in the literature (reviewed in refs. 12, 23, 33 and 34). This problem may be overcome by a combined electrochemical and electrophysiological procedure. In order to describe adequately the time course of the displacement photocurrents in biomembranes, it is necessary to introduce a novel capacitance that reflects transient charge separation and recombination (refs. 10 and 12). This capacitance, which has been shown to be physically distinct from the ordinary membrane capacitance, is called chemical capacitance (Fig. 2). The main characteristic is that chemical capacitance is charged by a current source from within the membrane whereas the membrane capacitance is charged by a current source external to the membrane. As a result, the chemical capacitance is in series with the photocurrent generator but the membrane capacitance is in parallel with the photocurrent generator. Thus, the photocurrent is ac-coupled via the chemical capacitance.

From the standpoint of making an electrical measurement, it is important to realize that the chemical capacitance causes a drastic reduction of the source impedance of the photocurrent generator at high frequency and that there is an interaction between the chemical capacitance and the ordinary membrane capacitance except under a strictly short circuit condition (ref. 34). As a consequence, the apparent relaxation time does not reflect the true intrinsic relaxa-

tion in most reported data. The extent of interaction depends on the magnitude
of the access impedance, which is the sum of the input impedance of the instrument, the electrode impedance and the impedance of the intervening electrolyte
solutions. Quantitative evaluation of this interaction can be made if the
access impedance is included in the equivalent circuit analysis of the electrical data. Deconvolution of the measured data allows the true photochemical
relaxation time to be recovered by the method of tunable voltage clamp measurements as described by Hong and Mauzerall (ref. 10). Such an approach permits
determination of the pseudo first order relaxation rate constant of the interfacial electron transfer reaction in a magnesium porphyrin-containing bilayer
lipid membrane (ref. 12). The pseudo first rate constant was found by this
method to be a linear function of the concentration of the aqueous electron
donor in accordance with the law of mass action governing a conventional bimolecular reaction.

The displacement photocurrent in a bacteriorhodopsin membrane is more complex. Clearly, there are at least two kinetically separable components: the B1
and the B2 components. However, the concept of chemical capacitance is still
applicable (refs. 22 and 23). Our analysis indicated that each component should
be associated with a separate set of equivalent circuit parameters. Therefore,
the first step toward a rigorous kinetic analysis requires a decomposition of
the two components. Decomposition was achieved by a special method of preparing
a dried multi-layered membrane with oriented bacteriorhodopsin, which invariably
gives rise to a pure B1 signal (ref. 22). In contrast, model membranes prepared
according to the method of Trissl and Montal (ref. 20) give rise to both components: B1 and B2. A pure B1 signal agrees with the equivalent circuit under a
wide range of experimental conditions. However, this agreement does not imply
that the B1 signal is generated by an interfacial process. Rather, the B1
signal reflects an intramembrane charge separation process, which is similar to
the primary light-induced charge separation process in the reaction center of a
photosynthetic membrane (oriented dipole mechanism).

Previously, two different models of light-induced charge separation processes
were described (Fig. 4): an intramembrane charge separation which leads to the
formation of a transient array of electric dipoles (oriented dipole mechanism)
and an interfacial charge transfer which leads to polarization of both aqueous
phases (interfacial charge transfer model). Application of the Gouy-Chapman
theory of the diffuse double layers yielded two slightly different equivalent
circuit representations of the charge separation processes (ref. 33). The two
equivalent circuits are macroscopically equivalent to the same irreducible
circuit in which there is a series capacitance, namely the chemical capacitance,
in the photocurrent generator. Thus, the electrical signal alone carries no
immediate clues as to which mechanism of charge separation is operating. How-

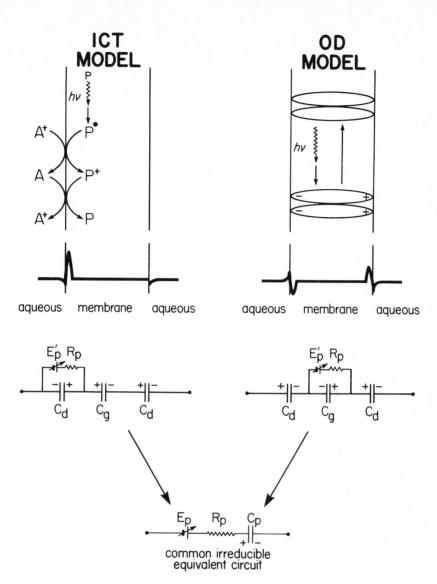

Fig. 4. Diagram explaining the difference between an interfacial charge transfer (ICT) mechanism and an oriented dipole (OD) mechanism in the generation of displacement photocurrents. The upper half of the diagram shows the two models and their charge distribution profiles across the membrane during the generation of the transient photocurrent. In the ICT model, P is the ground-state photopigment, and P^* is one of the intermediates. P^+ is a positively charged intermediate, and is a product of an interfacial charge transfer reaction: $P^* + A^+ \rightleftarrows P^+ + A$, where A^+ and A are charge donor and acceptor, respectively. The OD model is self-explanatory. The lower half of the diagram shows the two corresponding equivalent circuits obtained by a Gouy-Chapman analysis. Notice that the two circuits can be reduced to a common one with a series capacitance, C_p. C_p is a combination of three fundamental capacitances: a geometric (dielectric) capacitance, C_g, and two double layer capacitances, C_d. (Reproduced from the Bioelectrochemistry and Bioenergetics, ref. 33)

ever, differentiation between the two mechanisms follows a kinetic analysis.
Our analysis is based on the stipulation that the relaxation of intramembrane
charge separation is sufficiently decoupled from the interfacial processes as to
lose dependence on the conditions of the two aqueous phases whereas the relaxation of the interfacial process can be treated as a conventional bimolecular
reaction between the membrane-bound pigment and the aqueous proton donor or
acceptor. Furthermore, the interfacial reactions are expected to have a strong
dependence on the pH of the adjacent aqueous phase in accordance with the law of
mass action. This reasoning and the pH dependence of the ERP-like signal showed
that B1 is consistent with an intramembrane charge separation process, whereas B2
is consistent with the interfacial proton uptake at the intracellular surface.

The interpretation of pH data deserves a special comment. Previous reports
dismissed interfacial proton transfer as a possible mechanism of the B2 signal
generation (ref. 20). This latter conclusion was primarily based on the rather
modest pH dependence of the B2 component which showed virtually no pH dependence

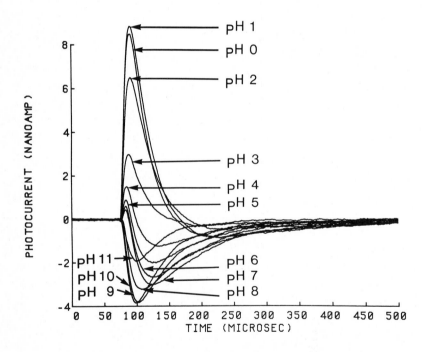

Fig. 5. The pH dependence of the two components of the ERP-like photosignal from
a bacteriorhodopsin model membrane reconstituted according to the method of
Trissl and Montal. The B1 component has a positive polarity as shown, and the
B2 component has a negative polarity. The experimental detail was described in
ref. 22. The photocurrent was measured with an access impedance of 40 kΩ and
the measurement condition is closer to a short-circuit than to an open-circuit
condition. See text for further explanation. (Reproduced from the Biophysical
Journal, ref. 22).

In the range between 5 - 8. On the contrary, our data shows a progressive change of B2 amplitude over the pH range from 0 to 11 (Fig. 5). This disagreement is a consequence of the kinetic distortion of data obtained under open circuit conditions due to infinite access impedance (ref. 34). Furthermore, the B1 component has no pH dependence between pH 0 and 11 (refs. 22 and 23). In contrast, a doubling in amplitude of the B1 component was reported to occur when the pH is shifted from 2 to 0 (ref. 35). This latter discrepancy may be due to an incorrect protocol for signal decomposition: part of the pH dependence of the B2 component had been attributed to that of the B1 component, owing to neglect of the effect of chemical capacitance (ref. 23). Preliminary evidence indicated existence of a third photosignal component (B2') representing interfacial proton release at the extracellular surface. The pH dependence of the B2' component is opposite to that of the B2 component as expected by the law of mass action (ref. 36). A rigorous kinetic analysis of the B2 and the B2' component is required to elucidate further the interfacial process of proton transfer.

EFFECT OF SURFACE CHARGES

The predominantly negative surface charges (due to the polar head groups of phospholipid) on the membrane surface generate negative surface potentials on both membrane surfaces. The interfacial charge transfer reactions also alter the surface charge densities and contribute to additional surface potentials. As a consequence, the interfacial concentrations of the aqueous charge donors or acceptors tend to be significantly different from their bulk concentrations. Since the relaxation of the interfacial charge transfer reaction depends on the interfacial concentration of the aqueous charge donor/acceptor rather than its bulk concentration, the surface charges can modulate the relaxation kinetics. That the surface charge is important in interpreting bacteriorhodopsin data has been demonstrated by Szundi and Stoeckenius (ref. 37). Furthermore, the variation of surface charges which leads to a change of electric fields inside the membrane may alter the second order rate constant of the interfacial charge transfer reaction via direct action on the membrane-bound pigment itself (ref. 31). An analysis based on this line of reasoning provides simple explanations for some anomalous behavior in light-activated spectral responses of bacteriorhodopsin preparations. Further discussion about the kinetic complexity arising from the surface charge effect can be found elsewhere (refs. 30 and 31).

CONCLUSIONS

It is quite possible that the above results concerning interfacial processes in the artificial magnesium porphyrin system and in the reconstituted bacteriorhodopsin system may be generally applicable to more complex systems such as bacterial chromatophores or the thylakoid membranes in higher plants. A better understanding of the ERP-like signal may also help to elucidate the possible role of the ERP in the visual phototransduction process (ref. 38). A detailed analysis of light-induced interfacial processes also allowed us to develop some design principles in molecular electronics (ref. 38-40). The following general conclusions may be stated.

1. Many important cellular functions in living organisms are performed by membrane-bound proteins. The interfacial events in pigment-containing biomembranes can be analyzed quantitatively by means of a relaxation method using pulsed light excitation.

2. Although the physiological functions of pigment-containing biomembranes are implemented primarily by integral protein components, kinetic analyses must be performed in the context of the membrane which serves as the supporting matrix for the functioning protein.

3. There are two interfaces in a membrane system that are separated by a distance on the scale of molecular dimensions. Some membrane-bound pigments actually span the thickness of the membrane and are exposed to two aqueous phases.

4. On the time scale of interfacial relaxation, the two interfacial charge transfer reactions are chemically decoupled. Nevertheless, the electrical signals accompanying the interfacial events remain coupled on the time scale of chemical relaxation because of the ultrathinness of the membrane and the rapid transit of charge separation effects.

5. A chemical reaction at a membrane-solution interface can be treated as a conventional heterogeneous bimolecular process that obeys the law of mass action. However, only the local reaction conditions in the adjacent aqueous phase are relevant.

6. The concept of chemical capacitance facilitates the interpretation of photoelectric signals accompanying an interfacial charge transfer process.

7. The surface charge density modulates interfacial bimolecular charge transfer reaction in a highly non-linear fashion.

8. A combined electrochemical and electrophysiological approach provides simple explanations of anomalous phenomena in pigment-containing biomembranes.

ACKNOWLEDGMENTS

The experimental work on magnesium porphyrin membranes was performed in the laboratory of Professor David Mauzerall, The Rockefeller University (NIH Grant GM 20729). The experimental work on bacteriorhodopsin was performed by Dr. T. L. Okajima (NIH Grants GM 25144 and EY 03334). This research is also supported in part by an Office of Naval Research Contract (N00014-87-K-0047).

REFERENCES

1 J. Deisenhofer, O. Epp, K. Miki, R. Huber and H. Michel, X-ray structure analysis of a membrane protein complex: electron density map at 3 Å resolution and a model of the chromophores of the photosynthetic reaction center from Rhodopseudomonas viridis, J. Mol. Biol., 180 (1984) 385-398.

2 J. Deisenhofer, O. Epp, K. Miki, R. Huber and H. Michel, Structure of the protein subunits in the photosynthetic reaction centre of Rhodopseudomonas viridis at 3 Å resolution, Nature (London), 318 (1985) 618-624.

3 E. Gorter and F. Grendel, On bimolecular layers of lipoid on the chromocytes of the blood, J. Exptl. Med., 41 (1925) 439-443.

4 J.D. Robertson, The molecular structure and contact relationships of cell membranes, Prog. Biophys. Biophys. Chem., 10 (1960) 343-418.

5 P. Mueller, D.O. Rudin, H.T. Tien and W.C. Wescott, Reconstitution of cell membrane structure in vitro and its transformation into an excitable system, Nature (London), 194 (1962) 979-980.

6 H.T. Tien, Bilayer Lipid Membranes (BLM): Theory and Practice, Marcel Dekker, New York, 1974.

7 S.J. Singer and G.L. Nicolson, The fluid mosaic model of the structure of cell membranes, Science, 175 (1972) 720-731.

8 A.N. Glazer and A. Melis, Photochemical reaction centers: structure, organization, and function, Ann. Rev. Plant Physiol., 38 (1987) 11-45.

9 P. Hinkle and R.E. McCarty, How cells make ATP, Sci. Am., 238(3) (1978) 104-123.

10 F.T. Hong and D. Mauzerall, Interfacial photoreactions and chemical capacitance in lipid bilayers, Proc. Natl. Acad. Sci. USA, 71 (1974) 1564-1568.

11 P.A. Millner and J. Barber, Hypothesis: plastoquinone as a mobile redox carrier in the photosynthetic membrane, FEBS Lett., 169 (1984) 1-5.

12 F.T. Hong, Charge transfer across pigmented bilayer lipid membrane and its interfaces, Photochem. Photobiol., 24 (1976) 155-189.

13 W. Stoeckenius, R.H. Lozier and R.A. Bogomolni, Bacteriorhodopsin and the purple membrane of Halobacteria, Biochim. Biophys. Acta, 505 (1979) 215-278.

14 W. Stoeckenius and R.A. Bogomolni, Bacteriorhodopsin and related pigments of Halobacteria, Ann. Rev. Biochem., 51 (1982) 587-616.

15 L. Stryer, Cyclic GMP cascade of vision, Ann. Rev. Neurosci., 9 (1986) 87-119.

16 L. Stryer, The molecules of visual excitation, Sci. Am., 257(1) (1987) 42-50.

17 K.T. Brown and M. Murakami, A new receptor potential of the monkey retina with no detectable latency, Nature (London), 201 (1964) 626-628.

18 R.A. Cone and W.L. Pak, The early receptor potential, in: W.R. Loewenstein (Ed.), Principles of Receptor Physiology, Handbook of Sensory Physiology, Vol. I, Springer-Verlag, Berlin, 1971, pp. 345-365.

19 W.J. Vredenberg and A.A. Bulychev, Changes in the electrical potential across the thylakoid membranes of illuminated intact chloroplasts in the presence of membrane-modifying agents, Plant Sci. Lett., 7 (1976) 101-107.

20 H.-W. Trissl and M. Montal, Electrical demonstration of rapid light-induced conformational changes in bacteriorhodopsin, Nature (London), 266 (1977) 655-657.

21 F.T. Hong and M. Montal, Bacteriorhodopsin in model membranes: a new component of the displacement photocurrent in the microsecond time scale, Biophys. J., 25 (1979) 465-472.

22 T.L. Okajima and F.T. Hong, Kinetic analysis of displacement photocurrents elicited in two types of bacteriorhodopsin model membranes, Biophys. J., 50 (1986) 901-912.

23 F.T. Hong and T.L. Okajima, Electrical double layers in pigment-containing biomembranes, in: M. Blank (Ed.), Electrical Double Layers in Biology, Plenum Press, New York, 1986, pp. 129-147.

24 P. Läuger and B. Neumcke, Theoretical analysis of ion conductance in lipid bilayer membranes, in: G. Eisenman (Ed.), Lipid Bilayers and Antibiotics, Membranes, Vol. 2, Marcel Dekker, New York, 1973, pp. 1-59.

25 G. Eisenman (Ed.), Lipid Bilayers and Antibiotics, Membranes, Vol. 2, Marcel

Dekker, New York, 1973.

26 S. McLaughlin, Electrostatic potentials at membrane-solution interfaces, Curr. Top. Membr. Transp., 9 (1977) 71-144.

27 P. Läuger, Kinetic properties of ion carriers and channels, J. Membr. Biol., 57 (1980) 163-178.

28 E.J.W. Verwey and J.Th.G. Oberbeek, Theory of the Stability of Lyophobic Colloids, Elseveir, Amsterdam, 1948.

29 H.T. Witt, Energy conversion in the functional membrane of photosynthesis: analysis by light pulse and electric pulse methods, the central role of the electric field, Biochim. Biophys. Acta, 505 (1979) 355-427.

30 F.T. Hong, Effect of local conditions on heterogeneous reactions in the bacteriorhodopsin membrane: an electrochemical view, J. Electrochem. Soc., 134 (1987) 3044-3052.

31 F.T. Hong, Internal electric fields generated by surface charges and induced by visible light in bacteriorhodopsin membranes, in: M. Blank and E. Findl (Eds.), Mechanistic Approaches to Interaction of Electric and Electromagnetic Fields with Living Systems, Plenum Press, New York, 1987, pp. 161-186.

32 H.T. Tien, Light-induced phenomena in black lipid membranes constituted from photosynthetic pigments, Nature (London), 219 (1968) 272-274.

33 F.T. Hong, Mechanisms of generation of the early receptor potential revisited, Bioelectrochem. Bioenerg., 5 (1978) 425-455.

34 F.T. Hong, Displacement photocurrents in pigment-containing biomembranes: artificial and natural systems, in: M. Blank (Ed.), Bioelectrochemistry: Ions, Surfaces, Membranes, Advances in Chemistry Series, No. 188, American Chemical Society, Washington, D.C., 1980, pp. 211-237.

35 L.A. Drachev, A.D. Kaulen, L.V. Khitrina and V.P. Skulachev, Fast stages of photoelectric processes in biological membranes: I. bacteriorhodopsin, Eur. J. Biochem., 117 (1981) 461-470.

36 F.T. Hong and T.L. Okajima, Rapid light-induced charge displacements in bacteriorhodopsin membranes: an electrochemical and electrophysiological study, in: T.G. Ebrey, H. Frauenfelder, B. Honig and K. Nakanishi (Eds.), Biophysical Studies of Retinal Proteins, University of Illinois Press, Urbana-Champaign, Illinois, 1987, pp.181-198.

37 I. Szundi and W. Stoeckenius, Effect of lipid surface charges on the purple-to-blue transition of bacteriorhodopsin, Proc. Natl. Acad. Sci. USA, 84 (1987) 3681-3684.

38 F.T. Hong, Relevance of light-induced charge displacements in molecular electronics: design principles at the supramolecular level, manuscript submitted for publication in J. Molecular Electronics.

39 F.T. Hong, The bacteriorhodopsin model membrane system as a prototype molecular computing element, BioSystems, 19 (1986) 223-236.

40 F.T. Hong and M. Conrad, The bacteriorhodopsin membrane as a prototype molecular electronic device, in: F.L. Carter and H. Wohltjen (Eds.), Proc. Third Int. Symp. on Molecular Electronic Devices, Arlington, Virginia, October 6-8, 1986, North-Holland, Amsterdam, in press.

Interfacial Phenomena in Biotechnology and Materials Processing,
edited by Y.A. Attia, B.M. Moudgil and S. Chander
Elsevier Science Publishers B.V., Amsterdam, 1988 — Printed in The Netherlands

THE UTILIZATION OF ELECTRICAL DIFFUSE DOUBLE LAYER THEORY IN UNDERSTANDING
TRANSPORT PHENOMENA IN SYNTHETIC AND IN BIOLOGICAL MEMBRANE SYSTEMS

M. Bender
Chemistry Department, Fairleigh Dickinson University, Teaneck, New Jersey
07666 U.S.A.

SUMMARY
   Whereas transport in the polymeric non-living system is of chaotic origin,
i.e. due to a concentration gradient, this is not necessarily a complete
story in biology, where living system evidence indicates oriented structured
purposeful flow of electrically charged particles. In either situation, vitro
or vivo, the flow is influenced by the extant state of the electrical diffuse
double layer in the system. The dependence of degree of flow and electrical
potential generated, on particular double layer conditions, is considered.

INTRODUCTION

   The usefulness of electrical diffuse double layer theory for interpretation
of properties of both synthetic membranes and physiological systems is discussed. Observations not appropriately analyzed by given classical relationships which were developed from too simplified a model, seem to be better
understood when the electrostatics is taken into account, both for homogeneous
and interfacial systems. In vivo transport,with the appropriate potential it
generates, is greatly influenced by the double layer ambiance, as is the transport due to a thermodynamic gradient.

DISCUSSION

   Synthetic membranes such as cellulose and cellulose acetate (CA) are essentially homogeneous in their absorption of water and electrolyte (refs. 1-4).
Thus, one is dealing with solutions in this sense, rather than interfaces within the membrane. There is molecular intimacy between the polymer chain network and the water molecules and the electrolyte ions. Transport properties
of these membranes then should be looked at in this light.

   These membranes manifest electrokinetic properties such as streaming potential, indicating the presence of an internal electrical diffuse double layer
network. This network is necessarily imposed on the homogeneity of the membrane.
Gouy-Chapman-Stern double layer theory helps explain the observations with cellulose and CA to be described. These observations in turn lend strong support
to utilization of the theory, recognizing it as an important factor of membrane
properties and extending to biological membranes.

The (inanimate) synthetic membrane is under thermodynamic chaotic randomness conditions, and diffusional transport depends on a concentration gradient. With the physiological system, not only are there thermodynamic fluctuations, but there is a "living"* "intelligent" transport of molecules and particles in cytoskeletal pathways, which travel through membranes and cell media to definite destinations fulfilling growth, maintenance of the system, and defense against inimical agents. They are necessarily electrically charged, travelling with energy much greater than the Boltzmann kT, and run over microtubules which presumably are also charged and supply energetics. This kind of motility was discussed by J.W. Griffin, R.J. Lasek, and T.S. Reese (ref. 5). Relevant papers are by R.D. Allen (ref. 6), S. Cohn (ref. 7), R.H. Miller, R.J. Lasek and M.J. Katz (ref. 8), and R.E. Buxbaum, T. Dennerll, S. Weiss and S.R. Heidemann (ref. 9), among others.

Conclusions on sorption homogeneity and electrical diffuse double layer characteristics in the synthetic membranes stem from the results of several mechanism studies (ref. 1-4, 10):

Simple salts in cellulose (Fig. 1) (ref. 3), and in CA (Fig. 2) (ref. 4), and the cationic compound diisobutylphenoxyethoxyethyl dimethyl benzyl ammonium chloride (DDBAC) in cellulose (Fig. 3) (ref. 10), are sorbed to the extent only of a few ions per 100 cellobiose units. Thus, it seems unlikely that there are specific sorption sites.

For both cellulose and CA, K^+ is sorbed to a greater extent than the more hydrated Na^+. This result is in line with the lyotropic series where ions are listed according to their electrokinetic effects.

The sorption of simple salts was seen to be irreversible, with unleachables according to the lower curves in Fig. 1 and in Fig. 2. It being observed that ion exchange replaced these ionic residues, H^+, K^+, Na^+ etc. membranes were produced. Interpretation is that the (sorbed) ions essentially make up the double layer network, ions of one sign of charge being preferentially held by the membrane molecular chains over the other sign of charge. At higher sorption concentrations, electrostatic interactive forces were weak, and solute is leached readily. But with decreased concentration, double layer forces of attraction become stronger and the stage of non-leachable ions is reached. Ion exchange occurs when the excess of the exchanging ion weakens the double layer interactions. The original sorbed ion is now out-populated so that subsequent leaching leaves essentially the newly introduced ion behind.

Ionic strength, an essential parameter in diffuse double layer theory is

---

* In this discussion, tissue is considered as alive when it generates an electrical potential between it and the medium it is in contact with, across their interface. This potential no longer exists when the tissue dies.

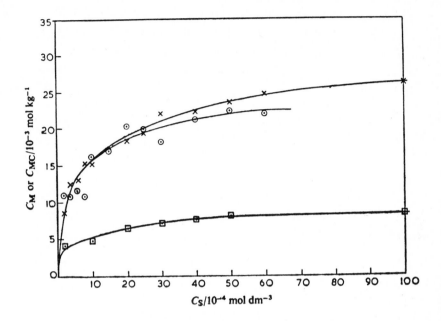

Fig. 1. Absorption equilibrium isotherms on hydrogen cellulose: KCl C_M(X) and C_{MC} ($\boxdot$) pH = 5.0; NaCl C_M ($\odot$) pH = 5.2 (ref. 3).

seen to be an important factor in sorption definition. An anionic system, the Na salt of reduced 16, 17 dimethoxy violanthrone (I) in NaOH and $Na_2S_2O_4$, with high ionic strength, the double layer consequently being swamped, showed essentially reversible sorption.

In further substantiation of the diffuse electrical double layer theory approach to understanding properties of the membranes, OH^- and H^+ appear as potential determining ions. These ions have strong effects on the sorption (ref. 1-4, 10). For instance OH^- accounts for the high solubility of complex anionics (ref. 1, 2) in cellulose. The take-up of simple salts in cellulose and CA decreases with lower pH (ref. 3,4), and Fig. 4 (ref. 10) shows that the sorption of DDBAC in cellulose is greatly influenced by H^+ and OH^- concentrations.

The diffusional transport observations made on these synthetic membrane systems are interpretable in terms of electrical double layer theory. Reciprocally, the results are a strong argument for the theory.

In all cases transport is highly concentration dependent, the dependency increasing with decreasing ionic strength. Fig. 5 (ref. 1) in the rightmost curve shows the steady state diffusion concentration gradient for anionic I in

108

cellulose at high ionic strength. While the diagonal should be the gradient if
the diffusion coefficient did not vary with concentration according to Fick's
first law, the curve, through interpretation of its slope, indicates that the
coefficient decreases continuously and significantly as the concentration decreases. This result implies increased interaction in the system when less of
the anionic solute is present. The interaction interferes with the freedom of
the chaotic diffusing motion of the solute molecules.

The steady state diffusion concentration gradient measured for KCl in cellulose is shown in Fig. 6 (ref. 3) by the curve marked "x". Note that the concentration in the membrane drops only by a small amount with distance over most
of the membrane. Then concentration changes suddenly, decreasing rapidly to
"zero".

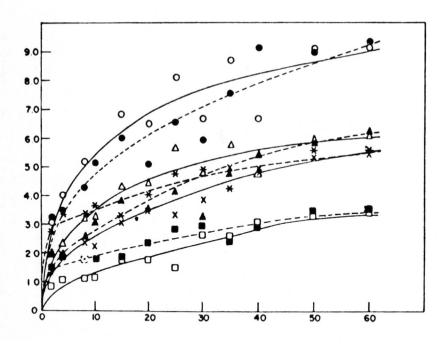

Fig. 2. Total equilibrium absorption and nonleachables for NaCl, NaI, KCl and
KI on hydrogen cellulose acetate: NaCl C_M(X) and C_{MC} (□); NaI C_M(*) and C_{MC}
(■); KCl C_M (⊙) and C_{MC} (△); KI C_M (●) and C_{MC} (▲). Chlorides (———),
iodides (-----). abscissa = C_S (mole liter^{-1} X 10^4), ordinate = C_M, C_{MC} (mole
kg^{-1} X 10^3) (ref. 4).

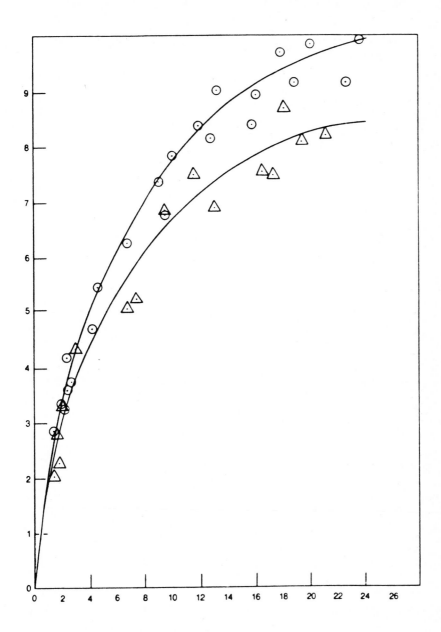

Fig. 3. Equilibrium sorption and desorption of DDBAC by cotton. pH = 6.50. (⊙), sorption; (△) desorption: abscissa = equilibrium solution concentration (moles DDBAC/liter x 10^4); ordinate = sorption concentration (moles DDBAC/g cotton x 10^6) (ref. 10).

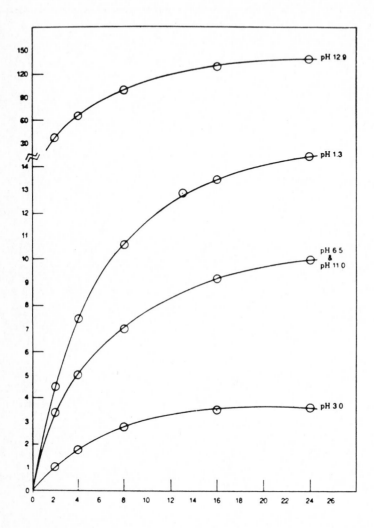

Fig. 4. Equilibrium sorption of DDBAC at different pH; abscissa = equilibrium solution concentration (moles DDBAC/liter x 10^4); ordinate = sorption concentration (moles DDBAC/g cotton x 10^6( )ref. 10).

With low ionic strength, especially at the low concentration side of the membrane, this system shows a large diffuse double layer in contrast to the swamped ionic strength conditions of the complex anionic system. Comparing the two gradients in their concentration dependence as deviations from the diagonal, the KCl-cellulose gradient deviates extremely in the curvature of the bow of Fig. 5, going to a radius so small as to be essentially a break. This result points to the presence in the KCl-cellulose of an electrostatic barrier which slows the diffusion process tremendously, the ions interacting strongly in the double layer.

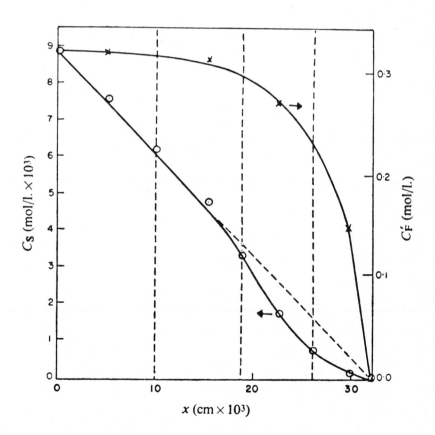

Fig. 5. Steady state concentration gradients in the four-layer membrane. X, C_F' against x; ⊙, C_S against x (ref. 1).

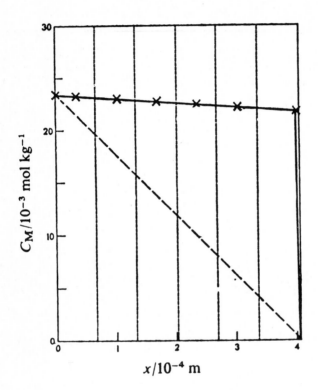

Fig. 6. Steady state concentration gradient in the six-layer membrane (ref. 3).

In the case of simple salts in CA, the ionic strength is even lower (much) than in the cellulose, and also the overall dielectric constant is lower. Both of these factors should bring on more extreme double layer interaction to interfere with the diffusion. Actually, more than one barrier to diffusion is found according to the concentration gradient curves in Fig. 7 (ref. 4) for NaCl, the result being more striking in the lower concentration runs. This correlates with the increased semi-permeability of CA over cellulose as in reverse osmosis where CA is much more effective than cellulose.

Conceptually, these data indicate that electrostatic barriers to transport must be an important aspect in biological systems which are comprised of a complex array of interfaces and particles and molecules including proteins and lipids, highly amphipathic in nature and electrically charged. Considering that electric currents (wherever generated) and measurable potentials exist, then

there is a conduit system, and electrostatic barriers should be a feature. Actually, for biological systems, a variety of observations reported in the literature may be interpreted as the result of electrical diffuse double layer
effects (ref. 11).

With respect to the nature of the physiological electric currents, researchers refer to Na^+, K^+, Ca^{++}, Cl^- etc. ionic flow, especially through specific
protein molecular channels across membranes, the molecular configuration of the

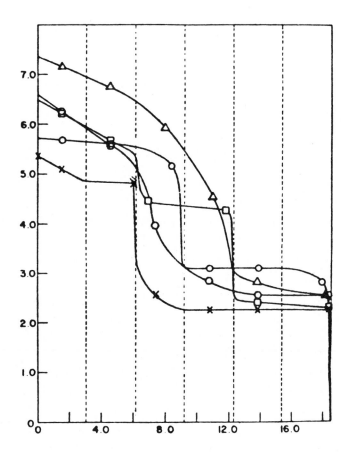

Fig. 7. NaCl diffusion concentration gradients at different C_s: 0.005 M (X),
0.010 M (⊙), 0.030 M (⊡), 0.050 M (⊙) and 0.10 M (△). Abscissa = x (m x
10^5), ordinate = C_M (mole kg^{-1} x 10^3) (ref. 4).

channel being essentially that with a polar "lining" and a hydrophobic exterior. Also reference is made to proton transfer including the Grotthuss type flow. A recent paper by Kuki and Wolynes (ref. 12) discusses electron tunneling in proteins. Not to be neglected are the living charged particles already mentioned, which in their flow must represent net electric current transfer. Associated with this flow there should be a potential of electrokinetic nature, i.e., the fall potential.

In terms of the energetics of electrical flow, much consideration is given in the literature to the thermodynamic concentration gradient as a source of power, particularly with ionic flow through protein channels. The metabolic pumping of ions is strongly meaningful. Apparently related to metabolism is the energy of motility of the "alive" particles and network referred to early in this paper.

Origin of charge of the "in vivo" network should be at least partly from preferential ionic sorption. But much of the "live" material is of protein nature, and therefore charged due to its state of ionization, there being little likelihood of isoelectric point conditions. Thus, diffuse electrical double layer theory should apply, and the amount and sign of charge on particles and surfaces in general, influencing electric current and potential, ought to be very much a function of species and concentration of bathing ions (ref. 11) as well as of the nature of the physiological substance. This has specifically been demonstrated with frogskin (ref. 11) including to the extent of reversal of the sign of the generated potential by appropriate ionic composition of the bathing solutions.

The electrostatic potential energy of double layer barriers is also a factor to consider in live transport of charges. Barriers affect the degree and orientation of the motility, their own sign and magnitude being a function of the ionic species and concentrations of which they are constituted.

One question on which there has been much thought is how a polar particle passes through the supposed non-polar areas of a membrane. Under "alive" conditions, transfer takes place with no such difficulties. Not only is there the energy for transfer but also for configuration change whether in the material under transport or the membrane through which the electricity is transferred. Polar-nonpolar incompatibilities are thereby alleviated since the protein can show hydrophobic or hydrophilic interface according to the necessary transfer conditions.

For the understanding of the "intelligence" or recognition properties manifest under these "in vivo" circumstances, application of "inanimate" physics concepts is a very difficult proposition. A recent letter to Nature (ref. 13) reports on the relaxation of DNA depending on its recognition of topoisomerase

I, suggesting that this configurational change might be important in a fundamental biological process. Papers given at the Whitehead Institute Symposium
held at the Massachusetts Institute of Technology, Oct. 13-15, 1985 (ref. 14),
address themselves to receptor structure and function, the cell's ability to
sort proteins, and biochemical pathways through which activated receptors
induce cellular responses. Reference is also made to Travers, A. and Klug, A.
on DNA wrapping and writhing (ref. 15). Actually there have been many (recent)
papers where the observations lend credence to the configurational changes suggested, creating interfacial compatibility for transport to take place.

The paper discusses the usefulness of electrical diffuse double layer theory
for interpretation of properties of both synthetic and biological membranes.
While the concept of the thermodynamic gradient is often utilized in energetics
explanations of electrical (ionic) currents in biological systems, acknowledgement must be made with respect to the electrical energy associated with the in
vivo oriented transport processes, in which, especially, proteins play a strong
part. Since pathway surface and transporting particle electrical charges depend on the species and concentration of charges in the ionic atmosphere
through which the transport takes place, then electrical diffuse double layer
theory principles have strong bearing on the magnitude and sign of electrical
potentials developed and the degree of electrical current flow.

REFERENCES

1   M. Bender and W.H. Foster, Jr., Diffusion and sorption in cellulose, Trans.
    Faraday Soc., 61 (1965) 159-169.
2   M. Bender and W. Foster, Jr., Diffusion and sorption in cellulose, Part 2,
    Trans. Faraday Soc., 64 (1968) 2549-2554.
3   M. Bender, J.K. Moon, J. Stine, A. Fried, R. Klein, and R. Bonjouklian,
    Diffusion and sorption of simple ions in cellulose: Ion exchange, J. Chem.
    Soc., Faraday Transactions I, 71 (1975) 491-500.
4   M. Bender, B. Khazai and T.E. Dougherty, Diffusion and sorption of simple
    ions in cellulose acetate-semipermeability, J. Colloid and Interfacial
    Science, 63 (1978) 346-352.
5   Symposium "Frontiers of Neuroscience" c/o AAAS Annual Meeting, 15-17 Feb.
    1987, Chicago, IL, U.S.A.
6   R.D. Allen, The microtubule as an intracellular engine, Scientific American,
    Feb. (1987) 42-49.
7   S. Cohn, Microtubule activation of kinesin AtPase activity, Nature, 326
    (1987) 16-17.
8   R.H. Miller, R.J. Lasek, M.J. Katz, Preferred microtubules for vesicle
    transport in lobster axons, Science, 235 (1987) 220-222.
9   R.E. Buxbaum, T. Dennerl, S. Weiss and S.R. Heidemann, F-Actin and microtu-
    bule suspensions as indeterminate fluids, Science, 235 (1987) 1511-1513.
10 M. Bender and R. Carmello, Sorption of diisobutylphenoxyethoxyethyl
    dimethyl benzyl ammonium chloride by cotton, J. Colloid and Interface
    Science, 86 (1982) 266-273.
11 M. Bender, K. Hillman, L. Tirri, M. Graben, D. Fenster and M. Rosensaft,
    The influence of the composition of the bathing solutions on frog skin
    membrane potential, in: M. Blank (Ed.), Bioelectrochemistry: Ions, Surfaces,
    Membranes, Advances in Chemistry Series 188, Amer. Chem. Soc., Washington,

D.C., 1980, pp. 411-443.

12 A. Kuki and P.G. Wolynes, Electron tunneling paths in proteins, Science, 236, (1987) 1647-1652.

13 H. Busk, B. Thomsen, B.J. Bonven, E. Kjeldsen, O.F. Nielsen and O. Wester-gaard, Preferential relaxation of supercoiled DNA containing a hexadecamer-ic recognition sequence for topoisomerase I, Nature, 327 (1987) 638-640.

14 J.L. Marx, A potpourri of membrane receptors, Science, 230 (1985) 649-651.

15 A. Travers and A. Klug, Nucleoprotein complexes - DNA wrapping and writhing, Nature 327 (1987) 280-281.

Interfacial Phenomena in Biotechnology and Materials Processing,
edited by Y.A. Attia, B.M. Moudgil and S. Chander
Elsevier Science Publishers B.V., Amsterdam, 1988 — Printed in The Netherlands

ADSORPTION OF POLYMERS ON CONTACT LENS SURFACES IN RELATION TO BIOLUBRICATION

K. Kumar and D. O. Shah, Center for Surface Science & Engineering,
University of Florida Gainesville, Florida 32611, U.S.A.

SUMMARY
     For the past few years, the Center for Surface Science and Engineering has
carried out extensive research on adsorption/desorption of polymers on contact
lens surfaces. The main objective of this research is to quantify the
adsorption/desorption properties of ophthalmic polymers using a wide variety
of lenses such as polymethylmethacrylate, silicron acrylate, etc.
     We have   carried   out   well coordinated research using a variety of
ophthalmic polymers. Solutions of these polymers were prepared in 0.9% saline
at pH 7. The concentrations of solutions were in the order of 0.1% - 2%,
maintaining viscosity up to 100 cp. The lenses were dipped in these solutions
for different time variations and were rinsed 0-50 times in saline and air
dried in horizontal position.   Adsorption of these polymers on these polymer
treated contact lenses was measured by using contact angle measurements and
ESCA. The silicon acrylate lens treated with polyvinyl alcohol showed
correlation of contact angle and ESCA measurements. It was found that the
degree of adsorption is inversely related to contact angle and coefficient of
friction measurements.

INTRODUCTION

     Nature has   designed   a very effective lubrication mechanism for the
blinking process.Generally a person blinks 4000 to 5000 times per day (réf. 1).
The fact that we do not experience a significant discomfort or wear due to the
blinking process is in part due to a very effective lubricating mechanism
operating in the eyelid-cornea system.   Surface phenomena   such   as the
spreading of meibomian oil at the air/tear interface, the kinetics of thinning
of tear film the rate of evaporation of water from the tear film, and the
lubrication of corneal surface and eyelid are pertinent to the normal blinking
process. Most of the processes occur every time we blink (ref. 2) (Figure 1).
     There are more than 20 million contact lens users in the United States
today. When one wears contact lenses, the structure of the   sliding surfaces
in the eye is modified. In contrast to the natural eyelid-cornea system, the
eyelid slides against the contact lens surface and a small movement also
occurs between the contact lens and the corneal surface (Figure 2). Thus, it
is expected that the surface properties of the contact lens material would

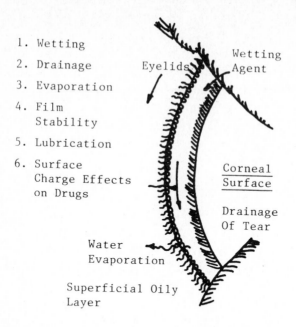

1. Wetting
2. Drainage
3. Evaporation
4. Film Stability
5. Lubrication
6. Surface Charge Effects on Drugs

Fig. 1 Surface phenomena in the eye.

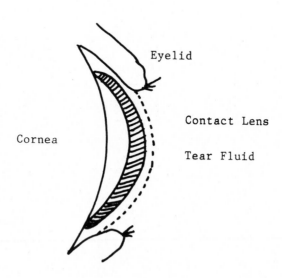

Fig. 2 Sliding of the eyelid against the contact lens surface during the blinking process.

significantly modify the friction and lubrication in the eye. The eye produces mucin, a mucopolysaccharide which is a surface active macromolecule and is likely to adsorb on the corneal as well as eyelid surfaces (refs. 3,4). Adsorbed mucopolysaccharide will stabilize on aqueous boundary layer on the sliding surface. During the blinking process, the friction is minimized presumably by the adsorbed film of macromolecule. Although we recognize that the mechanism by which macro-molecules provide lubrication in the cornea-eyelid system is unknown, we propose that the adsorption of polymers to the sliding surfaces has to be a very important and required step in the mechanism of lubrication in the eye. Figure 3 schematically illustrates the adsorption of polymer molecules on corneal and eyelid surfaces. It is obvious that the structure of polymers molecules, their hydrodynamic radii, the conformation of the adsorbed molecules, and the interaction of polymer molecules with the surface (i.e., the strength of adsorption) would influence the frictional forces generated during the sliding of the surfaces.

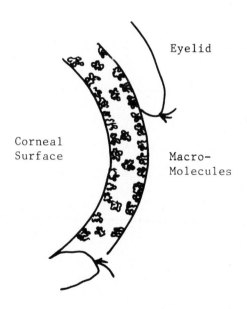

Eyelid

Corneal
Surface

Macro-
Molecules

Fig. 3   Adsorption of polymers and biolubrication of the corneal surface.

The subject of lubrication is concerned with the process of reducing frictional resistance occurring between two sliding solid surfaces. Any substance inserted between two sliding surfaces for the purpose of reducing the friction is called a "lubricant". There are two types of operating conditions in lubrication, namely boundary lubrication and hydrodynamic lubrication (ref. 5). In the former case, the lubricant film cannot support the load and contact occurs between the two surfaces (ref. 5). In this case, the coefficient of friction decreases with viscosity and speed and increases with load. In the case of hydrodynamic lubrication, the two sliding surfaces are separated with a thin film of lubricant. The frictional drag is entirely due to the rheological properties of the lubricant film. The coefficient of friction in this region increases with viscosity and speed and decreases with load. The friction between cornea and lid has been assumed to be of the boundary type. The surfaces, in this case, are likely to contact each other (ref. 6).

This paper focuses on the adsorption of polymers and commercial ophthalmic solutions on contact lens surfaces in relation to biolubrication.

MATERIALS, METHODS, INSTRUMENTS AND MEASUREMENTS

Materials

* Ophthalmic Polymers - Polyvinyl Alcohol (different molecular weights), Hydroxy methyl cellulose, Hydroxy ethyl cellulose, Chondroitin sulfate, Dextran, etc. have been used on Polymethylmethacrylate and Silicon acrylate lens surface for adsorption studies. These polymers were bought from Polyscience, Inc. or Aldrich Chemical Company, Inc.

* Commercial Ophthalmic Solutions - The following commercial ophthalmic solutions have been used for studies.

    1. AdsorboTear (Alcon) 2. Tears Plus (Allergan) 3. Lens Mate (Alcon)
    4. Adapettes (Alcon) 5. Tear Gard (Bio Products) 6. Tears Naturale
    (Alcon) 7. Muro Tears (Muro Pharmaceutical) 8. Neo-Tears (Barnes Hind)
    9. Hypotears (CooperVision) and 10. Liquifilm Tears (Allergan)

Methods

* Preparation of Polymer Solutions: Polymers were weighed based on their required concentration in the solution and added to a beaker containing 100 ml saline having a pH of 7. The mixture was stirred using a magnetic stirrer. The high molecular weight polymers had to be heated between 35° - 50° C until the polymer was completely dissolved.

* Preparation of Lens Material for Contact Angle, Coefficient of Friction and ESCA measurements: Three different size plates from big pieces of lens material were cut:

    a. 1/2 cm x 1/2 cm for ESCA measurements

    b. 2.5 cm x 2.5 cm for Adsorption/desorption studies and contact angle measurements

    c. 12.5 cm x 12.5 cm for coefficient of friction measurements

Lens plates for ESCA and contact angle measurements were dipped in polymer solutions for different time frames (3 hours, 8 hours and 24 hours).

Rinsing and drying processes

The plates were rinsed in saline. Depending on the experiment, the lens plates were rinsed in the following sequence: First plate was not rinsed while the second, third, fourth, fifth and sixth plates were rinsed 10, 20, 30, 40 and 50 times and dried in a horizontal plane laid on the tracks.

Instruments and Measurements

Contact Angle Measurements: A Rame-Hart goniometer was used to measure the advancing contact angle (θ) of solutions at equilibrium (approximately 3-5 minutes after a saline drop was deposited). The contact angle is the angle formed between the edge of a drop of the liquid and the surface upon which the drop is placed and is a measure of a solution's wetting ability. For example, water completely wets glass which is a hydrophilic surface and consequently the contact angle is zero. Water will not completely wet plexiglass and forms a contact angle of 65°. A solution is said not to wet at all if its contact angle is greater than 90°. Solutions with contact angles below 90° are said to wet incompletely. Under dynamic conditions, solutions with contact angles less than 90° can form a continuous liquid layer on some solid surfaces.

Surface Tension Measurements: The surface tension of polymer solutions and ophthalmic solutions was measured using Wilhelmy plate method. A platinum plate is suspended from one area of a balance and dipped into the liquid. One then gradually lowers the container holding the liquid and notes the pull on the balance when detachment occurs. A pressure transducer is used to determine the surface tension. The force is then converted into surface tension upon calibration of the instrument (ref. 7).

Viscosity Measurements: Viscosity measurements were taken by a Brookfield viscometer.

Coefficient Of Friction Measurements: A device has been developed in our
lab (Figure 4) which measures the coefficient of friction and the scuff load
of opthalmic solution for polymeric surface (ref. 8). A strain gauge bridge.
employed to measure the stability of the lubricant film.

Horizontal Force Measurements: The measuring circuit is stabilized by
running the machine with its cover in place for about 30 minutes. The stylus
is displaced horizontally in the direction of the plate rotation applying
definite force (0-1 gm, etc.) by means of a spring scales when it is out of
contact with the plate. This vertical force at a given sensitivity produces a
corresponding horizontal component that can be registered and measured on a
recorder chart. Coefficient of friction (μ) at a given vertical load and
speed (under a set of experimental conditions) can be computed for a lubricant
film from this calibration.

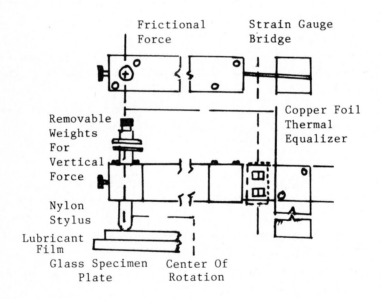

Fig. 4 Friction test apparatus, stylus and specimen geometry.

Vertical Force Adjustment: After installation of a clean specimen plate and stylus, the stylus is adjusted for a minimal clearance from the plate.   In a dry condition, this can be achieved by lowering the stylus until no contact sound is heard when the stylus is depressed towards the plate.   When tear substitute is introduced, it will form a film whose upper surface will contact the stylus, and because of its viscosity will produce a horizontal component increase initially in a nonlinear fashion up to a certain load.   Beyond this minimum load (1 or 2 gm is required to bring the stylus in contact with the specimen plate), the horizontal force begins to increase linearly as a function of load.   The coefficient of friction is determined as the slope of the linear portion of the plot between the measurements of vertical force (ΔFv) and horizontal force ΔF_H, i.e.

$$\Delta F_H / \Delta F_V \tag{1}$$

ESCA measurement: Electron spectroscopy for chemical analysis (ESCA) utilizes a monochromatic source of low energy x-ray to produce a core level ionization (ref. 9). The energies of the ejected photoelectrons are measured and subtracted from that of the incident photon to obtain their binding energies. These energies identify the elements present and from slight but readily detectable shifts in energy also contain information on the chemical states of the atoms.   The collected electrons derive only from the sample surface, thus ESCA reflects the characteristics of the sample surface.   This technique is expected to probe 5 to 30 $\overset{\circ}{A}$ thick surface zone of materials.

SPECIFIC EXPERIMENTS AND RESULTS

1. PMMA and silicon acrylate lens plates were treated with (dipped in) 1% hydroxypropyl methyl cellulose for 24 hours.   Solutions was prepared in saline.

2. The lens plates were rinsed in saline.

3. Rinsed lens plates were air dried in horizontal position, laid on tracks.

4. Contact angle on above treated plates were studied with a drop of saline and distilled water.

Table 1 represents the data of contact angle and coefficient of friction measurements on silicon acrylate and PMMA lens plates treated with Hydroxypropyl methyl cellulose.

TABLE 1

Contact Angle and Coefficient of Friction Measurements on Silicon Acrylate and PMMA Lens Plates Treated with 1% Hydroxypropyl Methyl Cellulose (HPMC) (Solution Prepared in Saline)

* pH of 1% HPMC solutions - 7.45
* Surface Tension - 72.695 dyne/cm
* Coefficient of Friction was measured between PMMA plate and PMMA ball of 0-25 inches in radius at a stylus velocity of 400 mm/sec with a vertical load of 10 gm.

Serial No.	Number of Rinses in Saline	Contact Angle with saline Silicon Acrylate Lens plate	Contact Angle with saline PMMA lens plate	Contact angle with distilled water on PMMA plate	Coefficient of friction Measurement on PMMA plate
1	Control (Clean plate)	59°	56°	66°	0.220
2	0	20°	21°	30°	
3	10	25°	23°	31°	
4	20	30°	23°	31°	
5	25	35°	23°	31°	0.12825
6	30	40°	23°	31°	
7	40	40°	23°	32°	
8	50	44°	25°	32°	0.1325
9	75	45°	39°	---	0.180
10	100	48°	56°	---	0.222

HPMC used for the experiment is from Aldrich Chemical Company, Inc. Cat #20,032-8, Lot #0112AL

Correlation of ESCA and Contact Angle Measurements of Polyvinyl Alcohol Adsorbed on Silicon Acrylate Surface

As shown in Figure 5, we have established that for a silicon acrylate surface, both the ESCA studies and contact angle measurements show that almost 50 rinses are required to remove the adsorbed polymer molecules from the surface.

POLYVINYL ALCOHOL (M.W. 25000, 88% HYDROLYZED)

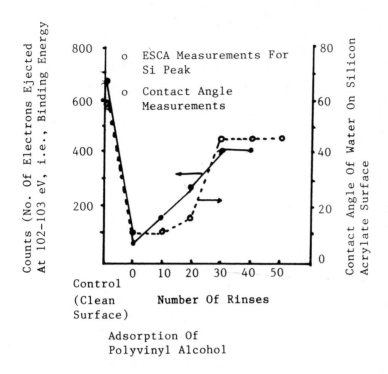

Fig. 5 A correlation of ESCA and contact angle measurements for polyvinyl alcohol adsorbed on a silicon acrylate surface.

Coefficient of Friction Measurement of Commercial Ophthalmic Solutions

   Table II represents the data of Coefficient of Friction, Viscosity, surface tension and contact angle for ten commercial solutions (ref. 10), Figures 6, 7, 8, 9.

TABLE 2

Surface Chemical and Lubrication Properties of Various Ophthalic Solutions

Commercial Tear Substitute	Coefficient of Friction*	Viscosity (cp)	Surface Tension (dyne/cm)	Contact Angle of the solution on clean PMMA
Adsorbotear (Alcon)	0.112	10.03	47.9	42°
Tears plus (Allergan)	0.121	4.52	46.4	50°
Lens mate (Alcon)	0.152	23.61	34.8	20°
Adapettes (Alcon)	0.153	2.13	60.0	52°
Tear gard (Bio Products Op.)	0.161	21.53	41.7	45°
Tears Naturale (Alcon)	0.164	7.03	30.8	26°
Muro tears (Muro Pharm.)	0.172	2.34	35.5	29°
Neo-tears (Barnes-Hind)	0.184	15.51	48.8	39°
Hypotears (CooperVision)	0.191	2.43	38.6	40°
Liquifilm tears (Allergan)	0.213	3.91	45.0	46°

* Coefficient of friction was measured between PMMA plate and PMMA ball of 0.25 inch in radius at a stylus velocity of 400 mm/s (80.4 revs/min) with a vertical load of 5 g.

† PMMA = polymethyl methacrylate

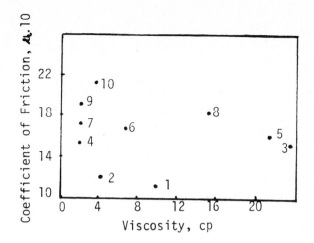

Fig. 6 Coefficient of friction as a function of viscosity of various ophthalmic solutions between polymethylmethacrylate with a vertical load of 5g at a speed of 400 mm/sec (stylus velocity). 1) Adsorbotear, 2) Tears Plus, 3) Lens Mate, 4) Adapettes, 5) Tear Gard, 6) Tears Naturale, 7) Muro Tears, 8) Neo-Tears, 9) Hypotears and 10) Liquifilm.

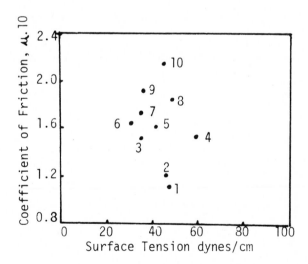

Fig. 7 Coefficient of friction for polymethylmethacrylate surfaces as a function of surface tension for various ophthalmic solutions: 1) Adsorbotear, 2) Tears Plus, 3) Lens Mate, 4) Adapettes, 5) Tear Gard, 6) Tears Naturale, 7) Muro Tears, 8) Neo-Tears, 9) Hypotears and 10) Liquifilm.

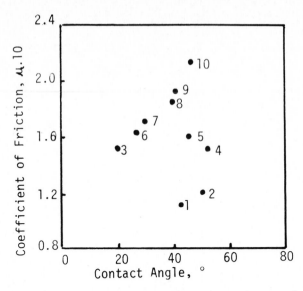

Fig. 8 Coefficient of friction of polymethyl methacrylate surfaces as a function of contact angle for various opthalmic solutions: 1) Adsorbotear, 2) Tears Plus, 3) Lens Mate, 4) Adapettes, 5) Tear Gard, 6) Tears Naturale, 7) Muro Tears, 8) Neo-Tears, 9) Hypo-tears and 10) Liquifilm.

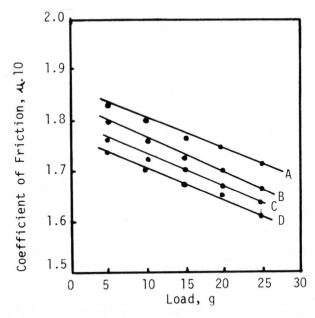

Fig. 9 Coefficient of friction between polymethyl methacrylate/ nylon at different loads and speeds for Tears Naturale.
      A. 400 mm/s
      B. 500 mm/s
      C. 600 mm/s
      D. 700 mm/s

DISCUSSION

The rationale behind studying adsorption of polymers on contact lens surfaces in relation to its biolubrication is to develop better understanding of biolubrication occurring between polymer molecules and corneal or contact lens surfaces. It is better to use low energy model surfaces such as PMMA, etc. to simulate a situation similar to that of the corneal surface with its critical surface tension of dyne/cm (ref. 11). The strongly adsorbed polymer on the lens surface will reduce friction of the sliding surfaces because of the increased thickness and stability of the tear film. Different polymers can adsorb at air/tear interface and the cornea/tear interface or both.

The ophthalmic solutions, comfort drops, cushioning agents and wetting agents are made with the blends of different ophthalmic polymers and other components. The interaction among the ingredient of these formulations will influence the surface properties, adsorption characteristics and coefficient of friction.

The solution with low surface tension and low contact angle will have the best wetting properties on contact lens and ocular surfaces (ref. 12). The ophthalmic solutions should have strong affinity for ocular surface, prolonged reaction time and low viscosity (ref. 13) since high viscosity could have undesirable side effects, such as blurring or vision and tendency to pull of epithelial filament.

In summary, the results presented in this paper establish that ophthalmic polymers have a wide range of coefficient of friction. Adsorption of ophthalmic polymers on lens surfaces is inversely related to contact angle and coefficient of friction measurements. Therefore, the polymers which are adsorbed strongly can be used for preparation of improved eye care products.

Acknowledgments

Alcon laboratories, Inc., Fort-Worth, Texas, provided the research grant that enabled the authors to undertake this project.

REFERENCES

1   Lerk, J.R., The Eye in Contact Lens Wear, Butterworth Co., 1985, p.6.
2   Brauninger, G.E., Shah, D.O., and Kaufman, H.E., Direct Physical
    Demonstration of Oily Layer on Tear Film Surface, American Journal of
    Ophthalmology, 1972, 73, pp 132-134.
3   Holly, F.J., and Lemp, M.A., Wettability and Wetting of Cornea Epithelium,
    Experimental Eye Research, 1971, 11, pp. 239-250.
4   Lemp, M.A. and Holly, F.J., Ophthalmic Polymers as Ocular Wetting Agents,
    Annals of Ophthalmology, January 1972, pp 15-20.

5   Appeldoorn, J.K., and Barnett, G., Frictional Aspects of Emollience, Proceedings of the Scientific Section of the Toilet Goods Assoc., December 1963, No. 40, pp 28-35.

6   Ehlers, N., the Preocular Tear Film and Dry Eye Syndromes, Acta Ophthalmology, 1965, 81, pp 111-113.

7   Adamson, A.W., Physical Chemistry of Surfaces, Wiley Interscience, New York, 1982, p 24.

8   Kalachandra, S., and Shah, D.O., Friction Tester for the Lubrication Properties of Ophthalmic Solutions, Rev. Sci. Instrum., June 1984,55, p 6.

9   Mullendore, A.W., Nelson, G.C., and Holloway, P.H., Surface Sensitive Analytical Techniques:An Evaluation, Sandia Laboratories,Alburquerque,N.M.

10 Kalachandra, S., and Shah, D.O., Lubrication and Surface Chemical Properties of Ophthalmic Solutions, Annals of Ophthalmology, Vol. 17, 11, Nov 1985, pp. 709-711.

11 Holly J.F., Lemp, M.A., Surface Chemistry of Tear Film: Implications for the Dry Eye Syndrome, Contact Lenses and Ophthalmic Polymers, J. Contact Lens Soc. Am., 1971, 5: 12-19.

12 Holly, F.J., and Lemp, M.A., Wettability and Wetting of Cornea Epithelium, Experimental Eye Research, 1971, 11, pp 239-251.

13 Lemp, M.A., Design and Development of An Artificial Tear. Presented at 80th Annual Meeting of the American Academy of Ophthalmology and Otolaryngology, Dallas, Texas, September 21-25, 1975.

PART 2

INTERFACIAL PHENOMENA IN ADVANCED MATERIALS

Interfacial Phenomena in Biotechnology and Materials Processing,
edited by Y.A. Attia, B.M. Moudgil and S. Chander
Elsevier Science Publishers B.V., Amsterdam, 1988 — Printed in The Netherlands

GENERATION OF AEROSOL PARTICLES BY BUBBLES

Richard Williams and Jordan Roy Nelson

David Sarnoff Research Center, SRI international CN-5300
Princeton, New Jersey (USA) 08543-5300

SUMMARY

A bubble in water rises to the surface and bursts, giving rise to a shower of small droplets. The droplets quickly evaporate and leave behind any dissolved solids as aerosol particles. Such bubble-generated particles can be a source of airborne contamination in clean rooms. During semiconductor processing, for example, racks of silicon wafers are routinely immersed in aqueous solutions and then removed. Bubbles generated in these operations would contribute to the salt aerosol in the room. In other work environments, bubble generation makes a similar contribution to respirable dusts in the work area.

We have generated aerosol particles by controlled bubbling of air through an $(NH_4)_2SO_4$ solution. Using a particle counter, we measured the number density and size distribution of the aerosol. The particles are solid, not liquid droplets, and small enough to be a true aerosol. We will show our data and discuss the particle generation process, together with the significance of bubble-generated aerosols.

INTRODUCTION

Water often contains dissolved solids, colloidal particles or microorganisms. Without giving it much thought, we usually assume that these materials stay in the water. When bubbles are present, this assumption is not correct. Bubbles rise to the surface and burst, giving rise to showers of fine droplets. The droplets quickly evaporate and leave behind solid aerosol particles. Bubbles in ocean surf generate enormous numbers of salt aerosol particles. The particles act as cloud condensation nuclei and, for this reason, meteorologists have studied the generation mechanism in detail.[1,2] A bubble 2-mm in diameter forms about 100 fine droplets that evaporate within seconds, leaving behind micron-sized salt particles. Colloidal particles can be injected into the air in the same way. It has been shown that radioactive clay sediment particles in the Irish Sea near a nuclear fuel reprocessing plant are carried into the atmosphere by bursting bubbles.[3] Similar water-air transfer may affect a proposed operation to boil off 2.1 million gallons of water containing dissolved radioactive material left after the accident at Three Mile Island.[4] Clearly, bubbles could also carry microorganisms from water into the air.

In our work, we have looked at possible effects of bubble-generated aerosol particles on a clean room environment. In semiconductor processing, effective filtering practices have been used to reduce the aerosol content of incoming air to very low levels. Most of the troublesome particles are now believed to be generated by clean room workers or to be suspended in the liquids used in the processing. A careful watch is now kept on the suspended-particle count of the liquids used. Bubble generation of aerosol particles might take place in aqueous solutions of several non-volatile materials that are now commonly used in semiconductor processing. These include NH_4F, $NaOH$, H_2SO_4, and H_3PO_4.

EXPERIMENTAL

We generated aerosol particles by passing nitrogen gas through a glass frit bubbler into a 0.1 molar aqueous solution of $(NH_4)_2SO_4$, using a gas flow rate of 5.7 liters/min. (Figure 1). The bubbles burst at the surface, forming a salt-particle aerosol. To count the particles, we used an aerosol particle counter, Hiac/Royco, Model 247, Pacific Scientific Co. A sampling head, mounted about 20 cm above the surface of our liquid, draws in air at a rate of 1 cu ft/min. The air passes through about 3 meters of tubing to a counter. Along the way, the liquid droplets evaporate, leaving only the salt particles in the airstream that reaches the counter. We confirmed this by experiments the particles over pure water blanks, with and without bubbling. In this case, the particle count was not enhanced by bubbling. Water droplets generated by the bubbles are not counted unless there is enough dissolved salt present to give rise to a solid particle when the droplet evaporates. We stress this point to emphasize that the data we show refer to a permanent aerosol made up of solid particles and does not refer to a transient fog of water droplets. When we use a salt solution, turning on the bubbler greatly enhances the particle count. When we use pure water, turning on the bubbler does not enhance the particle count.

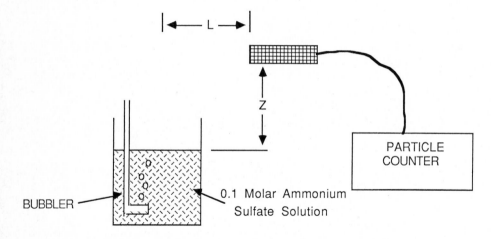

Figure 1- Experimental arrangement used to count the aerosol particles generated by bubbles passing through a solution of $(NH_4)_2SO_4$. In our experiments, the intake port of the counter was placed such that $L \sim 10$ cm and $Z \sim 15$ cm.

We did two kinds of experiment. In the first of these (Figure 2), the experiment was done in an ordinary laboratory where the air is not especially treated to remove airborne particles. The top graph shows the number of particles having sizes $> 5\mu$ for several different runs. In a run, the counter operates for a period of 1 minute. The bottom graph shows the count for particles having sizes $> 1.5\mu$.

SALT AEROSOL (5 μ Particles)

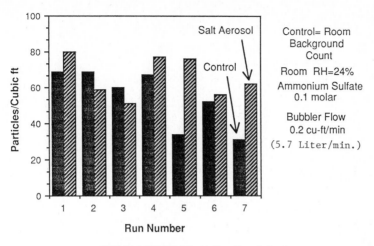

SALT AEROSOL (1.5 μ Particles)

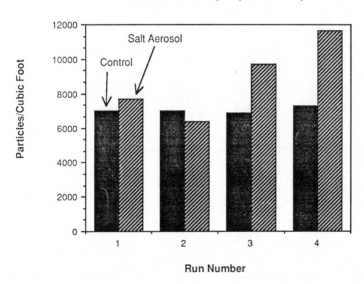

Figure 2- Data taken in ordinary laboratory air. The background count is relatively high, but the enhancement due to bubble-generated particles can be clearly seen. (100 `particles/cubic ft.` `= 3.5 particles/liter`).

In each case, the background count is with the bubbler shut off. The graphs show that bubbles in a salt solution increase the aerosol particle count significantly, even in ordinary room air. Figure 3 shows similar data taken in a connecting room between ordinary laboratory space and a semiconductor processing area. Some clean room practices are observed in this area, making it significantly cleaner than an ordinary room. Our background count for this room falls between the limits[5] for Class 10,000 and Class 1,000 rooms. The main difference between these data and

136

those of Figure 2 is the lower background count in Figure 3

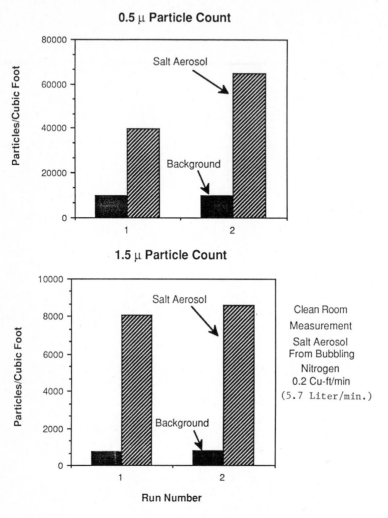

0.5 μ Particle Count

1.5 μ Particle Count

Clean Room
Measurement
Salt Aerosol
From Bubbling
Nitrogen
0.2 Cu-ft/min
(5.7 Liter/min.)

Run Number

Figure 3- Data taken in the connecting room to a semiconductor clean laboratory. Here the background particle count is reduced, and bubble-generated particles predominate.
(100 particles/cubic ft. = 3.5 particles/liter).

DISCUSSION

Our results show that bubbles in a salt solution generate high concentrations of aerosol particles. These particles did not exist, as such, in the water. They are created by the bubbles bursting at the surface. Water may also contain stable particles, such as colloidal materials or bacteria . Bubbles would carry these particles into the air in the same way[3].

In our experiments, when the bubbler is operating, the air quality in the vicinity corresponds to that in a Class 100,000 room. For semiconductor processing this would be a catastrophe. Of course, no operation in processing generates as many bubbles as we have used in our experiment. However, the immersion of racks of wafers into a liquid and the subsequent removal inevitably generates some bubbles and we would expect a corresponding increase of the concentration of aerosol particles.

In salt water fish tanks, it is common practice to bubble air continuously. We would expect this to cause a significant aerosol of salt particles as well as any microorganisms that are in the water.

REFERENCES

1 H.R. Pruppacher and J.D. Klett, Microphysics of Clouds and Precipitation", D.Reidel Publishing Co., Dordrecht, Holland, (1980) pp. 194-7.
2 D.E. Yount, "On the Evolution, Generation, and Regeneration of Gas Cavitation Nuclei", J.Acoust.Soc.Am., 71 (1982) pp. 1473-81.
3 M.I. Walker, W.A. McKay, N.J. Pattenden and P.S. Liss, "Actinide Enrichment in Marine Aerosols", Nature, 323 (1986) pp. 141-3.
4 New York Times, April 18, 1987.
5 "Clean Room and Work Station Requirement, Controlled Environment", U.S. Federal Standard No. 209a (1966).

Interfacial Phenomena in Biotechnology and Materials Processing,
edited by Y.A. Attia, B.M. Moudgil and S. Chander
Elsevier Science Publishers B.V., Amsterdam, 1988 — Printed in The Netherlands

THE SURFACE CHEMISTRY OF HIGH-Tc SUPERCONDUCTING $YBa_2Cu_3O_{7-x}$ MATERIAL

Mohamed E. Labib and Peter J. Zanzucchi

David Sarnoff Research Center, CN 5300
Princeton, New Jersey, (USA) 08543-5300

SUMMARY

The surface chemistry of $YBa_2Cu_3O_{7-x}$ was studied by electrophoresis, Fourier-Transform infrared spectroscopy and other techniques. Suspending 2% (by weight) powder in water increased the pH from 6 to 12. The specific conductance of the supernatant increased by a factor of 1000 over a 12-hour period. Electrokinetic studies indicated that the surface of the superconducting material is basic and highly unstable. Analysis of the residue, which remained after a prolonged soaking in water, showed a reduction of about 30% of barium. The formation of $Ba(OH)2$ was confirmed by wet chemical analysis. Data obtained by infra-red spectroscopy support the concept that formation of a $BaCO_3$ surface layer occurs.

INTRODUCTION

The discovery of the new ceramic high Tc yttrium-barium-copper oxide superconducting material (YBC) has resulted in an unprecedented interest in the properties of rare earth containing ceramics. Applications of the new material are likely to have a great impact on several technologies. The stability of this new ceramic material is a necessary requirement for developing such new applications. In comparison, the success of the silicon technology is attributed to the unique properties of silicon-silicon dioxide interface and to the surface stability.[1] This paper presents some aspects of surface chemistry and stability of the yttrium-barium-copper oxide material. Implications of the sensitivity of this material to ambient conditions will be discussed with regard to proposed applications.

In 1911, Onnes[2] discovered the phenomenon of superconductivity in metals. Thereafter, extensive research lead to the discovery of Niobium metal with a Tc of 17K and of Nb_3Ge with a Tc of 23K.[3] More recently, the discovery by Bednorz and Muller in 1986[4] of a mixed-valence copper oxide perovskite with a Tc of 33K has fueled research efforts and generated new interest in superconductivity. The subsequent discovery of the 90K-Tc-$YBa_2Cu_3O_{7-x}$ by Chu and coworkers[5] has made this material the center of attention for possible applications. Although the problem of stability in ambient has not been a major problem for metal and alloy superconductors, it is apparent that the new ceramic material suffers from inherent chemical instabilities.

Since the synthesis of high Tc ceramic materials involves grinding, milling,and other processes that require the use of solvents, the knowledge of surface chemistry is essential. In this paper, we report our results on the surface chemistry of the YBC material and on its stability , especially with respect to exposure to moisture and suspension in water or other solvents. Wet chemical, electrophoretic, and spectroscopic techniques were used in this investigation.

EXPERIMENTAL

Synthesis Procedure: The YBC material was prepared by solid-state synthesis. Barium carbonate, yttrium oxide, and cupric oxide were mixed and ground in a mortar. The mixture was placed in a platinum boat and fired at 950C for 24 hours in oxygen. The resultant material was reground and fired at 950C for another 6 hours, cooled to 500C over 6 hours, then annealed at 500C for 24 hours in an oxygen atmosphere. The material was cooled to room temperature, over 6 hours and then stored in a dry box under nitrogen.

Electrophoresis Techniques: The electrophoretic mobility of the YBC material was measured as a function of pH using two techniques, namely, microelectrophoresis and an acoustic-based technique. Laser-Z-Meter[R] Model 500 made by Pen-Kem Inc. was used for the microscope electrophoresis study. Acoustophoresis[R] Model 7000, made by the above company, was used to study stability and obtain surface chemical information on the YBC material at high solid concentration (10%).

Fourier-Transform Infra-red Spectroscopy: The infrared spectra of individual grains of YBC material were obtained by use of the microscope stage of a Digilab FTS-60 Fourier Transform Infrared Spectrometer. The maximum aperture for this microscope is about 400 microns and the minimum aperture is about 20 microns. Grains on the order of 50-200 microns were characterized. The infrared spectra of aggregates were obtained in two ways: by diffuse reflection using a Harrick diffuse- reflection attachment [Harrick Scientific Co., Ossining, New York] and a Digilab Model FTS-15C Fourier Transform Infrared Spectrometer or by the standard potassium bromide pellet technique using the same spectrometer. The pellet technique was used primarily for characterization of hydrolysis products, but not with untreated YBC material to avoid introducing possible unexpected surface chemistry.

RESULTS

<u>Microscopic Examination:</u> The material made by the above synthesis had a Tc of 90K as measured by susceptibility technique (Figure 1). When we placed a drop of de-ionized water on the surface of a sintered disc, a large amount of white substance formed, after drying the water droplet (Figure 2). Exposure of the YBC material to high relative humidity (70%) had a similar effect. SEM examination of the surface of the powder showed needle-like crystals growing out of the surface.[6]

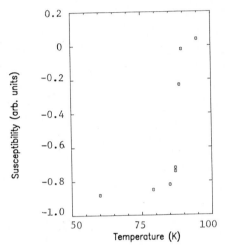

Figure 1: Susceptibility vs. temperature curve for the material used in this investigation.

(a) CONTROL
$YBa_2Cu_3O_{7-x}$

(b) AFTER CONTACT WITH WATER

Figure 2: Effect of water of the YBC material--a heavy white substance is formed after water evaporation.

<u>Wet-Chemistry Study:</u>   The pH of a dispersion made by suspending two grams of the YBC material in 100 ml of distilled water resulted in an increase in pH from 6.0 to 12.0--over a period of 12 hours (Figure 4).   In the same experiment, the specific conductance of the same suspension increased by a factor of 1,000 (Figure 5).   A plot of conductance versus square root of time gave a straight line which indicates that the   leaching process is mostly "diffusion-controlled."

Acid-base titration and emission spectroscopic analysis of the supernatant indicated that barium is leached out with subsequent   formation of barium hydroxide.   Further analysis of a dried residue showed the presence of   barium carbonate.

An important aspect of processing the YBC material is milling.   We found that milling in water destroys the superconductivity properties.[7]   Milling in isopropanol was safe and the resultant material was only slightly deteriorated, with respect to superconductivity.[7]   The specific conductance of a dispersion of YBC material in methanol or isopropanol increased by only a factor of 1.5 (Figure 6).   We believe that the increase in conductivity of the supernatant, in the case of alcohols, was due to residual water and to absorption of water from the   atmosphere.

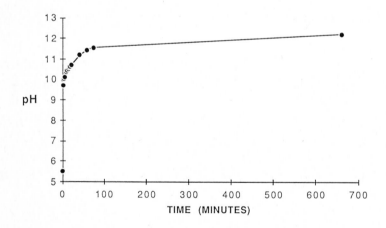

Figure 3: pH increase of a suspension of YBC material in water.

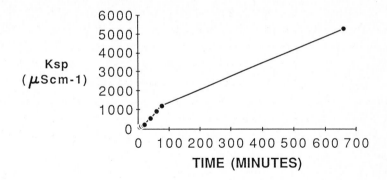

Figure 4: Specific conductance of YBC suspension in water.

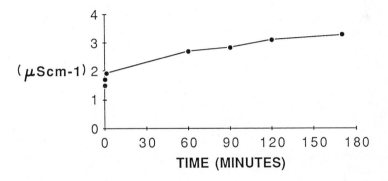

Figure 5: Specific conductance of a suspension of YBC in methanol.

Electrokinetic Studies: A 10% suspension of YBC material was investigated by the acoustophoretic technique. This technique is based on the fact that sonic waves disturb the electrical double layers around particles in a periodic fashion and that measurement of the potential drop across a known distance is a measure of the electrokinetic properties of the suspended particles. In concept, this technique is analogous to measurement of sedimentation potential. Due to the instability of the YBC material, we started the measurements quickly. When we titrated with KOH, the surface of the material had an iso-electric point of about 12. This indicates a highly basic surface.

144

hydroxyl and carbonate containing salts. The hydroxyl modes, 3650-3710 cm-1, are free hydroxyl modes characteristic of metal hydroxide[9] indicate the presence of Ba(OH)2 and an unidentified second metal hydroxide phase, possibly a complex hydroxide-carbonate barium salt. The strong absorption mode near 1400 cm-1 is characteristic of carbonate salts. By comparison, diffuse reflectance spectra of the same water-treated material are characterized by featureless hydroxyl absorption and carbonate anion absorption which suggest surface reaction with water and carbon dioxide to form surface carbonates.

Infrared spectra of white residue on water treated YBC material indicate that the hydroxides which initially form convert, in a short period of time, to carbonate salts and the hydroxyl modes do not appear in the spectrum (Fig. 9).

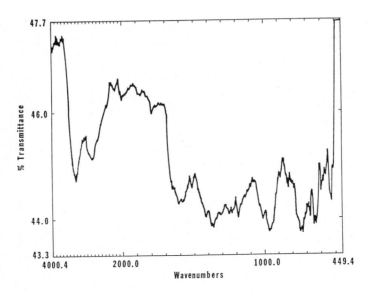

Figure 7:   Infra-red   spectrum of grain cluster of the YBC material.   Complex organic and inorganic infrared absorption characteristic of material: Organic salts (3000-2800 and 1650 cm^{-1}) ; Metal oxide (1000-450 cm^{-1}, see Reference 10).

The results of several acid-base titrations were not reproducible because of the reactivity of the material. Repeated measurements showed that the surface becomes positively charged even in highly alkaline media (pH >11).

The electrophoretic mobility was also measured using microelectrophoresis. At ionic strength of 10^{-3}, we identified that an iso-electric point at pH 9.0 (Figure 8). Again, the surface of the material suffered from instabilities as in the case of acoustophoretic measurements.

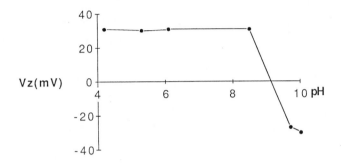

Figure 6: Zeta potential results of YBC particles in an aqueous system.

Fourier-Transform Infrared Study: By use of the microscope stage of a Digilab FTS-60 Fourier Transform Infrared Spectrometer, the effect of moisture on small grains of superconducting YBC ceramic could be studied. For untreated grains, which were approximately 50-200 microns in size, the infrared spectrum is characterized by: a large number of very weak, overlapping bands in the region of 1700-500 cm-1; the absence of strong dispersion effects (Christiansen effect) which suggests a low refractive index; and the absence of significant hydroxyl, or bound water, in the bulk of the grain(Fig. 7). By comparison, the diffuse reflectance spectrum of the same powder is essentially featureless with very weak structure in the 1200-400 cm^{-1} region which suggests the surface of the powders is comparable in composition to that of the bulk. The surface is not significantly hydrated nor do carbonates readily form on exposure to relatively dry air as occurs with some rare earth oxides.[8]

After addition of de-ionized water to the grains and its evaporation, a residue remains which surrounds individual grains. The infrared spectrum of the residue, Fig. 8, shows a complex spectrum which represents inorganic

146

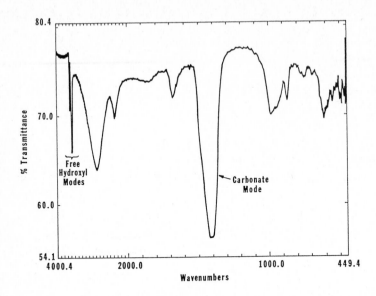

Figure 8: Infrared spectum of white residue as formed: Carbonate and hydroxyl containing salt.

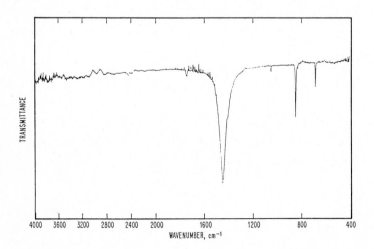

Figure 8: Infrared spectrum of white residue, aged: Carbonate, but hydroxyl free, salt.

DISCUSSION

The $YBa_2Cu_3O_{7-x}$ material is sensitive to moisture. The superconductivity properties deteriorate as a result of exposure to moisture. Immersing the powder in water, for 24 hours, distroys all the superconductive crystalline phase as judged by susceptibility and by independent x-ray diffraction measurements. Needle-like crystals are formed due to exposure to moisture. This material has been identified by Fourier-transform infra-red spectroscopy as barium carbonate.

Water interacts with the barium component of the YBC material to produce barium hydroxide which in turn forms barium carbonate as a result of reaction with atmospheric CO_2. This is the mechanism by which the moisture caused the deterioration of the YBC superconductivity. The increase in the pH and specific conductance of YBC-water suspension is attributed to the leaching out of barium. Barium seems to be highly reactive towards water in the perovskite structure. These data are consistent with the barium reactions proposed by Yan et al.[11] based on the x-ray study of water treated YBC material.

The linear increase in conductivity with the square root of time is attributed to a diffusion-controlled leaching process. We have estimated from x-ray fluorescence analysis that as much as 30% of the barium in YBC is lost during 12-hours the soak in water. Therefore, milling in water is not appropriate to process this material.

The deterioration of the surface layer due to exposure to ambient conditions has been demonstrated in a previous paper using x-ray photoelectron spectroscopy[6].

During applications, the reaction with moisture due to condensation causes serious limitations for applications of the new materials. The formation of a "dead layer" in the case of thin film YBC material would likely be an additional serious problem. We believe that use of passivation or encapsulation technologies will be required using this material in practical applications.

REFERENCES

1   G. Barbotten and A. Vapaille, Editors, "Instabilities in Silicon Devices -- Silicon Passivation and Related Instabilities", Volume 1, North-Holland, Amestrdam, New York, Oxford, Tokyo (1986).

2   H. Kamerlingh-Onnes,   "Disappearance of the Electrical Resistance of Mercury at Helium Temperatures",   Leiden Comm. 122b (1911).

3   CRC Handbook of chemistry and Physics (1983-1984).

4   J.G. Bednorz and K.A. Muller, "Possible High Superconductivity in the Ba-La Cu- O System", Z. Phys. B - Condensed Matter $\underline{64}$ (1986) pp. 189-93.

5   M.K. Wu, J.R. Ashburn, C.J. Torng, P.H. Hor, L. Gao, R.L. Meng, Z.J. Huang, Y.Q. Wang, and C.W.Chu"Superconductivity at 93K in a New mixed-phase Y-Ba Cu-o Superconducting Compound System at Ambient Pressure",   Phys. Rev. Lett. 58 (1987) pp. 908-10.

6   J.H. Thomas, III, and M.E. Labib,   "The Effect of water on the Surface Chemistry of Superconductive $YBa_2Cu_3O_{7-x}$", Proccedings of the Topical Conference on Thin Film Processing and Characterization of High Temperature Superconductors,   Americn Vacuum Society,   American Institute of Physics Symposium Series (1988).

7   M.E. Labib, and J.R. Matey, "Effect of Water on the Superconducting Properties of Y-Ba -Cu-O Superconducting Material,   to be published.

8   F.A. Cotton, and G. Wilkinson, "Advanced Inorganic Chemistry", Interscience Publishing Co. (1962) p.879.

9   K. Nakamoto,   "Infrared Spectra of Inorganic and Coordination Compounds", John Wiley and Sons, New York, London (1970) pp. 81-2.

10   S. Sugai,   "Effect of Oxygen Deficiency on the Infrared Spectra in $YBa_2Cu_3O_{7-x}$", Phys. Rev. B $\underline{36}$ (1987) pp. 7133-6.

11   M.F. Yan, R.L. Barns, H.M. O'Bryan, P.K. Gallagher, R.C. Sherwood, and S. Jin,   "Water Interaction with the Superconducting $YBa_2Cu_3O_7$ Phase",   Appl. Phys. Lett. $\underline{51}$ (1987) pp.532-4.

Interfacial Phenomena in Biotechnology and Materials Processing,
edited by Y.A. Attia, B.M. Moudgil and S. Chander
Elsevier Science Publishers B.V., Amsterdam, 1988 — Printed in The Netherlands

DYNAMICS OF GROWTH OF SILICA PARTICLES FROM ALKOXIDES

T. MATSOUKAS and E. GULARI
Department of Chemical Engineering, University of Michigan, Ann Arbor, MI 48109

SUMMARY
    Light scattering and Raman spectroscopy were used to study the growth of
silica particles from hydrolysis of tetraethylorthosilicate in the presence
of ammonia. Under conditions of excess water the growth is limited by the
slow hydrolysis. The number of particles reaches a steady-state early in
the process and after this point particles grow without any significant
nucleation. A simple model based on monomer addition growth offers a semiquantitative description of this process.

INTRODUCTION

    The hydrolysis and polymerization of lower silicon alkoxides, primarily

tetramethylorthosilicate (TMOS) and tetrethylorthosilicate (TEOS), have

received a lot of interest recently because of the potential applications

in ceramic processes. In the presence of water, alkoxides hydrolyze and

polymerize through a condensation scheme to produce a network of siloxane

bonds according to the following scheme:

$$Si(OR)_4 + x\ H_2O \rightarrow Si(OR)_{4-x}(OH)_x + x\ ROH \quad \text{(hydrolysis)}$$

$$Si - OH + OH - Si \rightarrow Si - O - Si + H_2O \quad \text{(condensation)}$$

(1)

The structure of the resulting network depends primarily on the pH. At low pH

and water-to-orthosilicate ratio chain-like polymeric units are formed (ref. 1)

with a strong tendency to gel. On the other hand, the presence of a base pro-

motes the formation of a sol whose structure is compact and highly cross-linked.

Under sufficient concentration of the base the sol is stable against aggregation

(ref. 2) and the size of the particles increases depending on the concentration

of the base. Ammonia has been shown (refs. 3 & 4) to promote formation of

spherical particles of narrow spread and of diameters that are sensitive to the

amount of ammonia. The particle size, as well as the spread of the distrib-

ution, depend upon the dynamic competition between nucleation and growth. In

addition to that, hydrolysis has an important part since it controls the release

of the active (hydrolyzed) monomer. The purpose of this work is to study the

factors that affect the dynamics of formation and the size of the particles in

the presence of ammonia. We performed intensity measurements to follow the

overall extent of growth and dynamic light scattering to obtain the time evolution of the particle size. Information about the kinetics of hydrolysis was obtained by Raman scattering. We also made use of the Rayleigh scattering of the plasma lines. These lines which are normally found in the beam provide a convenient way to follow the extent of growth simultaneously with the hydrolysis despite the fact that the siloxane network does not have characteristic Raman bands.

LIGHT SCATTERING MEASUREMENTS

Light scattering is particularly useful in the study of growth processes. The intensity probes the second moment of the particle distribution and the hydrodynamic radius provides a direct measure of the size.

The intensity scattered at an angle θ is given by (ref. 5)

$$I = M_2 P(qr)$$

$$q = \frac{4\pi n}{\lambda} sin\frac{\theta}{2}$$

(2)

where M_2 the 2nd moment of the particle distribution defined as $\sum i^2 n_i$, n_i the number concentration of particles of mass i (in units of the monomer), r a mean particle radius and $P(qr)$ the form factor which for spherical particles is given by (ref. 6)

$$P(qr) = \left(3\frac{sin(qr) - (qr)cos(qr)}{(qr)^3}\right)^2$$

(3)

This correction is important only for values of qr higher than approximately 1. For particles of low polydispersity we can relate the second moment to the first (the mass of the particles)

$$M_1 \sim \frac{M_1^2}{M_0}$$

(4)

where M_0 the total number of the particles, so that M_1 is related to the scattered intensity through

$$M_1 \sim \sqrt{\frac{IM_0}{P(qr)}}$$

(5)

If the number of the particles remains constant, the intensity is a measure of the particles. The low polydispersity of the resulting particles suggests that M_0 reaches a steady state soon, otherwise, prolonged nucleation should yield broad distributions. This we can experimentally test: Since $M_1 \sim M_0 r^3$ the second moment scales as

$$M_2 = \frac{I}{P(qr)} \sim M_0 r^6$$

(6)

If M_c is constant a log-log plot of M_2 vs r must yield a straight line with a slope of 6. A slope higher than 6 indicates ongoing nucleation (an increase in the number of particles) while a slope less than 6 indicates aggregation. An alternative interpretation of slopes less than 6 is that of particles with a fractal structure, namely particles for which the mass scales with the size as (ref. 7)

$$M_1 \sim M_0 r^{D_f}, \quad D_f < 3 \tag{7}$$

In this case the plot is still linear but the slope is equal to $2D_f$.

EXPERIMENTAL

Light scattering experiments were performed using an Argon ion laser operating at 514.5 nm wavelength. To avoid turbid suspensions of large particles we used low concentration of the orthosilicate. Typical composition of the sample was 0.0087 mol TEOS/l, 1.4 mol NH_4OH/l, and 3.2 mol water/l. Hydrodynamic radii were calculated from the Stokes-Einstein equation

$$r = \frac{kT}{6\pi D_z} \tag{8}$$

and the z-average diffusion coefficient was computed from a second order cumulant expansion of the autocorrelations function.

Raman experiments were performed at higher TEOS concentration because of detectability limitations. Also less ammonia was used in order to suppress the formation of large particles and the water concentration was kept well above the stoichiometric ratio (4/1 for complete hydrolysis). The spectra were obtained at an excitation wavelength of 488 nm.

The experiments reported here were all done at 20° C.

RESULTS
Light scattering

The particle size depends on the concentration of ammonia, as well as on the alcohol used as a solvent as shown in Fig. 1. In both alcohols, increasing the concentration of ammonia results in larger particles. Particles in ethanol grow to considerably larger sizes than those in methanol, but this difference eventually becomes smaller at higher ammonia concentration.

Fig. 2 shows the scattered intensity and the particle diameter as a function of time for particles grown in ethanol. Initially we observe an induction period during which the intensity remains constant at the solvent/background level. This period is followed by a slow increase of the intensity until the

152

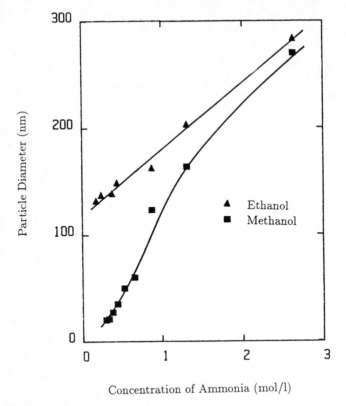

Fig. 1. The effect of ammonia and solvent on the size of the particle

reaction is over. The particle size follows a much faster transient. During
the induction time we cannot measure any meaningful autocorrelations and, as
shown in Fig. 2, the smallest size that we can detect is already within about
50% of the final size. Fig. 3 shows the log-log plot of the intensity
(corrected by the form factor) vs the radius. There is a linear dependence
from which we obtain a slope of 6.1. This shows that the number of the particles has reached a steady state, at least after the particles are within
50% of final size, which corresponds to about 13% conversion of the monomer.

Similar is the behavior when the solvent is methanol with the notable
difference that the growth is considerably faster (Fig. 4). Here we also show
the effect of ammonia. Increasing the concentration of NH_4OH results in a
shorter incubation time and faster growth rate.

We analyzed the intensity assuming first order kinetics and the linearized
plots are shown in Fig. 5. There is a remarkable linearity which apart from
the incubation time describes the growth very well. The fact that the

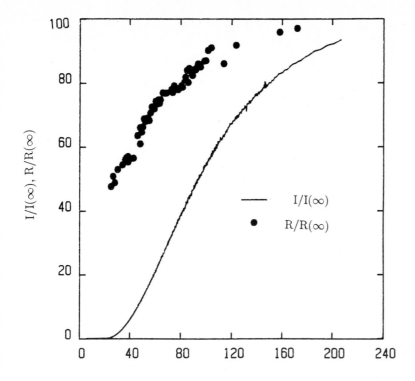

Fig. 2. The scattered intensity and the hydrodynamic radius as a function of time. (Measurements made at $\theta = 120°$)

equation of growth is of the form

$$\frac{dM_1}{dt} = k(c - M_1) \tag{9}$$

indicates that the rate depends on the concentration of the unreacted monomer $c - M_1$ but not explicitly upon M_1 as we would expect for growth which occurs by a surface reaction. Therefore the growth is limited by the availability of the hydrolyzed monomer.

Raman spectra

To verify this conclusion we followed the hydrolysis using Raman scattering. If hydrolysis is the rate limiting step the rate of growth should match the rate of hydrolysis. The practical difficulty is that the siloxane network does not have Raman bands that can be used for a quantitative analysis. In order to overcome this difficulty we made use of the Rayleigh scattering of the plasma lines. These lines are usually found in the laser

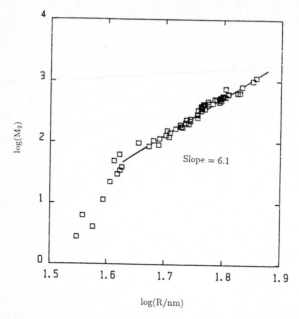

Fig. 3. The second moment vs the radius of the particle. The slope of 6.1 indicates that the number of particles remains constant.

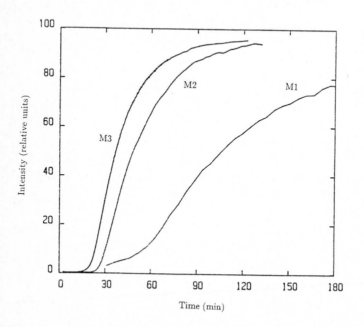

Fig. 4. Intensity measurements from particles grown in methanol with different concentrations of ammonia: (M1) 0.8 mol/1, (M2) 1.2 mol/1, (M3) 1.6 mol/1.

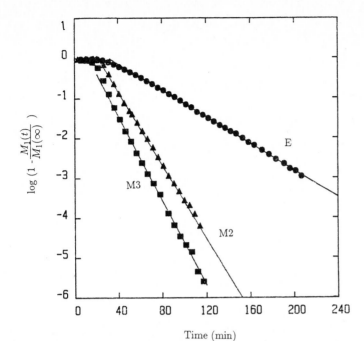

Fig. 5. Linearized plot of the particle mass assuming first order kinetics. Curve (E) is from particles grown in ethanol with the same concentration of ammonia as (M3).

beam and in normal operation they are removed with a grating. If they are left in the beam though, they will show up in the spectrum through Rayleigh scattering (their intensity is very low to produce Raman scattering). Their position is fixed and it does not depend upon the chemical character of the scatters but their intensity is a function of the concentration and size of the particles that scatter them. The intensity of the scattered plasma lines provides the same information that we would get in a light scattering experiment and the advantage is that these measurements can be performed from the same sample and simultaneously with the Raman measurements.

Fig. 6 shows Raman spectra from particles growing in methanol. Due to the low concentration of the orthosilicate only the two bands at 807 and 664 cm^{-1} are visible. The growing band at 880 cm^{-1} is due to the ethanol released during the hydrolysis. The sharp lines at 740, 563 and 531 cm^{-1} are all plasma lines and they increase in intensity as the particles grow. In Fig. 7 we plot the integrated intensity of the Raman bands and the 740 cm^{-1} plasma line. The plasma line shows the characteristic incubation time but no such feature is observed with the orthosilicate or the ethanol. In Fig. 8 we plot the linearized form of the intensity assuming first order kinetics. The y ordinate is

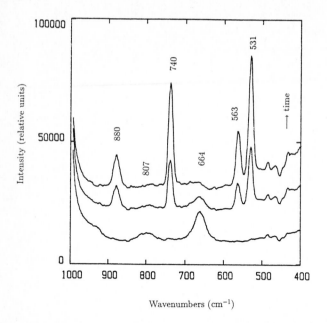

Fig. 6. Raman spectra of TEOS in methanol taken at 45-minute intervals and off-set for clarity. The 664 and 807 peaks are due to TEOS. The 880 peak is due to ethanol. The sharp peaks at 531, 563 and 740 are plasma lines.

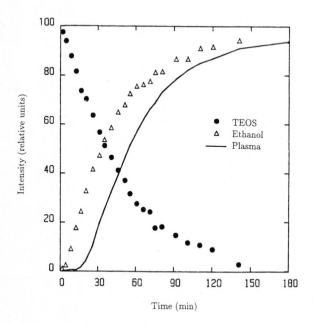

Fig. 7. The integrated intensity of the Raman and plasma peaks as function of time.

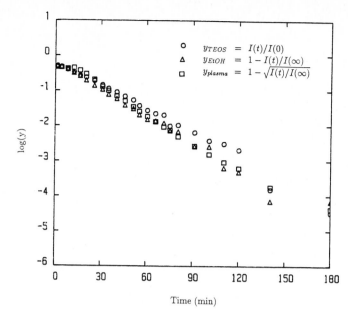

Fig. 8. Linearized plot of the intensity vs time assuming first order kinetics.

calculated for various bands as follows:

$$y_{TEOS} \quad = \quad I(t)/I(0)$$

$$y_{EtOH} \quad = \quad 1 - I(t)/I(\infty)$$

$$y_{plasma} \quad = \quad 1 - \sqrt{I(t)/I(\infty)} \tag{10}$$

We see that all the points fall on a single line with a slope equal to the first order hydrolysis constant. A similar plot is shown in Fig. 9 for particles grown in ethanol.

A GROWTH MODEL

The Raman spectra have confirmed the fact that hydrolysis is the rate limiting step and we can use this information in order to attempt a more quantitative description of the process and gain some insight about the effect of hydrolysis on the particle size. The proposed scheme of chemical equations is simplified to the following steps:

$$C_0 \xrightarrow{k_h} C_1 \quad (Hydrolysis)$$

$$2C_1 \xrightarrow{k_1} C_2 \quad (Nucleation)$$

$$C_1 + C_i \xrightarrow{k_i} C_{i+1} \quad (Growth) \tag{11}$$

158

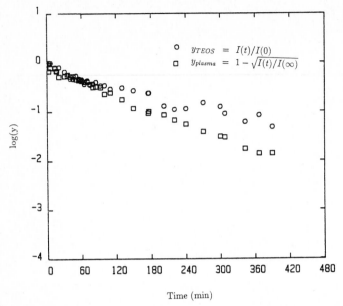

Fig. 9. Linearized plot of the intensity for particles grown in ethanol.

In this description hydrolysis is a first order process in the concentration of the orthosilicate (C_0) and produces the active monomer (C_1). C_1 reacts with either a monomer to produce a dimer C_2 (a "nucleus") or with a particle of mass i to produce a new one of mass $i + 1$. Assuming elementary reaction steps and that the rate of monomer-particle reaction is related to the particle size as $k_i = i \cdot k$, we can show (ref. 8) that the number of particles M_0 is proportional to the square roote of the dimensionless hydrolysis constant κ_0

$$M_0 \sim \sqrt{\kappa_0} = \sqrt{\frac{k_h}{k_p c}} \tag{12}$$

In qualitative terms this result shows that under conditions that don't affect the nucleation constant, the size of the particles depends on the ratio of the hydrolysis and the growth constants. In order to test this prediction experimentally, we first note that it is enough to measure the particle size under varying initial concentration of the orthosilicate, since from the mass balance $M_0 r^3 \sim 1$ the model predicts that the size of the particles should scale as the 1/6th power of the initial concentration of the initial concentration of the orthosilicate. Fig. 10 shows the log-log plot of the particle size vs concentration of TEOS for various concentrations of ammonia. The lines were drawn with a slope of 1/6 and describe the dependence well. Least squares fit produces slopes somewhat higher (between 1/6 and 1/5). Given the simplicity

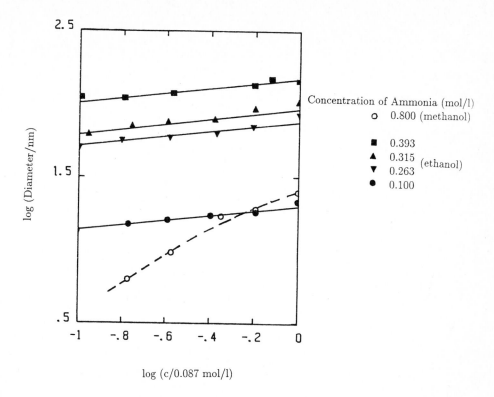

Fig. 10. The size of the particles as a function of the initial concentration of TEOS.

of the model we think the agreement is satisfactory. It must also be noted that slopes larger than 1/6 are consistent with fractal structure of the particles. Our light scattering results don't seem to support this view. We must observe, however, that light scattering cannot resolve this issue for particles in this diameter range and a more definite picture for the particle structures must come through SAXS. We also show in Fig. 10 a similar plot from particles grown in methanol. In this case the 1/6 power law does not seem to apply. But it must be taken into account that the small sizes at the low concentration of TEOS exhibit rather high polydispersities.

DISCUSSION

Hydrolysis is the rate limiting step, even in the presence of excess water. At low ammonia concentration and especially in ethanol, hydrolysis takes hours to reach completion. It is not surprising then, as it has been reported in the literature, that ammonia-catalyzed systems reach the gelation point before the completion of hydrolysis. The role of ammonia is apparently to accelerate the rate limiting step. Ammonia also promotes the formation of larger particles (fewer number of particles) and this in relation to our model suggests that the polymerization rate must also increase in such a way as to result in a smaller dimensionless hydrolysis constant κ_0 and therefore larger particles. Even with increased ammonia concentration the growth remains hydrolysis limited, as manifested by the first order dynamics and this has been the case in all the experiments that we performed. It is conceivable of course that as we increase the ammonia concentration hydrolysis may eventually cease being the rate limiting step. If this is the case, it must happen at such high concentrations that light and Raman scattering experiments cannot be performed because of the large particle size.

Nucleation is limited to the early stages of the process and once some critical concentration of sizable particles has been formed the process becomes essentially a monomer-particle reaction. Unfortunately, this is as much as light scattering can tell about nucleation. The actual nucleation process is inaccessible because of the weak scattering of small particles. As an order of magnitude example, particles at 35% of their final size scatter only 0.02% of the intensity of the fully grown particles. This is the reason why we cannot determine particle sizes less than ~50% of their final size. The slope of 6 which is obtained from the correlation between the 2nd moment and the particle size is indicative of the high degree of cross-linking and the compact structure of the particles (ref. 9) although light scattering is not very sensitive to the internal structure of particles of this size.

The factors that determine the intra-particle structure must be sought in the relative importance of particle-monomer diffusion and particle-monomer reaction. Both these processes are faster than hydrolysis and therefore our experiments cannot discriminate between diffusion limited and reaction limited growth.

The effect of the solvent is not easy to explain, but we speculate that the reverse hydrolysis reaction is responsible since the reaction takes place in excess concentration of alcohol. If mixed orthoesters are formed (ref. 1) methoxy groups must react faster than ethoxy groups and this can explain why the hydrolysis rate is higher in methanol than it is in ethanol. And if the condensation reaction has a comparable rate in both solvents, then larger particles must be formed in the solvent where hydrolysis is faster, i.e. methanol, which is what we actually observe. However our Raman spectra did not show any Si-O-Me bands during the hydrolysis of TEOS in methanol, but we must note that the presence of this band would be difficult to detect because of the low concentrations of TEOS that we used and the relatively high reactivity of this bond.

CONCLUSIONS

We have shown that the ammonia-catalyzed growth of silica particles is controlled by the slow hydrolysis of TEOS even under conditions of excess water concentration. The nucleation is limited to the early stage of the process as manifested by the fact that the number of particles remains constant for the most part of the growth. Ammonia accelerates both hydrolysis and condensation rates but the effect on the condensation rate must be much stronger. A simple monomer addition model provides a good basis for understanding the growth and describes correctly the correlation between the particle size and the concentration of the orthosilicate, at least for growth in ethanol. And finally, we have demonstrated the use of the plasma lines as a means of obtaining information about particle growth simultaneously with the chemical information that we obtain from the Raman spectrum.

ACKNOWLEDGEMENTS

Financial support for this project by the National Science Foundation grant # CBT 830350 is gratefully acknowledged.

REFERENCES

1 C.J. Brinker, K.D. Keefer, D.W. Schaeffer, R.A. Assink, B.D. Kay, C.S. Ashley, Sol-Gel transistors in simple silicates II, Journal of Non-Crystalline Solids, 63 (1984) 45.
2 R. Iler, The Chemistry of Silica, John Wiley & Sons Inc., New York City, New York, 1979, 213.
3 W. Stoeber, A. Fink, E. Bohn, Controlled growth of monodisperse silica spheres in the micron size range, Journal of Colloid and Interface Science, 26 (1968) 62.
4 A.K. Van Helden, J.W. Jansen, A. Vrij, Preparation and characterization of spherical monodisperse silica dispersants in aqueous solvents, Journal of Colloid and Interface Science, 81 (1981) 354.
5 J. Feder, T. Jossang, A reversible reaction limiting step in irreversible immunoglobulin aggregation, in R. Pynn and A. Skjeltrop (Eds.), Scaling Phenomena in Disordered Systems, Plenum, New York, 1985, 99.
6 P. Kratochvil, Particle scattering functions, in M.B. Huglin (Ed.), Light Scattering from Polymer Solutions, Academic Press, New York City, New York, 1972, 340.
7 J.E. Martin, B.J. Ackerson, Static and dynamic scattering from fractals, Physical Review A, 31 (1985) 1180
8 T. Mastoukas, E. Gulari, Monomer addition growth with slow initiation step, submitted to Journal of Colloid and Interface Science.
9 K.D. Keefer, Growth and structure of fractally rough silica colloids, in E.J. Brinker, D.E. Clark, D.R. Ulrich (Eds.), Better Ceramics Chemistry, Vol. 73, Materials Research Society, Pittsburgh, Pennsylvania, 1986, 79.

Interfacial Phenomena in Biotechnology and Materials Processing,
edited by Y.A. Attia, B.M. Moudgil and S. Chander
Elsevier Science Publishers B.V., Amsterdam, 1988 — Printed in The Netherlands

SUSPENSION PROPERTIES OF ALUMINA AND TITANIA SYSTEM: EFFECT OF ADDED SURFACE
ACTIVE AGENTS

A. Srinivasa Rao
Department of Ceramics
Rutgers University
Piscataway, NJ 08855-0909

SUMMARY

   The electrophoretic mobility, rheology and sedimentation behavior of alumina and titania powders (ratio alumina:titania 1:2) was investigated in aqueous
dispersions in presence of three cationic surface active agents in the pH range
3-10. The results suggest that all three additives adsorb on to the surface of
alumina and titania powders in suspension above pH 5.6 and the addition of additives lowers both the slip viscosity and the sedimentation rate of the particles from suspension. The increase in stability of the suspensions also remains independent of the electrophoretic mobility of the dispersed powders in
presence of additive solutions at high concentration (1-2%(wt)).

INTRODUCTION

   It has been well established that a better homogenization of heteroparticulate systems such as alumina and zirconia or alumina and titania powders used
in ceramic processing can be achieved from suspension phase. Therefore it will
be useful to optimize the properties such as the stability of slip prior to the
slip casting operation. In heteroparticulate systems the stability of the
suspension often depends upon the degree of mutual flocculation and different
approaches were suggested in the literature in order to overcome flocculation.
For example while some investigators [1-3] have emphasized the control of pH
of the system, others have suggested the use of an additive [4]. This project
was undertaken in order to investigate the stability of aqueous dispersions of
alumina and titania dispersions in presence of three cationic surface active
agents. In this paper we report our results obtained from alumina and titania
suspensions following electrophoresis, sedimentation and rheology experiments.

EXPERIMENTAL

   Commercial alumina and titania powders that were obtained from Alcoa and
Tioxide International companies and three commercial addtivies Jeff Amines (Jeff
Amine M-302, Jeff Amine M-310 and Jeff Amine M-320 manufactured by Jefferson
Chemicals) were used in this investigation. All three additives are ethoxylated
amines, however, they differ in their chemical composition. It was suggested by
the manufacturer that these agents exhibit both the characteristics of cationic
and non-ionic additives depending upon their concentration in the suspension and

also upon the activity of the adsorbing sites. While Jeff Amine M-302 and M-310 contain 2 and 10 ethylene oxide chains respectively, Jeff Amine M-320 contains 20 chains. The total solids concentration in our slips used in this investigation was 30%(wt) and the ratio of alumina:titania was 1:2 respectively. The additive concentration investigated here range from 0-2%(wt). The pH of the suspensions was adjusted by adding either 1 mol dm^{-3} HCl or NaOH solutions followed by thorough mixing. The slips used for the electrophoresis experiments were prepared in 10^{-3} mol dm^{-3} KCl solution in order to maintain the electrical double layer thickness constant in the pH range 3-10.

From the particle size measurements made on these powders using Coulter Counter we found that both alumina and titania are polydispersed with a median diameter of ~6.3 and 3.2 microns respectively. In addition, it was also noticed that the titania powders contain a larger fraction of sub micron particles than alumina and the details of those measurements are given elsewhere [5]. The electrophoretic mobility of these powders was measured using Micromeritics Mass Transport Analyzer and the viscosity of the suspensions was measured using Brookfield Digital Viscometer provided with a small sample adapter. The sedimentation experiments were carried out in 100 ml graduated cylinders by following the rate of fall of the settling particle interface from suspension over the first two hour period.

RESULTS

The electrophoretic mobility of alumina and titania powders (ratio 1:2) in 10^{-3} mol dm^{-3} KCl and 0.2, 0.5, 1.0 and 2.0%(wt) additive solutions (Jeff Amines A-302, M-310 and M-320) is shown in Figure 1. These results suggest that all three additives are adsorbed on to the surface of alumina and titania powders in solution above pH 5 and are very effective in changing the electrophoretic mobility of the mixture in concentrated dispersion. Figure 2 shows the viscosity of the alumina and titania slips in water and the additive solutions measured at 9 s^{-1} (shear rate). From these results shown in Figure 2 it can be concluded that with few exceptions, that the viscosity of the 30%(wt) slips in water is higher than in additive solution and an increase in the concentration of the additive decreases the slip viscosity. It also appears that the viscosity maximum in water at pH 5.6 tend to shift to a higher pH value in additive solutions.

Figure 3 shows a typical rate of fall of the interface profiles of alumina and titania powders from suspension. The sedimentation rate of the settling powder interface was calculated from the above plots by measuring the intial linear portion of each sedimentation curve. Figure 4 shows the sedimentation rate of alumina and titania powders from aqueous suspensions with or without the presence of additive. These results suggest that the sedimentation rate of

alumina and titania particles decreases with an increase in the additive concentration and the sedimentation rate of the settling particles is higher in
water than in additive solutions. It also appears that the rate of sedimentation in water remains independent of the suspension pH in the range pH 5-9.

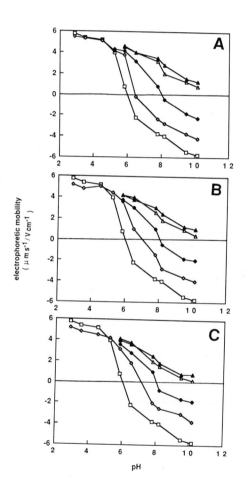

Figure 1. Electrophoretic mobility of alumina and titania powders in 30%(wt)
dispersions prepared in (□) 10^{-3} mol dm^{-3} KCl, (◇) 0.2%(wt), (◆) 0.5,
(△) 1.0 and (▲) 2.0%(wt) additive and 10^{-3} mol dm^{-3} KCl solutions.
(A) additive Jeff Amine M-302, (B) additive Jeff Amine M-310 and (C) additive
Jeff Amine M-320.

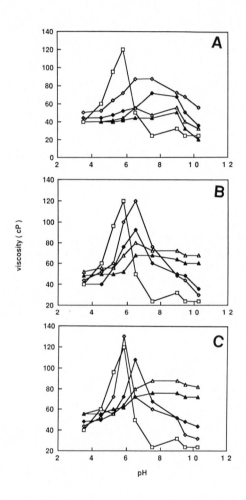

Figure 2. Viscosity of 30%(wt) alumina and titania (ratio alumina:titania 1:2) slips prepared in (A) Jeff Amine M-302, (B) Jeff Amine M-310 and (C) Jeff Amine M-320 solutions. Additive concentration (□) 0, (◇) 0.2, (◆) 0.5, (△) 1.0 and (▲) 2.0%(wt). Shear rate 9s^{-1}.

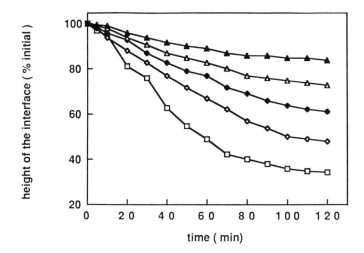

Figure 3. Rate of fall of the interface of alumina and titania particles from (□) 10, (◇) 15, (◆) 20, (△) 30 and (▲) 40%(wt) aqueous suspensions pre-prepared in 1.0%(wt) Jeff Amine M-310 at pH 5.6.

DISCUSSION

The rapid decrease in the electrophoretic mobility of a dispersed oxide due to the addition of suitable surface active agent to the suspension is well known [6]. Such a behavior determines the optimum concentration of the additive required for the processing of suitable slip. A similar analogy based on the suspension rheology can also be derived. However, if one compares the results of the suspension stability and the rheology and electrophoresis results shown above a different picture emerges. For example, the viscosity of slips prepared in water shows a sharp increase in the slip viscosity at pH 5.6 and above pH 6.0 the viscosity decreases. Both sedimentation and electrophoresis results suggest that the behavior of the slip remains independent of pH in the range pH 5.6-10 and the addition of additive in the concentration 1%(wt) eliminates these differences.

It is possible that the differences observed in the electrophoresis, sedimentation and rheology results may be due to the difference in the particle-particle interaction. For example, in suspensions of either alumina or titania powders the particle-particle interaction arising from the surface charge are nearly the same; however, in heteroparticulate suspensions such as alumina and

titania while alumina particles are positively charged at pH 7, titania particles are negatively. It is possible that at pH 7, these particles may tend to
flocculate due to heteroflocculation. In order to unequivocally determine and
attribute the observed differences, we estimated the equivalent flocc diameter
of the settling particles from suspension using the modified Steniour equation
[7] which is given as:

$$d = 2 \left[\frac{9Q\ (1-\varepsilon)}{2g\ (\rho_s - \rho_1)\ \varepsilon^3\ \theta(\varepsilon)} \right]^{0.5}$$

where d is the diameter of the settling particle, Q the sedimentation rate, ρ_s
and ρ_1 are the density of dispersed solid and liquid respectively. $1-\varepsilon$ is the
volume fraction of the dispersed solid and $\theta(\varepsilon)$ is called the shape factor.

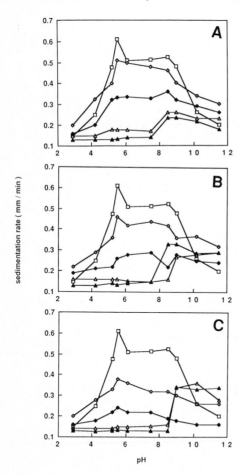

Figure 4. Sedimentation rate of alumina and titania powders from aqueous dispersions prepared in (□) 0, (◇) 0.2, (◆) 0.5, (△) 1.0 and (▲) 2.0%(wt)
(A) Jeff Amine M-302, (B) Jeff Amine M-310 and (C) Jeff Amine M-320.

Figure 5 shows the normalized flocc size (represented as the ratio of the mean diameter of the settling particle in suspension to the mean diameter of the primary particle) as a function of electrophoretic mobility of the particle in suspension. These results suggest that alumina and titania particles undergo heteroflocculation in water and in additive solutions (Jeff Amine M-302, M-310 and M-320) at low concentrations ($\leq 0.5\%$(wt)). At higher concentrations ($\geq 1.0\%$(wt)) the flocculation of the particles is at a minimum.

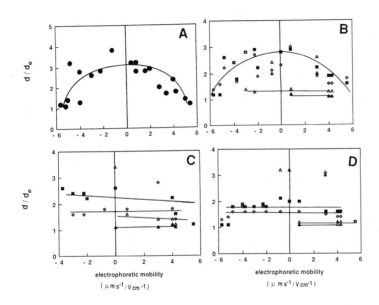

Figure 5. (d/d_o) versus electrophoretic mobility of alumina and titania powders from (A) water, (B) Jeff Amine M-302, (C) Jeff Amine M-310 and (D) Jeff Amine M-320. Additive concentration (■) 0.2, (◇) 0.5, (△) 1.0 and (▲) 2.0%(wt).
d = mean diameter of the flocc estimated from sedimentation analysis.
d_o= mean diameter of the primary particle estimated from particle size analysis.

CONCLUSIONS

From the present investigation the following conclusions can be derived:

1. Both alumina and titania in 30%(wt) suspension in water behave electrophoretically positive below pH 5.6 and above pH 5.6 their mixture behaves as negatively charged powder. All three additives adsorb on to the surface of the powder mixture and the addition of the surface active agents to the slip shifts the point of zero charge of the mixture towards the higher pH scale.

2. The slip viscosity in water is maximum at pH 5.6 and the addition of the additives not only tends to shift the viscosity maximum towards the higher scale of the pH, but also lowers the slip viscosity.

3. The sedimentation rate of alumina and titania particles in water and additive solutions appears to be independent of pH in the pH range 5.6-9.0. Addi-

tion of the surface active agents appears to improve the stability of the suspensions and an increase in the amount of additive increases the stability of
the slip.

4. In water and additive solutions at concentrations ($\leq 0.5\%$(wt)) alumina and
titania particles tend to flocculate and remain independent of the surface
charge. However, in additive solutions at high concentrations investigated
here ($\geq 1.0\%$(wt)) the dispersed powders remain as primary particles and also
behave independently of the surface charge.

REFERENCES

1   I. A. Aksay, F. F. Lange and B. L. Davis, Uniformity of Al_2O_3-ZrO_2 com-
    posites by colloidal filtration, Comm. Amer. Ceram. Soc., 66, C-190 (1983).
2   A. S. Rao, Effect of pH on the suspension stability of alumina, titania
    and their mixtures, To be published in Ceramics International.
3   S. Baik, A. Bliar and P. F. Becher, Properties of Al_2O_3-ZrO_2 composites by
    surface chemical behavior, To be published in MRS Symp. proceedings.
4   E. M. DeLiso, The dispersion of an aqueous heteroparticulate oxide system:
    alumina/zirconia, PhD Thesis, Rutgers University, Piscataway, NJ (1986).
5   A. S. Rao, Electrokinetic behavior of alumina, titania and their mixtures
    in aqueous dispersions, To be published in Ceramics International.
6   G. D. Parfitt, Dispersions of powders in liquids, 3rd ed., Applied Science
    Publishers, Inc., Englewood, NJ (1981).
7   D. Dollimore and K. Karimian, Sedimentation of suspensions: Factors effec-
    ting the hindered settling of alumina in a variety of liquids, Surface
    Technology, 17, 239 (1982).

Interfacial Phenomena in Biotechnology and Materials Processing,
edited by Y.A. Attia, B.M. Moudgil and S. Chander
Elsevier Science Publishers B.V., Amsterdam, 1988 — Printed in The Netherlands

COLLOIDAL STABILITY OF OXIDIZED SILICON PARTICLES IN ETHANOLIC AND AQUEOUS MEDIA

Evelyn M. DeLiso[*] and Alan Bleier[†]

[*]Center for Ceramics Research, Rutgers University, P.O. Box 909, Piscataway, NJ 08855

[†]Metals and Ceramics Division, Oak Ridge National Laboratory, P.O. Box X, Oak Ridge, TN 37831

ABSTRACT

The stability of an oxidized silicon powder was studied in ethanolic and aqueous suspension. Properties of 10^{-3} M LiCl (ETOH) suspensions were studied as a function of the addition of an ethoxylated amine additive denoted herein as KD-2. Potentiometric titrations were used to characterize the acid-base properties of the powder and KD-2 and to develop a calibration curve for measuring the adsorption of KD-2 from ethanolic solutions. Evaluation of zeta potential and sedimentation volume were made for suspensions in 0.15 M NaCl (Aq) over the pH range 3 to 9.

Adsorption of KD-2 onto the silicon powder surface in 10^{-3} M LiCl (ETOH) follows a Langmuirian model. Stability in suspensions containing KD-2 is essentially steric and the comparison of rheological and adsorption behavior can be explained by examining the degree of surface coverage of KD-2 and applying theories of polymer flocculation and heterocoagulation.

Stability of the suspension studies is influenced by the magnitude of the van der Waals forces. Long-term stability of oxidized silicon in 10^{-3} M LiCl (ETOH) suspensions occurs because these forces are small. In 0.15 M NaCl (Aq), somewhat larger van der Waals forces operate and stability requires a sufficiently large repulsive force, which in our case derives from the acid-base character of silanol-type sites and is Coulombic.

INTRODUCTION

This work examines suspensions of partially oxidized silicon metal in ethanol and aqueous media in the presence and absence of an amine additive. Colloidal stability of the silicon powder in ethanol was found to be influenced by the concentration of the additive, whose stabilizing mechanism appears to be essentially steric and whose adsorption is governed by the interaction of the basic amine functionality with the acidic oxidized silicon surface. Detailed analysis of the adsorption isotherm suggests that the configuration of the additive in the adsorbed state is closely related to that in bulk solution, as interpreted from pycnometry data.

Stability in aqueous media was principally assessed in the absence of the additive and relates to the zeta potential of the powder which varies slightly with pH. Maximum zeta potential and surface charge density values are considerably less than those found for particles of quartz and amorphous silica under similar conditions. The charging mechanism responsible for this behavior and the corresponding suspension stability derived from acid-base properties of the silicon surface.

CHARACTERIZATION OF MATERIALS

Silicon

A commercial silicon metal powder (Komatsu Electronic Metals Co.) was used. The powder has an oxidized surface and was fabricated (ref. 1) from silane gas by a vapor-phase reaction using a process similar to that reported by Gregory and Lee (ref. 2). Various physical characteristics of the silicon powder were evaluated and are summarized in Table 1.

The composition of the surface was determined by electron spectroscopy for chemical analysis (ESCA; Perkin Elmer, Model 560) using Mg Kα-radiation as an excitation source operated at 15 kV and 300 watts. The morphology of the silicon particles, as ascertained from transmission electron microscopy (TEM; JEOL, Model 100CX), is spherical but the particles form snake-like chains of as many as 300 primary particles.

The crystallite size giving rise to x-ray diffraction (XRD) was evaluated by analyzing the degree of line broadening. The diffractometer (Siemens, Model D-500) was equipped with a graphite-monochromatized Cu Kα-radiation source. The (220) reflection of Si was used to estimate the crystallite size at 8.9 nm by the Scherrer equation (ref. 3). The estimated cluster diameter of 500 nm from TEM micrographs (Table 1) agrees with the hydrodynamic diameter determined in ethanol using photon correlation spectroscopy (PCS; Malvern, Sub-Micron Particle Analyzer, Model 4700C). The specific surface area of the powder was measured using nitrogen gas adsorption (Quantachrome, Quantasorb Sorption System) and calculated using the Brunauer, Emmett and Teller (BET) equation (ref. 4). An equivalent spherical diameter for the primary particles was calculated from the BET data. Finally, potentiometric titration of the powder surface revealed weakly acidic sites relative to water.

Additive

A commercial additive, denoted here as KD-2 (ICI Speciality Chemicals, Hypermer KD-2, Lot No. 96615) was used. The Fourier transform infrared (FTIR; Perkin Elmer, Model 1750) spectrum in Figure 1 indicates that this material contains both aliphatic and ether functionalities.

These features are respectively indicated by the double adsorption peak in the vicinity of 2900 cm^{-1} and the C-O-C stretching vibration at 1110 cm^{-1}. An amine group is also present in the molecular structure (ref. 5), and appears (ref. 6) responsible for the stretching vibration at about 3496 cm^{-1}. However, it is not presently possible to distinguish unequivocally amongst primary,

TABLE 1

Physical Characteristics of Silicon Powder

	Parameter	Method of Analysis
Reaction system	$SiH_4(g)$	---

Composition,
  (atomic % normalized)
    Si
    0
    C
	
	51.7
	30.3
	18.0
		ESCA

Shape	Chains of spheres	TEM
Crystal type	Amorphous; cubic	TEM, XRD
Crystallite size(nm)	8.9	XRD line broadening
Surface area (m^2/g)	59.9	BET

Particle diameter (nm)
    Primary particles
    Cluster size
    Hydrodynamic (ETOH)
	
	43
	500
	463.6
		BET
		TEM
		PCS

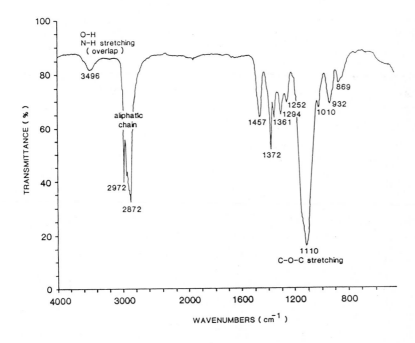

Fig. 1. FTIR spectrum of KD-2.

secondary, and tertiary amine characteristics of N-H species on the basis of
the IR data alone.

Potentiometric titration of the KD-2 in 0.15 M NaCl(Aq) exhibited an
equivalence point, consistent with the presence of a single type of amine and
an apparent pK_a-value of 8.9. This value is in reasonable agreement with those
for various fatty amines found in the literature (refs. 7, 8). The equivalent
weight of KD-2 found (ref. 9) by analyzing the first derivative of pH titrations is 1738 ± 4 (99 percentile). A 1.7-w/o loss was found, however, when
this material was irradiated with 0.25 watts of infrared heat (OHaus, Moisture
Determination Balance) for 60 minutes. Concentrations reported here are based
on the assumption of 100% activity (ref. 5), since it was not established
whether or not the material lost on heating contains basic amine.

Other Chemicals

Other chemicals were Fisher reagent grade. The water was doubly distilled
and pH-adjustments were made with HCl and NaOH; ionic strength was maintained
using NaCl. LiCl was used, on the other hand, to maintain a constant electrolyte concentration in absolute ethanol due to the salt's high solubility.

EXPERIMENTAL PROCEDURE

Potentiometric Titration

Potentiometric titration (Radiometer America, Model RTS 822) was used to
evaluate the acid-base properties of the Si powder and KD-2 and to determine
surface charge density, σ_0, of Si-particulates in aqueous media. A double
junction reference electrode and a glass electrode were used for pH
measurements. The concentration of electrolyte in the outer junction of the
reference electrode was matched to the salt level of the suspension being
titrated. All solutions were bubbled for 10 to 20 min with argon to ensure
removal of $CO_2(g)$. From the aqueous titration data, of which examples are
given in Figure 2, it is possible to calculate (ref. 10), as functions of pH,
the dissociation constant of the additive (K_a), the surface charge density of
Si (σ_0), and the fraction of KD-2 that is protonated (α_+). An aqueous soxhlet
extraction was performed for 48 h on the silicon powder prior to titration.
Potentiometric titration was also used to establish the adsorption isotherm
for KD-2 on silicon powder in 10^{-3} M LiCl (ETOH).

Pycnometry

The molal volume of KD-2 in 10^{-3} M LiCl (ETOH) was calculated from measurements of solution density obtained at 25 °C. (ref. 11) See Figure 3. Linear
regression analysis of the pycnometric data gave the following equation with a
correlation coefficient of 0.9999, where ρ is the measured density and C is the
molality of KD-2.

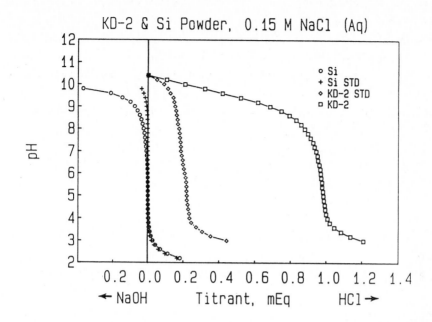

Fig. 2. Potentiometric titration of Si powder and KD-2 along with corresponding data for the suspension medium, 0.15 M NaCl (Aq).

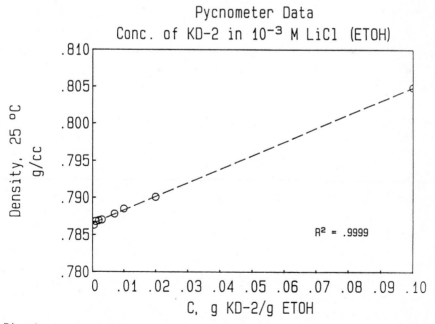

Fig. 3. Density of KD-2 as a function of the additive concentration in 10^{-3} M LiCl (ETOH) at 25 °C.

$$\rho = 0.2031\ C + 0.7864 \tag{1}$$

where ρ = the measured pycnometric density and c = concentration of KD-2.

Rheology

Viscosity, η, was measured on selected suspensions with a parallel plate viscometer (Rheometrics, Fluid Spectrometer) operating in the thixotropic loop mode at 24 °C. The thixotropic loop was divided into four zones as shown in Figure 4. These are: (1) 30 s at 50 s^{-1}, (2) 30 s ramp down to 0 s^{-1}, (3) 180 s ramp up to 200 s^{-2}, and (4) 180 s ramp down to 0 s^{-1}. Samples were prepared by ultrasonically dispersing 8 v/o silicon powder in 10^{-3} M LiCl(ETOH) with desired concentrations of KD-2. Samples were then agitated for 2 d to allow adsorption to reach equilibrium. In order to ensure sample homogeneity and proper sampling of the systems, the suspensions were sheared at 14,750 rpm for 30 s in a blender (Waring, Commercial Blender) prior to evaluation.

Adsorption

The adsorption of the additive onto Si powder was studied in ethanol suspensions containing 10^{-3} M LiCl. Amount adsorbed was determined by measuring the quantity of KD-2 removed from solution after exposure to the powder. Suspensions for the adsorption study were prepared according to the procedure just described for the rheological measurements and agitated for 2 d using a reciprocal shaker.

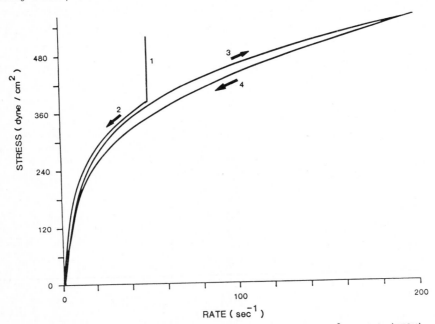

Fig. 4. Flow curve for 8 v/o Si powder suspensed in 10^{-3} M LiCl (ETOH). The concentration of KD-2 is 0.003 g KD-2/g ETOH. See text for details of the thixotropic loop.

A small portion of each suspension was then centrifuged, after which 5 g of supernatant were combined with 25 g of 10^{-3} M NaCl (Aq). This 30-g solution was subsequently titrated using standardized HCl. The amount of KD-2 remaining in solution after exposure to Si was determined from a calibration curve developed for this 3.52 M C_2H_5OH (Aq) system. Regression analysis of the linear calibration curve, µeq HCl vs mg KD-2, yielded a slope of 0.582 and a correlation coefficient was 0.9914. The amount adsorbed was taken as the difference between initial concentration of KD-2 in solution and that found remaining after centrifugation.

Electrokinetics

The electrophoretic mobility, μ_e, of Si was measured (Rank Brothers) in 0.15 M NaCl (Aq) over the pH range 3 to 10 and in 10^{-3} M LiCl (ETOH). Stock suspensions with these electrolyte concentrations were aged for 2 d. The aqueous suspension was stored at pH 5.1 and adjusted to desirable values several hours prior to measurement. The mobilities of aqueous and ethanolic suspensions were measured with cells having cylindrical and flat rectangular cross-sections, respectively. Each mobility value was calculated from the average of 20 velocity measurements.

Zeta potential, ζ, expressed in mV, was calculated from the μ_e-values using Henry's equation (ref. 12) following the guidelines given by Smith (ref. 13). The primary particle size given in Table 1, 43 nm, was used for determining the appropriate Henry correction factor. The ultimate equations used for calculating the zeta potential at 25 °C are $\zeta = 14.77 \mu_e$ for 0.15 M NaCl (Aq) and $\zeta = 83.85 \mu_e$ for 10^{-3} M LiCl (ETOH).

Sedimentation

The sedimentation behavior of suspensions containing 0.71 w/o Si was evaluated at pH 3.2, 5.1, and 8.9 in 0.15 M NaCl (Aq) and in 10^{-3} M LiCl (ETOH). The height of the interface was measured at desired time intervals for two weeks and is expressed as a fraction of the total height of the solid-liquid mixture.

RESULTS AND DISCUSSION
Rheology and Adsorption

Rheological data were collected for Si suspensions in 10^{-3} M LiCl (ETOH) as a function of the concentration of KD-2. A typical flow curve is given in Figure 4. This example exhibits pseudoplasticity and hysteresis; thixotropy was also evident, although this feature is not depicted in the figure. Most suspensions exhibited these characteristics.

Curve a in Figure 5 demonstrates that small additions of KD-2 increase the viscosity of 8 v/o Si suspensions, indicating enhanced "structure"in the system and probably flocculation. Larger additions of the additive lead to values of viscosity less than the maximum observed, suggesting restabilization of the suspensions. However, the viscosity under these latter conditions is somewhat higher than that observed in the absence of KD-2. Curve b is the adsorption isotherm of KD-2 on Si-powder and appears knee-shaped, with a steep initial slope that indicates high affinity for equilibrium concentrations of KD-2 below 12 mg per mg solvent, corresponding to an initial concentration of 0.002 mgKD-2 per mg solvent. Surface coverage appears to approach a constant level above 85 mg KD-2 per mg solvent and no evidence of multilayer adsorption presents itself. Note that the increased viscosity of suspensions containing small concentrations of KD-2 coincides with approximately 50 % coverage and that ultimate viscosity for high concentrations of KD-2 corresponds to saturation adsorption.

The adsorption isotherm appears to be Langmuirian (ref. 14). Langmuir's theory for adsorption of gases and vapors on solids (ref. 15), when applied to dilute solutions of KD-2 in 10^{-3} M LiCl (ETOH)in equilbrium with Si-powder can be written

$$\Gamma = \Gamma_{max} \, KC_e/(KC_e + 1) \, , \tag{2}$$

where Γ is the adsorption concentration expressed as mg KD-2 per m^2 of Si; Γ_{max} is monolayer coverage corresponding to saturation of the surface; C_e is the equilibrium concentration of KD-2; and K is the equilibrium constant for the adsorption-desorption process. Eq. 2, rearranged as

$$C_e/\Gamma = (1/\Gamma_{max}) \, C_e + (1/\Gamma_{max})(1/K) \, , \tag{3}$$

and applied to the data in Figure 5 yields the plot given in Figure 6. Neglecting the two lowest concentrations that approach insensitivity of the potentiometric titration method described earlier, regression analysis yields a linear relationship for Ce/Γ vs Ce that has a correlation coefficient of 0.9983. The fitted equation describing the relationship in Figure 6 is

$$C_e/\Gamma = 4.9135 \, C_e + 14.875 \, . \tag{4}$$

This expression provides respective values for Γ_{max} and K of 7.05×10^{16} molecules KD-2 per m^2 and 0.574 per molality. These quantities and the coefficients in Eq. 4 were used to calculate Curve b shown in Figure 5. Finally, the values for Γ_{max} and K lead, respectively, to estimates of

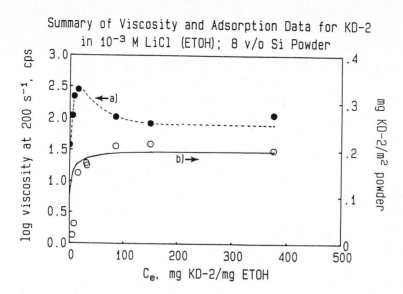

Fig. 5. Viscosity and adsorption data for KD-2 in 10^{-3} M LiCl (ETOH) suspensions at 8 v/o Si powder. Samples were exposed to KD-2 solutions for 2 d. Curve (a) viscosity data. Curve (b) adsorption isotherm.

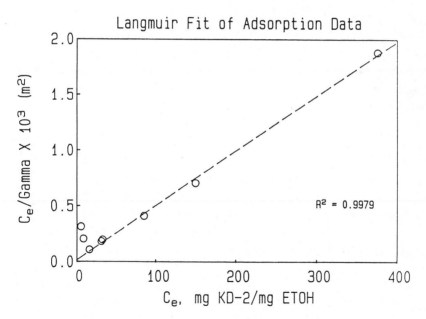

Fig. 6. Langmuirian analysis of the adsorption data in Figure 5 for KD-2 onto 8 v/o Si suspensions.

14.2 nm^2 for the area occupied by a KD-2 molecule in the adsorbed state and 0.55 kT per molecule for the change in Gibb's free energy between the adsorbed and solute states, ΔG_{ads}. The value for ΔG_{ads} indicates that adsorption is thermodynamically spontaneous and represents a small driving force for adsorption, a characteristic which is somewhat surprising in view of the apparently high affinity character of Curve b in Figure 5.

Pycnometry

The pycnometric data in Figure 3 were used to estimate the effective size of KD-2 molecules in 10^{-3} M LiCl (ETOH), assuming sphericity. The effective radius is 0.869 nm and corresponds to a cross-sectional area of 2.37 nm^2. Comparison of this value with the estimated area per molecule on the surface of silicon for saturated adsorption determined in the previous section indicates that KD-2 is probably interacting with specific sites on the particle surface and that the area-based concentration of these sites is low.

Sedimentation and Electrokinetics

Aqueous experiments are summarized in Figure 7. During the first hour after sample preparation, rate of settling at pH 3.2 is rapid indicating colloidal instability, but it is slow at pH 8.9 which suggests that these latter conditions lead to suspensions that are quite stable to agglomeration. Behavior at pH 5.1 is intermediate between these extremes. Additionally, note that acidic conditions correspond to the lowest values of ζ-potential, whereas basic conditions at pH 8.9 favor values notably higher. Thus, zeta potential apparently determines the general stability of these aqueous systems. Paradoxically, data gathered at significantly longer aging times of 72 to 240 h contradict this trend, and clearly indicate that sediment height increases with increasing pH and corresponds to a reduced concentration of solid in the sediment. See Table 2.

Behavior in ethanolic media differs from that just described. The ζ-potential in 10^{-3} M LiCl (ETOH) was found, in the absence of KD-2, to be -10.9 mV. However in contrast to the aqueous systems characterized by low values of ζ, ethanolic suspensions remained well-dispersed, even after long aging periods of two weeks. Little-to-no sediment could be detected under these conditions. This observed stability agrees with the rheological datum point in Figure 5 for systems free of KD-2.

In view of the greater electrostatic repulsion at pH 8.9 anticipated from the profiles over the range pH 3 to 9 given in Figure 7b, a possible explanation for this behavior may evolve from the following hypotheses. It appears that relatively more intracluster (chain) repulsion may exist in basic media, as compared to conditions below pH 7, and that this repulsion may generate, in

TABLE 2

Comparison of Sedimentation Behavior of 0.71 v/o Si
in 0.15 M NaCl (Aq) at Short and Long Aging Times

pH	Setting rate, $\Delta h / \Delta \log(t)$[a]	v/o Solids in sediment[b]
3.2	Most rapid	1.7
5.1	Intermediate between those at pH 3.2 and 8.9	1.3
7.3	c	~1.0[d]
8.9	Slowest	0.9

[a]Age is < 1 h; See Figure 7a; h = height of
  Si-containing portion of suspension, t = time.
[b]Estimated values based on ages of 72, 120, and
  240 hours for each pH value.
[c]Data are not included in Figure 7a for clarity;
  resembles data for pH 5.1.
[d]Similar to but greater than value obtained at
  pH 8.9.

turn, a greater hydrodynamic radius of the chain-like particulates described
in Table 1. An enhanced size may be detectable by PCS techniques and is,
therefore, a candidate property to be examined in future experiments.

Other Behavior and Observations

Suspensions of Si powder were difficult to prepare in 0.15 M NaCl (Aq).
Incorporation of the powder in suspension was not easy due to incomplete wetting
of the powder, behavior apparently manifested in the sedimentation studies. For
instance, three of the systems in Table 2, those at pH 3.2, 5.1, and 7.3,
exhibited bubbles permeating upward during the first hour. This combination of
rapid settling and expulsion of bubbles from the Si-rich portion of the suspension at extended age was greatest in pH 7.3. Moreover, expulsion of bubbles
apparently rearranges the structure of sediment, with most disruption occurring
under neutral pH conditions and best packing of the snake-like chains at pH 3.2.
Kinetics of bubble evolution were not investigated in this study and may need to
be examined in order to understand in detail the trend in Table 2 for systems
between pH 3.2 and 7.3.

Finally, these observations on possible incomplete wetting and bubbling may
ultimately underscore the stability data in Figure 7a and Table 2. The degree
of oxidation of Si is apparently limited and particles seem to have retained
some hydrophobic character (ref. 16). In view of the silane origin of our
powder, it is noteworthy that Popper and Ruddlesden (ref. 17a) report similar
gas evolution under acidic conditions, behavior which White (ref. 17b)

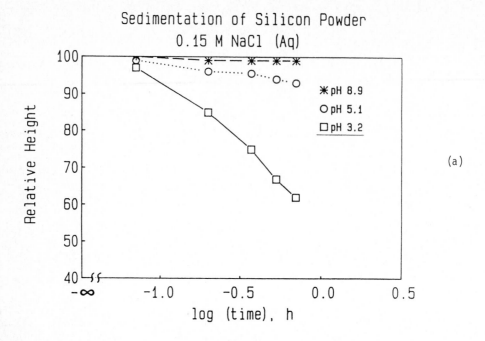

(a)

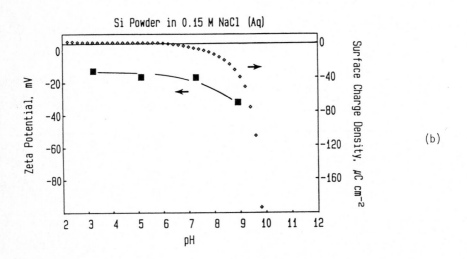

(b)

Fig. 7. Behavior of Si powder in 0.15 M NaCl (Aq). (a) Sedimentation data for relative total suspensions height as a function of time at pH 3.2, 5.1 and 8.9. (b) Zeta potential, ζ, and surface charge density, σ_0, as a function of pH.

attributes to oxidation of Si and production of $H_2(g)$.

CONCLUSIONS AND CLOSING REMARKS

The work presented here constitutes an initial effort to understand the broad range of surface and colloidal properties exhibited by oxidized silicon powder. Ultimately, this understanding will be used to predict and to control the behavior of Si powder in ceramic fabrication processes, whether they be predominately aqueous or nonaqueous methods.

Specifically, we conclude the following points from the data presented:

1. Colloidal stability of oxidized Si powder is of electrostatic origin in 0.15 M NaCl (Aq). Moreover, Figure 7 and Table 2 demonstrate that greatest stability exists at pH 9 owing to a sufficiently high ζ potential of -31.8 mV. Recent studies (ref. 18) for which the concentration of NaCl is 10^{-3} M (Aq) reveal high stability over the entire pH range 3 to 9. Thus, a critical concentration of electrolyte exists between 10^{-3} M and 0.15 M NaCl (Aq) and for the high level of electrolyte used in this study a critical value of ζ exists between -16.5 and -31.8 mV.

2. Stability in 10^{-3} M LiCl (ETOH) is high in the absence of KD-2 and is reduced by small additions of this material, as seen in Figure 5. These conditions correspond to submonolayer coverage. Adsorption of KD-2 onto oxidized Si appears to arise from the interaction of the basic amine functionality of the dispersant with the acidic oxidized silicon surface.

3. High concentrations of KD-2, relative to the degree of surface coverage, reduce the destabilizing effect noted in Conclusion 2 but do not yield maximum stability (lowest η).

4. Comparison of estimated molecular cross-sectional areas for KD-2, based on adsorption and pycnometric data, suggest that monolayer coverage corresponds to only about 17 % of the available surface. Also, indiscrimant adsorption resulting from reduced solute-solvent interactions is not considered probable.

5. Further work on FTIR, surface charge characterization and PCS is needed (a) to elucidate the chemical aspects of the interactions between KD-2 and the powder surface and those among molecules of KD-2, (b) to identify the orientation of KD-2 in the adsorbed state, (c) to understand why a monolayer corresponds to such a low coverage of available surface, (d) to resolve the apparent discrepancy between very low σ_0-values and low but discernable values of ζ below pH 7 (Figure 7b), and (e) to understand the comparison between short and long aging periods on the sedimentation behavior summarized in Figure 7a and Table 2.

Generally, the mechanisms of colloidal stability and destabilization of oxidized silicon particles appear to be founded in the classical DLVO theory of lyophobic colloids (ref. 19), modified to include the model of Healy and La Mer (ref. 20) for the ethanolic systems that contain KD-2. That is, electrostatic and van der Waals forces are present in both media investigated, 10^{-3} M LiCl (ETOH) and 0.15 M NaCl (Aq). In fact, excellent stability was observed in each case, viz. in the absence of KD-2 in the ethanolic medium (Figure 5) and under sufficiently basic conditions of pH 8.9 in an aqueous NaCl system (Figure 7a). However when KD-2 is present in ethanolic systems, conditions of destabilization and restabilization were additionally identified, indicating that KD-2 plays a specific role in the behavior of these suspensions.

Electrokinetic experiments described earlier demonstrated that the ζ-potential developed in ethanol by our Si powder is -10.9 mV, a value which is probably too low to establish a significant electrostatic barrier between neighboring particles. This situation seems to be confirmed by related research (ref. 18) and will be discussed fully when the work in Conclusion 5 is completed. Nonetheless, Bleier (ref. 21) has recently shown that the presence of electrostatic forces is an unnecessary condition for colloidal stability of 46.5-nm Si particles in ethanol since van der Waals forces between such particles generate an attractive energy of only ~1 kT when the separation distance corresponds to two molecular layers of solvent.

The status of van der Waals forces is similar in aqueous media but the analogous attractive energy is ~6 kT, a value which suggests that a repulsive force, albeit weak, may be required for colloidal stability, depending on particle size (ref. 21). In this regard, the data in Figure 7b support the conclusion that electrostatic forces enhance the stability of Si in 0.15 M NaCl (Aq).

The role of KD-2 in 10^{-3} M LiCl (ETOH) requires additional commentary. The experimental data in Figure 5 clearly demonstrate that maximum destabilization corresponds to approximately 0.5-monolayer coverage. Under these conditions, vis à vis Healy and La Mer (ref. 20), the probability of collision for neighboring particles that have degree of coverage, θ, and approach one another with their covered regions aligned is given by the term θ^2. The corresponding probability for collisions with uncovered areas aligned is $(1 - \theta)^2$ and, finally, the probability of two particles approaching one another with alignment of a covered area on one particle and an uncovered region on the other is given by $2\theta(1 - \theta)$. These expressions and the systems to which we apply them here address the process of heterocoagulation (ref. 22), i.e., the interaction of dissimilar particles or surfaces. Since systems without KD-2 and those for which $\theta \to 1$ when KD-2 is present are stable, as just described, we principally concern ourselves here with the term $2\theta(1 - \theta)$ which has a maximum value of 0.5 when $\theta = 0.5$. The destabilization inferred from the rheological data in

Figure 5 is greatest when θ is ~0.5. Consequently, we conclude that KD-2 can "bridge" particles of Si.

A final comment on the values of ζ in 10^{-3} M LiCl (ETOH) when $\theta = 0$ in the absence of KD-2 and when $\theta \rightarrow 1$ is appropriate. Calculations (ref. 18) based on the recent treatment of Green and Parfitt (ref. 23),when applied to our system, suggest that a significant (~20%) of the average intercluster spacing is occupied by adsorbed KD-2 when $\theta \rightarrow 1$. Such a high percentage would reasonably raise η, relative to the system without KD-2. More work is clearly needed to resolve this issue further.

ACKNOWLEDGMENTS

We thank C. G. Westmoreland for his laboratory assistance at Oak Ridge National Laboratory and F. E. Stooksbury for her help in compiling the manuscript. Also, we gratefully acknowledge the experimental help of L.Johnson and W. Symons of Rutgers University for the x-ray diffraction and TEM work, respectively. The ESCA analysis was done by R. Moore of Perkin-Elmer Physical Electronics Division, Edison, N.J. This work was sponsored by the Center for Ceramics Research at Rutgers University and by the Division of Materials Sciences, U.S. Department of Energy, under contract No. DE-AC05-840R21400 with the Martin Marietta Energy Systems, Inc. at Oak Ridge National Laboratory.

REFERENCES

1 F. Takao, Komatsu Electronic Metals Co., Hiratsuka, Japan, Aug. 1986.
2 O. J. Gregory, S. Lee and R. C. Flagan, Reaction sintering of submicrometer silicon powder, J. Am. Ceram. Soc., 70[3] (1987) C52-C55.
3 B. D. Cullity, Elements of X-ray Diffraction, 2nd Ed., Addison-Wesley Publishing Co., Inc., Reading, Mass., 1978, p. 102.
4 S. Brunauer, P. H. Emmett and E. Teller, "Adsorption of gases in multimolecular layers, J. Am. Chem. Soc., 60 (1938) 309-319.
5 Hypermer Polymerics, ICI Speciality Chemicals Brochure, Release 65-1760-012, April 1987.
6 R. T. Conley, Infrared Spectroscopy, 2nd Ed., Allyn and Bacon, Inc., Boston, 1972.
7 H. K. Hall, Jr., Correlation of the base strengths of amines, J. Am. Chem. Soc., 79 (1957) 5441-5444.
8 J. W. Smith, Basicity and complex formation, The Chemistry of the Amine Group, S. Patai, ed., Interscience, New York (1968) pp. 161-204.
9 J. S. Fritz and G. H. Schenk, Jr., Quantitative Analytical Chemistry, 2nd Ed., Allyn and Bacon, Inc., Boston, 1969, p. 584, p.172.
10 R. O. James and G. A. Parks, Characterization of aqueous colloids by their electrical double-layer and intrinsic surface chemical properties, Surface and Colloid Science, Vol. 12; edited by E. Matijević, Plenum Press, N.Y. (1982) pp. 119-216.
11 D. P. Shoemaker, G. W. Garland, J. I. Steinfeld and J. W. Nibler, Experiments in Physical Chemistry, 4th Ed., McGraw-Hill, Inc., 1962, p. 161.
12 D. C. Henry, Cataphoresis of suspended particles, Proc. Roy. Soc., A133 (1931) 106-129.

13 A. L. Smith, Electrical phenomena associated with the solid-liquid
    interface, Dispersion of Powders in Liquids, 3rd Ed.; edited by
    G. D. Parfitt, Applied Science Publishers, New Jersey (1981) pp. 99-147.
14 I. Langmuir, The adsorption of gases on plane surfaces of glass, mica and
    platinum, J. Am. Chem. Soc., 40 (1918) 1361-1403.
15 P. C. Hiemenz, Principles of Colloid and Surface Chemistry, 2nd Ed., Marcel
    Dekker, Inc., N. Y., 1986, p. 398.
16 A. Bleier, The role of van der Waals forces in determining the wetting
    and dispersion properties of silicon powder, J. Phy. Chem., 87[18] (1983)
    3493-3500.
17 (a) P. Popper and S. N. Ruddlesden, The preparation, properties and
    structure of silicon nitride, Trans. Br. Ceram. Soc. 60 (1961) 603-626;
    (b) J. White, Trans. Br. Ceram. Soc. 60 (1961) 625.
18 E. M. DeLiso and A. Bleier, unpublished work.
19 (a) B. V. Derjaguin and L. D. Landau, Acta Physicochim. URSS 14 (1941)
    633; (b) E. J. Verwey and J. Th. G. Overbeek, Theory of the Stability
    of Lyophobic Colloids, Elsevier, Amsterdam, 1948, 205 p.
20 T. W. Healy and V. K. La Mer, The energetics of flocculation and
    redispersion by polymers, J. Colloid Sci. 19 (1964) 323-332.
21 A. Bleier, Fundamentals of preparing suspensions of silicon and
    related ceramic powders, J. Am. Ceram. Soc. 66 (1983) C79-C81.
22 J. Gregory, J. Colloid Interface Sci. (a) 42 (1973) 448; (b) 55 (1976) 35;
    see also D. R. Kasper, Theoretical and Experimental Investigations of the
    Flocculation of Charged Particles in Aqueous Solutions by Polyelectrolytes
    of Opposite Charge, Ph.D. Thesis, California Institute of Technology,
    Pasadena, 1971, 201 p.
23 J. H. Green and G. D. Parfitt, Stability of concentrated pigment
    dispersions in p-Xylene/Aerosol-OT solutions, Poly. Mat. Sci. Eng. 52
    (1985) 420-425.

Interfacial Phenomena in Biotechnology and Materials Processing,
edited by Y.A. Attia, B.M. Moudgil and S. Chander
Elsevier Science Publishers B.V., Amsterdam, 1988 — Printed in The Netherlands

ELECTROKINETIC BEHAVIOR OF MOLYBDENUM, SILICON, AND COMBUSTION—SYNTHESIZED
MOLYBDENUM DISILICIDE POWDERS DISPERSED IN AQUEOUS MEDIA

S. C. Deevi[1], C. K. Law[1] and A. S. Rao[2]

[1]Department of Mechanical Engineering, University of California, Davis,
California 95616

[2]Department of Ceramics, Rutgers University, Pistcataway, New Jersey 08855

SUMMARY
To understand the electrokinetic behavior of combustion–synthesized
molybdenum disilicide powder in aqueous media, commercial silicon, molybdenum
and combustion—synthesized molybdenum disilicide powders were dispersed in
10^{-3} mol dm^{-3} KCl and the zeta potential of powders was measured in the pH
range 1.5 to 10 using microelectrophoresis technique. The results suggest
that while silicon powder is negatively charged in the pH range 4 to 10,
electrophoretic behavior of Mo is analogous to that of a basic oxide. The
zeta potential of molybdenum disilicide powder was positive below pH 10 and
negative above pH 10. Upon calcination at 500°C for two hours the isoelectric
points of both silicon and molybdenum disilicide powders shift toward the
lower pH scale.

INTRODUCTION

The recent surge of interest in the synthesis and processing of non-oxide
ceramic materials has led many investigators to develop novel methods for
synthesizing different refractory materials (1,2). Among the various methods,
combustion-synthesis technique has gained prominence to synthesize materials
such as transition metal borides, carbides, nitrides, and silicides (3-6). In
this technique, a strong exothermic reaction initiated at the surface by using
an ignition coil or laser liberates adequate heat to heat the adjacent layers
of reactants until the reactants are consumed. During this process, very high
temperatures are generated and the product phase at any instant during the
reaction is separated from the reactants by a reaction zone. Unless densified
during the reaction, the product is a fused porous solid and can be ground
easily to obtain powders of varying sizes. The combustion-synthesis process
requires low-cost equipment while reaction times in this process are
insignificant as compared to conventional processes. The expulsion of
absorbed gases and low boiling metals gives rise to a pure homogeneous product
thereby eliminating the homogenizing step.

High purity molybdenum disilicide ($MoSi_2$) has been successfully
synthesized using combustion-synthesis technique (7). Silicide coatings find

applications in high temperature oxidizing environments in places such as jet engines, guided missiles, radiative shields and as leading edges on aerospace vehicles (8). Slurry preparation is one of the steps involved in the coating process, and it is essential to understand the surface properties of the combustion—synthesized powders for optimizing the slurry properties. Also ceramic components with high degree of homogeneity can be produced from fine powders either by slip casting or by filter pressing well identified aqueous or non-aqueous concentrated slips of $MoSi_2$ powders. It is reasonable to assume that any poorly dispersed powder agglomerates in the slip reduces the reliability of the fabricated ceramic component. Consolidation of fine powders will be easier, provided the surface chemical properties such as the surface charge of powders is well optimized. This project was undertaken to understand the electrokinetic behavior of as-received molybdenum, silicon, and as-synthesized $MoSi_2$ powders and to compare the charge behavior of molybdenum and silicon with that of as-synthesized $MoSi_2$ powder in aqueous media. We present briefly some of the features of combustion-synthesis of $MoSi_2$ followed by the electrokinetic behavior of Mo, Si and combustion—synthesized $MoSi_2$ powders.

EXPERIMENTAL

As—received commercial Mo and Si powders with a minimum purity of 99% were used in the present investigation. While Mo particles were spherical, silicon particles were flaky with an irregular shape (Figure 1). Size distributions determined using a Quantamet image analyzer with more than 4000 particles are indicated in Figure 2. Surface areas (Table 1) were measured using standard N_2 BET analysis by carrying out the adsorption at 77 K. Chemical analysis of the elements (Table 1) were carried out using emission spectrochemical analysis technique. Although impurity content in Mo powder is less, iron and aluminum were the major impurities in silicon.

Combustion—synthesis of $MoSi_2$ was carried out in the laboratory by homogenizing the powders (ratio of Si to Mo is 2:1 (at%)) in an ultrasonic mixer using small quantities of an organic solvent followed by the removal of dispersion medium under vacuum. Cylindrical pellets were ignited in a static pressure of argon (1.2 atm) using an ignition coil. After the reaction was complete, the porous $MoSi_2$ compacts were ground in an agate mortar. The mean particle size of $MoSi_2$ was 2.3 μ, and the surface area of the powder was 0.37 m^2/g. Figure 1 also denotes the morphology of ground $MoSi_2$ powders.

Dispersions suitable for microelectrophoresis experiments were prepared by dispersing 0.1 gram of each powder in 10^{-3} mol·dm^{-3} KCl solution using an ultrasonic probe for five minutes. The pH of the suspension was adjusted by

Mo Powder

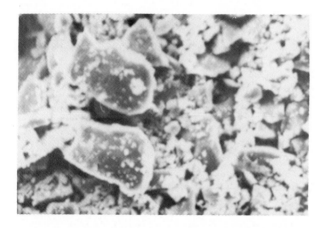

Si Powder

MoSi$_2$ Powder

Fig. 1   Surface Morphologies of Mo, Si and combustion–synthesized MoSi$_2$ powders.

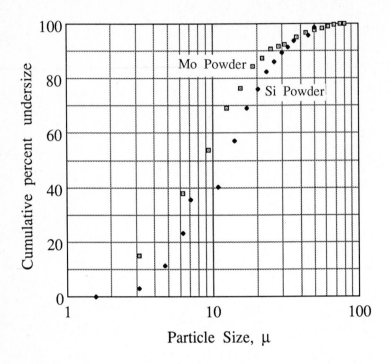

Fig. 2 Particle size distributions of Mo and Si powders.

TABLE 1

Surface areas and impurity contents of Mo and Si Powders.

	Mo Powder	Si Powder
Surface Area	0.20 m^2/g	2.15 m^2/g
Impurities, ppm		
Al	20	600
Ca	6	200
Cr	28	100
Cu	8	20
Fe	88	2000
Mg	1	20
Ni	73	40
W	100	--

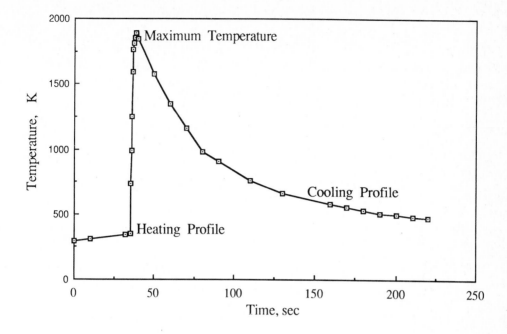

Fig. 3 Temperature profile of 63.08% Mo - 36.92% Si compact during combustion-synthesis of $MoSi_2$ in Argon.

adding a few drops of either 1 mol dm^{-3} HCl or NaOH solutions followed by thorough mixing. In all our electrophoresis experiments suspensions were prepared in 10^{-3} mol dm^{-3} KCl to maintain the electrical double layer thickness constant in the pH range investigated. Zeta potentials of the dispersed powders were determined using a Pen Kem Laser Zee meter and due care was taken in noting down the final pH of the suspension prior to the measurement.

RESULTS

Ignition of cylindrical pellets of Mo-Si mixture in argon caused an exothermic reaction on the surface of the compact raising the surface temperature to 1886 K and initiating a combustion wave. The combustion wave propagated slowly through the compact at a finite velocity converting the reactants to products. The thermocouple located at the centerline of the compact indicated the temperature profile of the compact during the reaction (Figure 3). The rate of heating involved (calculated from Figure 3) during the reaction corresponds to approximately 4.8 x 10^4 K/min. Note that only the compact temperature is raised to the combustion temperature and the surrounding atmosphere is closer to room temperature. This is one of the novel features of the combustion-synthesis technique, eliminating the use of industrial furnaces and long heating times to synthesize high temperature materials.

The following reaction takes place:

$$Mo + 2Si \rightarrow MoSi_2 \qquad \Delta H = -31.4 \ \frac{K \cdot cal}{g \cdot mole}$$

giving rise to $MoSi_2$. During the reaction liquid Si (melting point of Si $\approx$ 1683 K; temperature in the compact from figure 3 is $\approx$ 1900 K) reacts with solid Mo powders (melting point of Mo $\approx$ 2883 K) and the process can be termed as a liquid-solid reaction. An interesting aspect observed in combustion-synthesized samples is a local sintering phenomenon (Figure 4) which is explained in terms of the particle-particle contact and the spread of liquid Si.

X-ray diffraction analysis of the product indicates that the combustion-synthesized sample is highly crystalline with conversion corresponding to 100% (Figure 5). The product crystallizes in tetragonal structure and the lattice parameters (a_0 = 3.20, c_0 = 7.845) and x-ray density (6.29 g/cm^3) determined from the diffractogram correspond well with the reported values (9). Chemical analysis of the product indicated that the product is much purer than the original reactants (10). Fully characterized $MoSi_2$ with powder characteristics as described in experimental section has been used to study the electrokinetic behavior.

Figure 6 shows the zeta potential versus pH profiles of as received Mo and Si powders dispersed in 10^{-3} mol dm^{-3} KCl solution. The results suggest that the surface of silicon is negatively charged above pH 3.8 and the surface of Mo is positively charged in the pH range 1.5 to 10.4. Although the electrokinetic behavior of silicon is in agreement with the literature value for silica (11), the behavior of Mo is different from that of the earlier report on MoO_3 (12). The difference may be attributed to the fact that in the case of Mo, oxide layers are present only on the surface.

Figure 7 indicates the zeta potential versus pH profile of $MoSi_2$ powder. These results suggest that $MoSi_2$ is positively charged below pH 10 and above pH 10 the surface behaves as if it is negatively charged. Another interesting feature of Figure 7 is that the positive zeta potential drops sharply from pH 1.50 to pH 1.85 before it rises to a maximum at pH 4. There has been no reported work on the electrokinetic behavior of $MoSi_2$ and no comparison can be made to any previous work.

Fig. 4   Local sintering phenomena in the combustion—synthesized MoSi$_2$ compact.

DISCUSSION

It is reasonable to assume that a thin layer of the respective oxide is present as an impurity on the surfaces of Mo, Si and MoSi$_2$ powders. In fact, an examination of the powders using x-ray photoelectron spectroscopic technique indicated that the surfaces of powders are associated with some oxygen. It is well known that a surface charge is imparted to the dispersed particles by the ionization of surface hydroxyl groups in aqueous suspension (13). In pure substances and in oxides, pH has been known to control the

194

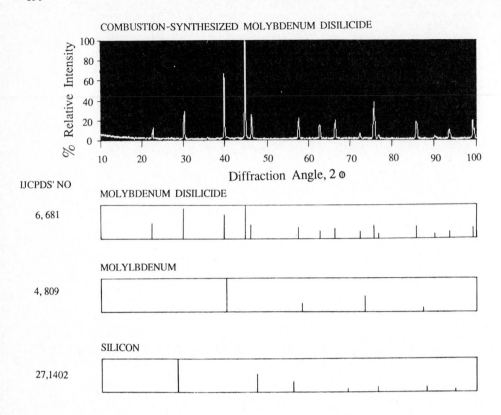

Fig. 5   X-ray diffractogram of combustion—synthesized MoSi$_2$ and IJCPDS patterns of MoSi$_2$ and Si.

surface ionization behavior (14). It has been suggested that a heterogeneous surface containing more than one type of surface functional group may exhibit a number of isoelectric points (IEP- isoelectric point value corresponds to the pH at which the net surface charge on the particle is zero) depending on the relative dissociation of surface groups.

Although there exist a number of papers on the stability of mixed oxide systems, only a few studies relate the surface heterogeneity to the net electrokinetic potential.   However, James and Healey (15) have proposed a mechanism for explaining the electrokinetic behavior of silica particles dispersed in a solution containing hydrolyzable cobalt ions which were adsorbed first onto the surface of silica and precipitated.   The authors indicated three points of zero charges corresponding to the isoelectric points of silica, cobalt hydroxide, and another more basic hydroxide of cobalt as a result of the adsorption of cobalt ion followed by the precipitation of cobalt

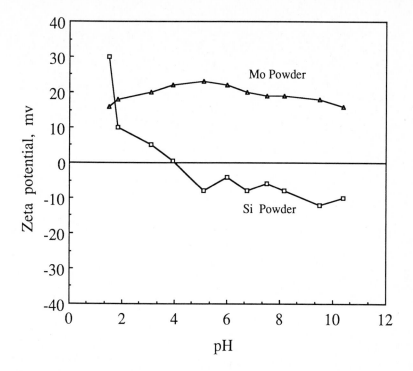

Fig. 6   Zeta potential versus pH profiles of Mo and Si powders dispersed in 10^{-3} mol dm^{-3} KCl solution.

hydroxide on the surface of silica.   Recently, Pugh and Tjus (16) have suggested from their electrokinetic study on copper (II) hydroxy coated zinc sulfide particles that a number of isoelectric points may arise from surface doping of the second phase into the first phase.

It has been suggested empirically from experimental evidence (17) that the mechanism of James and Healey can be extended to heterogeneous and/or heteroparticulate systems.   In the same manner, the electrokinetic behavior of $MoSi_2$ can be explained in terms of the electrokinetic behavior of a mixed oxide system.   A schematic diagram of the electrokinetic behavior of a mixed oxide system in suspension is shown in Figure 8.   Figure 8a shows the electrokinetic behavior of two pure oxides, oxide A and oxide B.   However, the shape of the profile and the intermediate IEP depends upon the concentration of the individual species in the mixture.   For example, Schwarz et al (18) indicated that in silica dispersions the point of IEP shifted due to the addition of alumina.   The IEP of 100% silica at pH 4 has shifted to 4.85, 5.85 and 6.28 due to the addition of 25, 64 and 75% alumina, respectively.

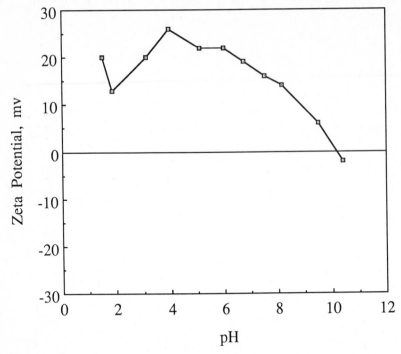

Fig. 7 Zeta potential of combustion-synthesized $MoSi_2$ powder dispersed in 10^{-3} mol dm^{-3} KCl solution.

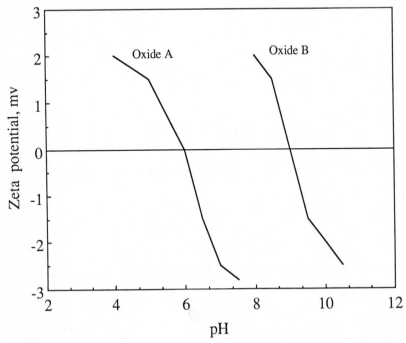

Fig. 8a A schematic diagram of the electrokinetic behavior of pure oxides.

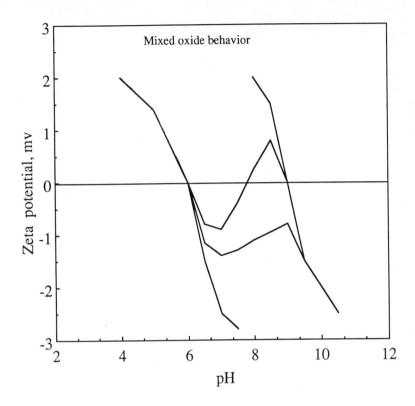

Fig. 8b A schematic diagram of the electrokinetic behavior of mixed oxides dispersed in water.

From the above, Figure 6 can be explained as representing typical shapes of pure silica and molybdenum oxide surfaces analogous to the behavior of pure oxide A and oxide B (Figure 8b). Figure 7 may represent the combined acidic silica surface and basic molybdenum oxide surface (as in the case of a mixed oxide surface behavior - Figure 8b). To prove that the shape of zeta potential versus pH curve for $MoSi_2$ represents the combined oxide surfaces of Mo and Si, we calcined Mo, Si and $MoSi_2$ in air at 500 $^{\circ}$C for two hours and measured their zeta potentials. Figure 9 shows the zeta potential versus pH profiles of calcined Si and $MoSi_2$ powders along with the profile of pure silica. It can be noticed from the figure that the IEP of Si shifted from pH $\approx$ 4 in Figure 6 to pH ~ 3 in Figure 9 due to calcination (Figure 9) and the silicon powder surface electrophoretically tends to behave to a greater extent as pure silica. This could be due to the fact that calcined Si can form stronger bonds in water suspensions due to the oxide layer present on the surface. The partial dissolution of MoO_3 above pH 5 prevented the accurate

198

determination of the zeta potentials of calcined Mo powders. Thermogravimetric curve of calcined Mo powders indicate that Mo powders are oxidized significantly to MoO_3 with a weight gain of 40% (Figure 10). During calcination, $MoSi_2$ is oxidized with a weight gain of 22% (Fig. 10.) The aqueous chemistry of MoO_3 indicates that MoO_3 is not attacked by acids but dissolves in basic media (19). Therefore, at a sufficient basicity (or pH) of the solution MoO_3 formed on the surface is dissolved, leaving behind a SiO_2 layer. The IEP shift from pH ~ 10 to pH ~8 due to calcination gives credence to the above explanation.

A IEP shift towards still lower pH values has been observed (20) at calcination temperatures of 700°C, 900°C, 1100°C and 1300°C. The shift towards low IEP values is explained as due to evaporation of MoO_3 at higher temperatures leaving behind $MoSi_2$ with a layer of silica. Volatilization of MoO_3 begins around 850°C and its vapor pressure at 1360°C is close to 700 mm of Hg.

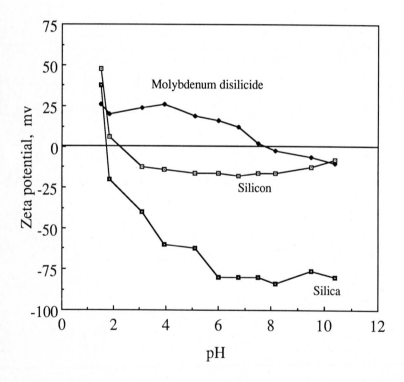

Fig. 9 Zeta potential of calcined Si, $MoSi_2$ and as-received pure silica powders dispersed in 10^{-3} mol dm^{-3} KCl solution. Calcination at 500°C in air for two hours.

Fine particles of $MoSi_2$ can be assumed to behave as colloids holding a film of water. They acquire a charge due to ionic dissociation and the magnitude of the charge (zeta potential) determines the repulsive power of particles. Based upon Anderson and Murray's (21) theoretical and experimental evidence, it is reasonable to assume that the maximum in the zeta potential versus pH curve corresponds to the minimum in the viscosity versus pH curve. The sign of the charge and its magnitude indicates that slip casting of $MoSi_2$ ceramic components can be carried out successfully using combustion-synthesized $MoSi_2$ powders.

We have studied and explained the complex nature of zeta potential versus pH behavior for slip casting of $MoSi_2$ ceramic components. It should be stressed that additional experiments need to be carried out to determine the influence of particle size and size distribution to optimize the slip casting parameters.

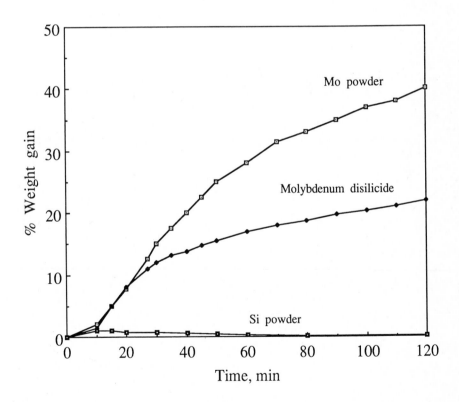

Fig. 10   Isothermal thermogravimetric curves of Mo, Si and $MoSi_2$ powders at $500^\circ C$ for 2 hours.

CONCLUSIONS

(a) Different electrokinetic behaviors have been observed for Mo and Si in dilute aqueous dispersions. For example, while Mo powders in suspension exhibited a positive charge in the pH range 1.5 to 10.5, Si powders showed negative electrophoretic behavior above pH ~4.

(b) The IEP of $MoSi_2$ is around pH 10 and the zeta potential decreased sharply around pH 2 before reaching a maximum at pH 4. IEP values of Si and $MoSi_2$ shifted towards lower pH values upon calcination.

(c) The sign of the charge and magnitude of the zeta potential of combustion-synthesized $MoSi_2$ indicate that slip casting technique can be used to produce ceramic components of $MoSi_2$.

REFERENCES

1   Ceramic Powder Science , Advances in Ceramics, 21 (1987).
2   Abstracts of 89th Annual Meeting of the American Ceramic Society, Pittsburgh, PA, April 26-30, (1987).
3   A. G. Merzhanov, Combustion Processes in Chemical Technology and Metallurgy, Chernogolovka, (1975).
4   A. G. Merzhanov and I. P. Borovinskaya, Dokd. Akad. Nauk SSSR, 204, 2 (1972).
5   DARPA/ARMY Symposium on Self-Propagating High Temperature Synthesis, Oct. 21-23, Daytona Beach, Florida (1985).
6   J. B. Holt and Z. A. Munir, J. Mat. Sci., 21, 251 (1986); also P. D. Zavitsanos and J. R. Morris, Jr., Ceram. Eng. Sci. Proc., 4 (7-8), 624 (1983).
7   S. C. Deevi and C. K. Law, Spring Meeting of Materials Research Society, Paper H1.5, Apr. 21-23, Anaheim, CA (1987).
8   G. V. Samsonov and A. P. Epik in Coatings of High-Temperature Materials , Ed. H. H. Hausner, Plenum Press, N.Y. (1966).
9   G. V. Samsonov and I. M. Vinitskii, Handbook of Refractory Compounds , Plenum Publishing Co., (1980).
10   S. C. Deevi and C. K. Law, to be published.
11   G. A. Parks, Chem. Reviews, 65, 177 (1965).
12   M. Escudey and F. Gil-Llambías, J. Coll. and Int. Science, 107(1), 272 (1985).
13   G. D. Parfitt, Dispersion of Powders in Liquids , 3rd ed., Applied Science Publishers, Inc., Englewood, N.J. (1981).
14   G. A. Parks, Adv. in Chem. Ser., 67, 121 (1967).
15   R. O. James and T. W. Healey, J. Coll. and Int. Science, 40(1), pp 42, 53, 65 (1972).
16   R. J. Pugh and K. Tjus, J. Coll. and Int. Science, 117(1), 231 (1987).
17   A. S. Rao and G. D. Parfitt, The Properties of Coal/Water Slurries Based on Lower Kittanning and Splash Dam Coals , Report of Investigation submitted to the Gulf Research and Development Corp., and DOE, Pittsburgh, PA. (1984).
18   J. A. Schwarz, C. T. Driscoll and A. K. Bhanst, J. Coll. and Int. Science, 97(1), 55 (1984).
19   F. A. Cotton and and G. Wilkinson, Advanced Inorganic Chemistry", Wiley Eastern Pvt. Ltd., New Delhi (1972).
20   S. C. Deevi, C. K. Law and A. S. Rao, to be published.
21   P. J. Anderson and P. J. Murray, J. Amer. Ceram. Soc., 42, 70 (1959).

Interfacial Phenomena in Biotechnology and Materials Processing,
edited by Y.A. Attia, B.M. Moudgil and S. Chander
Elsevier Science Publishers B.V., Amsterdam, 1988 — Printed in The Netherlands

ELECTROKINETIC BEHAVIOR OF SILICON CARBIDE, ALUMINUM NITRIDE, TITANIUM DIBORIDE, TITANIUM CARBIDE AND BORON CARBIDE DISPERSED IN DILUTE AQUEOUS DISPERSIONS

S. C. Deevi[1], C. K. Law[1] and A. S. Rao[2]

[1]Department of Mechanical Engineering, University of California Davis, CA 95616

[2]Department of Ceramics, Rutgers University, Piscataway, New Jersey 08855

SUMMARY

The electrokinetic behavior of powders of combustion synthesized titanium diboride and titanium carbide, and of commercial silicon carbide, aluminum nitride and boron carbide, was investigated in the pH range 1.5 to 11 using microelectrophoresis technique. The zeta potentials of the above materials were compared with the results obtained on pure silica, alumina and titania. The electrokinetic behavior of both silicon carbide and aluminum nitride appears to be simple and resembles that of pure silica and alumina, respectively. However, the zeta potentials versus pH profiles of titanium diboride, titanium carbide and boron carbide are complex. The complex electrokinetic behavior may be attributed to the different hydroxy complex species formed on the powder surface due to the chemical reaction between the dispersed powder and water at the solid/liquid interface.

INTRODUCTION

The final properties of ceramic components such as microstructure and composition depend on the methods employed in assembling the fine particles through various forming processes like slip casting, pressing, extrusion, injection molding, spraying, deposition, etc. Bowen (1) in a recent review demonstrated the necessity to understand and control the parameters influencing the morphology of high temperature ceramic materials. Processing of high temperature ceramic materials for optical and electronic components, rocket engines, turbine blades and as large structural components requires methods which yield homogeneous components with minimal porosity. This can be achieved with well dispersed suspensions of fine particles, since the surface charge acquired due to their colloidal behavior determines (a) how soon the dispersed particles agglomerate and (b) the density of shapes or compacts formed by particles under the influence of an electric field. It is well known that a high degree of reliability can be achieved in processing of ceramic materials by optimizing the particle packing structure based on a knowledge of the interfacial chemistry of dispersed ceramic powders (2-4).

Slip casting of ceramic materials is an attractive alternative method for the conventional injection molding technique. The basic theories concerned with the slip casting of oxides should be applicable to non-oxide ceramic materials. They are all based on the diffuse double layer theory, which has been used to understand the double layer properties such as electrophoretic mobility and zeta potential of ceramic powders (5-9). Although much information is available on the nature and behavior of ceramic oxides such as silica, alumina, titania (10-12), very little is known about the non-oxide ceramics such as carbides, nitrides and borides of titanium, silicon, zirconium, hafnium, etc. Since surfaces of non-oxide ceramic materials may be oxidized to a certain degree depending upon the processing, storage and handling of powders, it is possible that these powders may exhibit electrokinetic behavior similar to their oxide cousins.

Uncertainties associated with the preparation, handling and storage of powders may be eliminated by synthesizing the powders with combustion synthesis technique, otherwise known as self-propagating high temperature synthesis technique. The technique has generated great interest (13-16) due to the short reaction times coupled with the extremely high temperatures generated during the reaction. Electrokinetic behaviors of combustion synthesized titanium carbide (TiC), titanium diboride (TiB_2) and commercial silicon carbide (SiC), boron carbide (B_4C) and aluminum nitride (AlN) were compared with the oxides of silicon, aluminum and titanium. As expected, we noticed a similarity in the electrokinetic behavior of SiC and AlN with the corresponding oxides like silica (SiO_2) and alumina (Al_2O_3). But in the case of TiC, TiB_2 and B_4C, we observed a complex electrokinetic behavior which cannot easily be explained based on the behaviors of their corresponding pure oxides. In this paper we (a) discuss briefly the synthesis aspects of TiC and TiB_2, and (b) present preliminary results which call attention to the complex electrokinetic behavior of TiC, TiB_2 and B_4C.

EXPERIMENTAL

Commercial carbon, boron and titanium powders were used for synthesizing titanium carbide and titanium diboride powders. Particle size distribution of as-received powders were determined using a laser granulometer by dispersing the powders either in an aqueous medium or in an organic solvent with the aid of a dispersing agent. Figure 1 indicates that the powders of carbon, titanium and boron are polydispersed and their median diameters are in the range 11.5 to 15.5 μ (Table 1). Powders of carbon and boron assume the shapes of either flakes or platelets while titanium powders possess a smooth, spherical surface (Figure 2). Combustion synthesis of TiC and TiB_2 was carried out by obtaining cylindrical compacts (from homogenized powders) of

1.96 cm diameter by approximately 2.54 cm length.

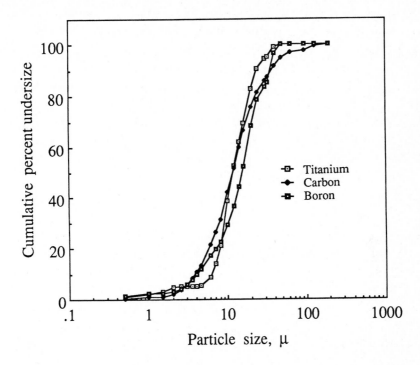

Fig. 1 Particle size distributions of titanium, carbon and boron powders

TABLE 1

The Median Diameters of Various Powders

Powder	Median Diameter, μ
Carbon	11.6
Titanium	11.5
Boron	15.5
Silicon Carbide	14.6
Aluminum Nitride	1.8
Alumina	2.7
Titanium Carbide	5.0
Titanium Diboride	6.0
Titania	6.3
Boron Carbide	18.5

204

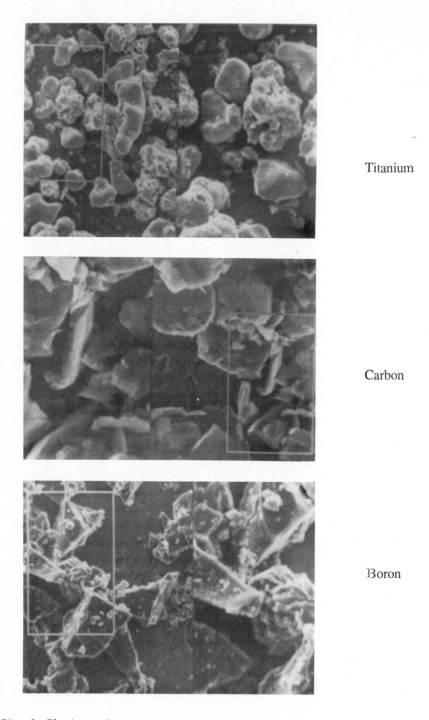

Titanium

Carbon

Boron

Fig. 2 Electron micrographs of titanium, carbon and boron powders

fine powders of commercial SiC, AlN, B$_4$C and oxides of silicon (SiO$_2$), aluminum (Al$_2$O$_3$) and titanium (TiO$_2$) were also used in the present set of experiments along with the synthesized powders. Particle size distributions of various ceramic powders used for electrokinetic measurements are polydispersed as shown in Figure 3 and their median diameters range from 1.8 to 18.5 μ (Table 1). The scanning electron micrographs (Figure 4) also indicate their polydispersed nature.

Dispersions suitable for microelectrophoresis were prepared in 10^{-3} mole/dm^3 KCl solution by dispersing 0.1 g of each powder using an ultrasonic

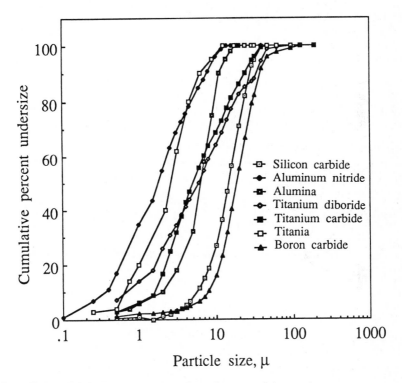

Fig. 3 Particle size distributions of various ceramic powders

probe for 5 minutes. Suspensions were prepared in 10^{-3} mole/dm^3 KCl solution to maintain the electrical double layer thickness fairly constant in the pH range investigated. The pH of the suspension was varied in the range 1.5 to 10.5 by adding either 1 mole/dm^3 HCl or NaOH followed by thorough mixing using an ultrasonic probe. The zeta potential of the dispersed powders was measured from these solutions after 24 hours and at each time due care was taken in noting down the final pH of the suspension prior to the measurement. Pen Kem Laser Zee meter was used to determine the zeta potential and velocities of at least 20 to 30 particles were measured to get an accurate value.

COMBUSTION SYNTHESIS OF TiC AND TiB$_2$

A strong exothermic reaction between the reactants is a prerequisite for combustion synthesis of ceramic materials. Solid-solid reactions are exothermic in nature and a significant amount of heat is liberated during the course of the reaction. If titanium and boron powder were to react to form TiB$_2$, the free energy change, ΔG, must be highly negative since the entropy change in most solids is assumed to be equal to zero (i.e. $\Delta S \approx 0$) or negligible. This dictates that the change in enthalpy must be negative for the reaction to be exothermic. The ΔH or heat liberated during the reaction raises the temperature of the compact by an amount which can be estimated based on

$$\Delta H = \int_{298}^{T} C_p dT$$

where Cp is the specific heat of the product.

As can be noted from Table 2, all the reactions are exothermic and the reaction leading to the formation of TiB$_2$ is most exothermic of all. Accordingly, the adiabatic combustion temperature is also very high. In the case of TiB$_2$, the adiabatic combustion temperature is much higher than the melting point of the product such that melting of the product takes place. This could also happen in the case of TiC. In all other cases, melting points of products are much higher than the respective adiabatic combustion temperatures. Synthesis of SiC and B$_4$C is difficult due to their low heats of formation, and external energy or other methods must be used to synthesize these materials.

To synthesize TiB$_2$ and TiC, the exothermic reaction is initiated on the surface of the compact by raising the surface temperature with a heated ignition coil. Due to the thermal conductivity of the compact, the heat liberated during the exothermic reaction is initially dissipated through the compact. A point is reached, however, when the reaction temperature rises appreciably accelerating the reaction rate and the rate of heat release. At this point, heat liberated is sufficient enough to heat the adjacent layers of reactants and the reaction becomes self-sustained in an uncontrolled manner propagating through the reactants. The reaction front, or combustion wave, propagated at a certain rate depending upon the atmosphere, diameter of the compact, initial particle sizes of reactants and density of the compacts (17). A typical temperature profile of TiC shown in Figure 5 indicates that the reactants were heated at a heating rate of about 10^5 $^\circ$C/min.

The reaction in the synthesis of TiB_2 is very fast and propagated at a rate of about 1.5 to 8.0 cm/sec. Very high rates were observed with amorphous boron powder and in some cases the compact reacted in an explosive manner. Combustion wave propagated smoothly in the synthesis of TiC and the velocities are in the range 0.5 to 1.9 cm/sec depending upon the particle sizes of reactants. Characterization of the products by x-ray analysis indicated the products to be pure TiC and TiB_2. SEM examination revealed that the products are porous with sponge like appearance.

Titanium diboride

Titanium carbide

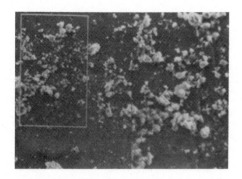

Boron carbide

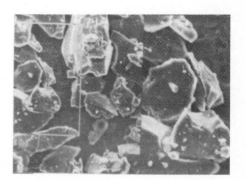

Aluminum nitride

Fig. 4   Electron micrographs of titanium carbide, titanium diboride, boron carbide and aluminum nitride powders.

Table 2
Thermodynamic Properties

Reaction	Heat of Formation ΔH^o_{298}, k·cal/mole	Adiabatic Combustion Temperature, K	Melting Points of the Products, K
Ti + 2B → TiB$_2$	-66.9	3190	3073
Ti + C → TiC	-44.1	3210	3210
Si + C → SiC	-15.6	1800	2970
4B + C → B$_4$C	-17.0	1000	2720

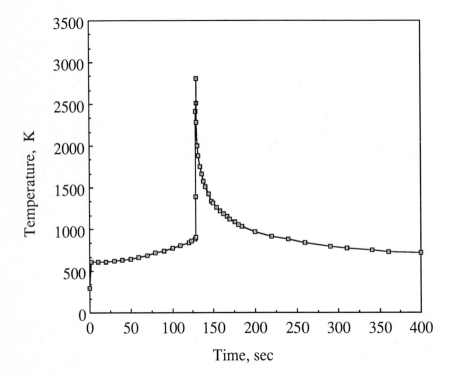

Fig. 5 Temperature profile obtained during combustion synthesis of TiC

RESULTS

   The zeta potential versus pH plot of SiC powder is shown in Figure 6.
Commercial SiC powders exhibited isoelectric point (IEP) values in the
range 2.5 to 3.0. Note that each point on the curves is a separate experiment
and pH is a log quantity. The results suggest that the SiC powders exhibit

an acidic character similar to that of pure silica, even though their magnitudes are much lower than pure silica. Also evident from Figure 6 is that isoelectric point of silicon carbide powders are higher than that of pure silica.

Figure 7 shows the zeta potential of AlN sample dispersed in 10^{-3} mole/dm^3 KCl solution. Also shown in Figure 7 is the zeta potential of commercial alumina (Alcoa A-16). The results indicate that the electrokinetic behavior of AlN is completely different from that of SiC exhibiting a basic character similar to that of alumina. However, the difference between the behavior of alumina and AlN in addition to the magnitude of their potential, appears to be a sudden decrease in the zeta potential of AlN at pH 6 followed by a slight increase at pH 7.

Figure 8 shows the zeta potential of TiB$_2$ and also pure titania. Titanium diboride exhibits a complex surface behavior, similar to that of a

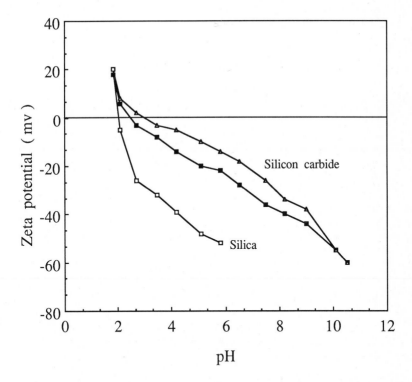

Fig. 6   Zeta potentials of silicon carbide powder and pure silica powder dispersed in 10^{-3} mole/dm^3 KCl solution

complex mixed oxide (18). The zeta potential versus pH behaviors of TiC and B$_4$C are indicated in Figures 9 and 10. As can be noted from Table 3, isoelectric points of TiC have shifted towards basic scale as compared to the isoelectric points TiB$_2$. Unlike titania (pure oxide of titanium), titanium carbide and titanium diboride exhibited four isoelectric points at the pH ranges indicated in Table 3.

DISCUSSION

Surface examination of the powders using x-ray photoelectron spectroscopy indicated that the surfaces of the powders are coated with a thin film of the oxide impurity. Due to the oxygen impurity, powders can easily form bonds with water when dispersed in an aqueous medium. The zeta potential behavior of silicon carbide resembles that of silica and the IEP of silicon carbide is in the pH range ~2.5 to 3.0 while the IEP of silica is at pH ~2. It is well known that silanol functional groups present on the surface of silica are responsible for the acidic behavior exhibited by silica (19,20). The acidic behavior shown by silicon carbide powders similar to that

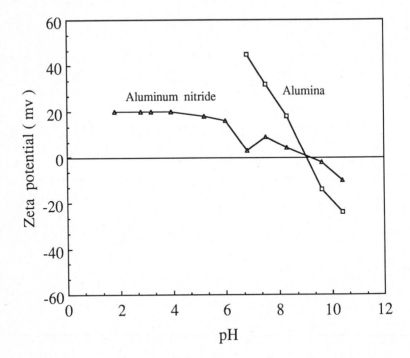

Fig. 7 Zeta potential of aluminum nitride and alumina particles dispersed in 10^{-3} mole/dm^3 KCl solution

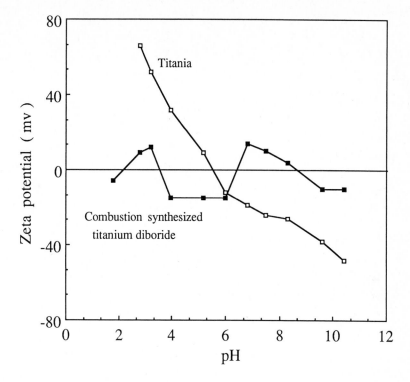

Fig. 8 Zeta potential of titanium diboride and titania powders dispersed in 10^{-3} mole/dm^3 KCl solution

of the silica surface implies the presence of silanols even on the surfaces of silicon carbide powders. Recently, the existence of a silica coating based on the surface chemistries of silicon carbide and silicon nitride powders has been reported without any apparent difference in their silanol densities. (21). Therefore, the observed electrokinetic behavior may be explained as follows:

Acidic Medium: $Si----OH_{(SiC)} + H^+_{aq} \rightarrow Si----OH^+_2 \ (SiC)$

Basic Medium: $Si----OH_{(SiC)} + OH^-_{aq} \rightarrow Si----O^-_{(SiC)} + H_2O(\ell)$

The IEP value of aluminum nitride is at pH ~9 agreeing well with the IEP value of alumina (at pH ~9). The oxide layer present on the surface of aluminum nitride can easily take up protons from aqueous solutions giving rise to a positive charge on the surface. With increase of pH neutrality arises depending on the acid-base equilibrium constants at the surface. Figure 2 implies that such a situation occurs at pH ~9 both for aluminum nitride and

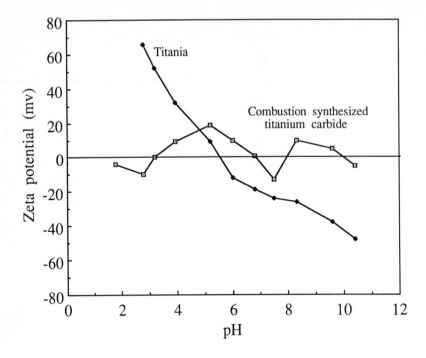

Fig. 9 Zeta potential of titanium carbide and titania powders dispersed in 10^{-3} mole/dm^3 KCl solution

alumina. Since the oxide layer determines the magnitude of the surface charge, it is possible that in the case of aluminum nitride the surface may not have been covered fully with an oxide layer.

Any explanation of the observed electrokinetic behaviors of titanium carbide, titanium diboride and boron carbide must consider the aqueous chemistry of titanium, carbon and boron. It is well known that the aqueous chemistry of titanium is complex due to its various oxidation states, -1, 0, +2, +3 and +4. Valences -1, 0, +2, +3 are oxidized to Ti^{IV} by water and Ti^{IV} may be considered to be the most stable state (22). Titanium^{IV} compounds readily undergo hydrolysis giving rise to Ti-O bonds, and in titania only Ti-O bonds exist. In the case of titanium carbide, both titanium and carbon can form bonds with oxygen complicating the surface structure. Carbon does not undergo ionization and is negatively charged in contact with water. It is widely accepted at the moment that surfaces of carbons and coals contain polar functional groups like hydroxyl, carboxyl and even lactone groups (23-25). The oxygenated surface complexes give carbon the negative charge due to the attachment of hydroxyl ions from the water and possibly also of anions from

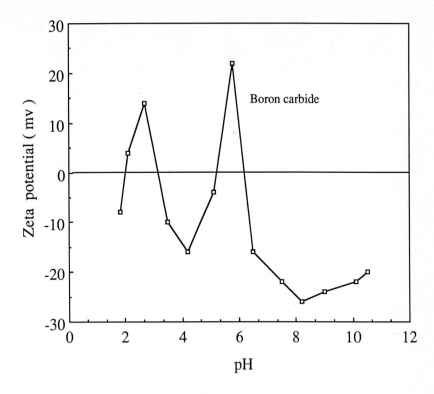

Fig. 10 Zeta potential of boron carbide powder dispersed in 10^{-3} mole/dm^3 KCl solution.

Table 3
Isoelectric Points of Various Ceramic Powders.

Ceramic Powder	Isoelectric Points
Titanium Diboride	2.18, 3.50, 6.50, 8.66
Titanium Carbide	3.20, 6.82, 7.95, 10.00
Boron Carbide	2.00, 3.16, 5.22, 6.22
Titania	5.50

the electrolyte. In contrast, as can be seen from Figures 8 and 9, titania is positively charged above pH ~5.5 and negatively charged below that value. It is opportune to mention at this point that titanium also forms bonds of the type Ti-O-C (19), thereby further complicating the surface behavior. Based on the above arguments, it can be surmised that charge and magnitudes of zeta potentials vary depending upon the relative strengths of the polar functional groups associated with titanium and carbon, titanium and boron, and boron and carbon with variation of pH.

It has been suggested recently that "a number of point zero charges" may arise due to adsorption or nucleation of simple hydrated ionic species, or surface nucleation of (hydrolyzed) polymeric species or due to the attachment of colloidal hydroxides formed in solution (26). Therefore, qualitatively it can be argued that the number of IEP's observed in Figures 8 to 10 may correspond to the different oxyhydrates formed on the powder surface due to chemical reactions at the solid/liquid interface as in the case of a complex mixed oxide behavior (18).

Although acid-base reactions at the interface have been considered above to explain the electrokinetic behavior of titanium carbide, titanium diboride and boron carbide, a more meaningful description of the zeta potential versus pH behavior is possible only after a thorough understanding of the surface reactions occurring at the interface. We intend to (a) examine the surfaces with FT-IR spectroscopy to determine the nature of surface groups present, (b) determine the electrokinetic behavior with different electrolytes at various concentrations and (c) obtain the acid-base equilibrium constants (pK_a and pK_B).

CONCLUSIONS

Silicon carbide powders and aluminum nitride dispersed in water electrophoretically behave similarly to the oxides of silicon and aluminum. The electrokinetic behavior of titanium carbide, titanium diboride and boron carbide is quite complex compared to the electrokinetic behavior of pure titania, carbon or boron and exhibits a number of zpc values. The origin of the multiple zpc values is not yet clear, however, it is possible that a chemical reaction between the dispersed carbide or boride and water may be taking place at the solid/liquid interface forming the surface coating of hydrolyzed species.

ACKNOWLEDGEMENTS

The authors would like to thank Ms. T. Gembala and Mr. W. Warren of University of California, Davis for their assistance in the experimental work, and Mr. J. Popp of Callery Chemical Company for supplying sub-micron amorphous boron powder.

REFERENCES

1   H. K. Bowen, Mat. Sci. Engg, 44, 1 (1980).
2   J. S. Reed, T. Carbone, C. Scott and S. Lukasiewitz, in Processing of Crystalline Ceramics, Materials Science Research, Vol. 11, (1978).
3   A. E. Pasto, C. L. Quackenbush, K. W. French, and J. T. Neil, in Ultrastructure Processing of Ceramics, Glass, and Composites, Ed. D. R. Ulrich and L. L. Hench, Wiley, New York (1984).
4   F. F. Lange, J. Am. Ceram. Soc., 67 (2) 83 (1984).
5   J. Th. Overbeek, "The Interaction Between Colloid Particles," in Colloid Science, H. R. Kruyt, Ed., Elsevier, Amsterdam, pp 245-277 (1952).
6   G. D. Parfitt, "The Dispersion of Powders in Liquids," Elsevier Pub. Co., New York (1969).
7   R. J. Hunter, "Zeta Potential in Colloid Science," Academic Press, New York (1981).
8   G. A. Parks, Chem. Rev., 65 177 (1965).
9   G. A. Parks, Adv. in Chem. Ser., 67 121 (1967).
10   F. Y. Wang, ed., "Ceramic Fabrication Processes," Academic Press, New York (1976).
11   R. O. James, Advances in Ceramics, 21 349 (1987).
12   E. M. DeLiso, A. S. Rao and W. R. Cannon, Advances in Ceramics, 21, 525 (1987).
13   A. P. Hardt and P. V. Chung, Combust. Flame, 21, 77 (1973), also Combust. Flame, 21, 91 (1973).
14   A. G. Merzhanov and I. P. Borovinskaya, Dokl. Akad. Nauk S.S.S.R., 204(2), 366 (1972), also Comb. Sci. Technol., 10, 195 (1975).
15   DARPA/ARMY Symposium on Self-Propagating High Temperature Synthesis, Oct. 21-23, Daytona Beach, Florida (1985).
16   K. V. Logan and J. D. Walton, Proc. Ceram. Eng. Sci., 712 (1984).
17   S. C. Deevi and C. K. Law, to be published.
18   S. C. Deevi, C. K. Law and A. S. Rao, "Electrokinetic Behavior of Molybdenum, Silicon and Combustion Synthesized Molybdenum Disilicide Powders Dispersed in Aqueous Media," Proceedings of this meeting.
19   J. Eisenlawer and E. Killman, J. of Colloid Interface Sci., 74 108 (1980).
20   R. K. Iler, The Chemistry of Silica, Wiley, New York (1979).
21   P. K. Whitman and D. L. Feke, Adv. Cer. Mat., 1 (4), 366 (1986).
22   F. A. Cotton and G. Wilkinson, "Advanced Inorganic Chemistry," Wiley Eastern Pvt. Ltd., New Delhi (1972).
23   R. N. Smith, D. A. Young and R. A. Smith, Trans. Farady Soc., 62, 2280 (1966).
24   F. H. Healey, Y. F. Fu and J. J. Chessick, J. Phys. Chem., 59, 399 (1955).
25   W. R. Kube, Ind. Eng. Chem., Chem. Eng. Data Ser., 2, 46 (1957).
26   R. J. Pugh and K. Tjus, J. Coll. and Int. Science, 117 (1), 231 (1987).

Interfacial Phenomena in Biotechnology and Materials Processing,
edited by Y.A. Attia, B.M. Moudgil and S. Chander
Elsevier Science Publishers B.V., Amsterdam, 1988 — Printed in The Netherlands

EFFECTS OF ADSORPTION OF POLYACRYLIC ACID ON THE STABILITY OF α-Al$_2$O$_3$, m-ZrO$_2$, AND THEIR BINARY SUSPENSION SYSTEMS

A. Bleier and C. G. Westmoreland

Oak Ridge National Laboratory, Metals and Ceramics Division, P.O. Box X, Oak Ridge, TN 37831

ABSTRACT
     The influence of polyacrylic acid having a molecular weight of 90,000 on the stability of aqueous suspensions containing either α-A$_2$O$_3$, m-ZrO$_2$, or both oxides was systematically studied over the range pH 3 to 11 using sedimentation, rheological, electrophoretic, adsorption, and pycnometric methods. The anionic polymer adsorbs on each type of surface, with the corresponding isotherms appearing Langmuirian. Acid-base phenomena giving rise to charging of the oxide particulate surfaces and the polymer appear ultimately responsible for adsorption and overall stability behavior.

INTRODUCTION
     This report describes recent efforts to understand the role of polyacrylic acid, PAA, on various colloidal and interfacial phenomena encountered in the presintering processing of ceramic composites that contain alumina and zirconia. This system represents important structural ceramic components, particularly for structural uses. (ref. 1) Though this study specifically focuses on incorporation of monoclinic zirconia, m-ZrO$_2$, into a matrix of alpha alumina, α-Al$_2$O$_3$, the general results should also apply to the processing of α-Al$_2$O$_3$ and t-ZrO$_2$ (tetragonal), since we emphasize the generic principles by which the ceramic composite's precursory binary suspensions behave.

     Following a description of behavior observed in unary suspensions containing 0.5-μm α-Al$_2$O$_3$ and 0.2-μm m-ZrO$_2$, obtained in the absence and presence of PAA having a molecular weight of 90,000, we describe experiments using binary suspensions of Al$_2$O$_3$ and ZrO$_2$. Experiments to understand colloidal and interfacial phenomena include rheological, sedimentation, adsorption, electrophoretic, and potentiometric titration methods. Density measurements were used to evaluate consolidated pellets prepared by pressure filtration.

CHARACTERIZATION OF MATERIALS
Alumina and Zirconia
     Table 1 summarizes some of the characteristics of the powders used in this study. The powders were fractionated by size according to the procedure

TABLE 1

Chemical and Physical Properties of Alumina and Zirconia Powders[a]

Property	Characterization		Analytical method(s)
Composition	Al_2O_3	ZrO_2	X-ray Diffraction
Phase	Alpha	Monoclinic	X-ray Diffraction
Density, g cm^{-3}	3.98	5.56	(ref. 2)
Surface Area, m^2 g^{-1}	7.92	13.5	BET Gas Adsorption
Diameter, nm	500, 710[b]	200	Laser Light Scattering
Isoelectric Point	8.1	6.5	Surface Titration[d], Microelectrophoresis

[a]Alumina: Sumitomo Chemical, Tokyo, HPS-40, Lot No. 40615; Zirconia: Toya Soda, Atlanta, TZ-0, Lot No. Z006291P.

[b]Lot No. 40313, used for initial titration experiments.

[c]Leeds & Northrup, Microtrac Particle-Size Analyzer, Model 7991-3, St. Petersburg, FL.

[d]Site density, N_s, is respectively assumed for calculation of surface charge density on Al_2O_3 and ZrO_2 to be 2.7 and 2.5 nm^{-2}. (ref. 3)

described by Baik et al. (ref. 4) X-ray diffraction analyses confirmed that no other crystalline phases were present in each sample.

The alumina was produced hydrothermally and is similar to that examined by Cesarano et al. (ref. 5); the zirconia is of the same origin as that studied by Baik et al. (ref. 4) Single point BET analysis of the surface area yields equivalent spherical diameters of 190 nm for Al_2O_3 and 79.9 nm for ZrO_2, values that are notably, namely 62 and 60 %, less than the ones in Table 1 derived by light scattering analysis. Stock suspensions were prepared and stored in polypropylene containers until stability, adsorption, and electrophoresis experiments were preformed.

Powder for titrimetric analysis was cleaned using soxhlet extraction by water for at least 72 h, a duration that is sufficient to remove all traces of soluble surface impurities. (ref. 6) Preliminary titration experiments indicated that the total concentration of soluble, reactive surface impurities was adequately minimized by soxhelation. Cleaned powder was ultimately dispersed in doubly distilled water for titration studies.

Polyacrylic Acid

This material (Aldrich Chemical Company, Lot. No. 0324BL) has the repeat unit structure, $CH_2CH(CO_2H)$, with a repeat unit mass of 72.1 g mol^{-1}, is linear, and has a molecular weight of 90,000. Individual sites on the polyacrylic acid (PAA) chain may be either neutral, as suggested by the structure, or bear a negative charge that develops via dissociation; see Eq. (1). Unless noted otherwise, concentration of polymer is expressed as g PAA per 100 g solid.

$$CH_2CH(CO_2H) \rightleftarrows CH_2CH(CO_2^-) + H^+ \tag{1}$$

The intrinsic acidity coefficient, K_a^{int}, that thermodynamically describes this reaction for the sample used in this study is 2×10^{-5}. The distribution of neutral and anionic sites depends on the solution pH, with pH = 5.9 being the condition for dilute solutions at which 50 % of the sites exist in each form; the pH at which this condition is met varies with only slightly ionic strength. (ref. 7) These acid dissociation properties were evaluated using potentiometric titrimetry; this procedure is described later.

Other Chemicals

HCl and NaOH were used to adjust pH and NaCl was used to maintain ionic strength. All reagents were of analytical grade (Fisher Scientific) and were used without further purification. Whereas doubly distilled water was used in most experiments, deionized water was used for some rheological and electrophoretic measurements. Prepurified nitrogen or argon was used as an inert atmosphere for all powder and polymer titrations. Finally, paraffin (Fisher Scientific) was used to coat pellets prior to determination of their density.

EXPERIMENTAL PROCEDURES

Potentiometric Titration

An automatic titration system (Radiometer America, Model RTS822) equipped with a double junction reference electrode, was used. The electrode was filled with 0.01- and 0.1-M NaCl for selected suspensions and PAA solutions respectively, having ionic strength in the ranges 0.001- to 0.01- and 0.1- to 1.0-M. (ref. 6) $N_2(g)$ or $Ar(g)$ saturated with water vapor was bubbled through each sample for at least 30 min prior to titration of powder with either acid or base; this procedure ensured removal of residual $CO_2(g)$ from the sample. Typical titration rates were 2 pH-units per h for powder and 3 to 4 pH-units per min for polymer. Data on the amount of titrant to adjust pH of powder suspensions and polymer solutions of known concentration and at fixed ionic strength were then compared to similar data gathered on the corresponding NaCl solutions. Thermodynamic evaluation of ionization and complexation occurring at the surface of powders was performed using a modified version of MINEQL (ref. 8), a computer program for calculating equilibrium speciation in complex aqueous systems. The version used for the oxide powders specifically considers the triple layer model of electrical double layers (refs. 3, 9) and ionic adsorption (ref. 10).

Sample titrimetric data for α-Al_2O_3, and m-ZrO_2 in 0.01 M NaCl are given in Fig. 1. The site-fractional charge, $\alpha_\pm$, and the corresponding surface charge density were determined over a wide pH range by analyzing plots (ref. 7) similar to those in Fig. 1. These analyses yielded the points of zero charge (p.z.c.) of α-Al_2O_3 and m-ZrO_2 respectively at pH 8.1 and 6.5; see Fig. 2.

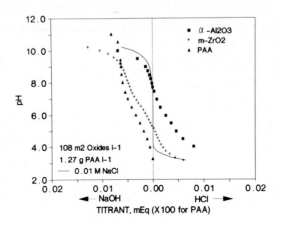

Fig. 1. Sample titration data for α-Al$_2$O$_3$, m-ZrO$_2$, and PAA (90,000 molecular weight) in 0.01 M NaCl. See text for details.

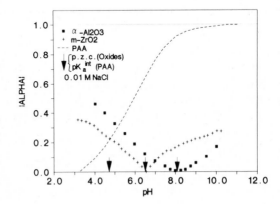

Fig. 2. Absolute value of site-fractional charges α$_\pm$ for oxides and α$_-$ for PAA as a function of pH in 0.01 M NaCl.

Similar data were gathered on the polyelectrolyte, owing to its weakly acidic character (ref. 11). Figure 1 also contains sample titration data for PAA and the dashed curve in Fig. 2 summarizes the relative charge characteristics of this material in 0.01 M NaCl. The profile of α_- vs pH, where α_- is the fraction of all carboxylic acid groups is charged, demonstrates that α_- is relatively small over most of the acidic pH range investigated. Also, quite basic conditions relative to those at which pH = pK_a^{int}, are required for α_- to approach unity. Data in Figs. 1 and 2, along with related ones (ref. 7) obtained at other values of ionic strength were analyzed by the procedures outlined by Overbeek and coworkers (ref. 12) and treatments of Tanford (ref. 11) and Molyneux (ref. 13).

Adsorption Isotherms

Potentiometric titration was used to determine the quantity of PAA remaining in solution after exposure to a known amount of oxide surface area. Each sample was prepared according to the following procedure: An aliquot of stock suspension containing 5 g of either solid was placed in a 50-cm^3 polypropylene centrifuge tube. This portion was then diluted with a predetermined amount of NaCl solution and water to yield the desired concentration of NaCl, usually 0.01 M, a concentration that is insufficient to induce rapid destabilization of the unary suspensions. (ref. 7) After adjustment to the desired pH, samples were allowed to approach equilibrium using a reciprocal shaker for 40 to 48 h, with pH being readjusted to the desired value after approximately 24 h, if necessary; this adjustment was typically small and within 0.5 pH-units. A desired amount of PAA solution of known concentration (C_i) and which had been previously adjusted to the same pH and ionic strength was then introduced. The final mass of each sample was 25 g and ultimate respective concentrations for α-Al_2O_3 and m-ZrO_2 were 5.88 and 4.26 % v/v.

Samples were then reagitated using a reciprocal shaker for 2 d, after which time they were subjected to centrifugation (Sorvall, Model RC-5) for 1 h at 6000 rpm (7000 G). A 15-cm^3 aliquot of clear supernatant was removed from each sample for measurement of pH and analysis by potentiometric titration, a procedure akin to that used by Cesarano et al. (ref. 5) for sodium polymethacrylate. For the PAA systems described herein, NaCl solution was added to each aliquot of supernatant such that the final volume was 30 cm^3. Each sample of diluted supernatant was adjusted to pH 3.5 with HCl and then titrated to the end-point with NaOH using the first-derivitive mode of the titration system described in the previous section; the added HCl did not significantly affect the initial, 30-cm^3 volume. The concentration of polymer in the titrated samples was then determined from the calibration curve given in Fig. 3. This value was then used to calculate for each original system C_e, the concentration of PAA in equilibrium with oxide powder, and Γ, the amount adsorbed.

Evaluation of Stability

Sedimentation and rheological studies were used to evaluate the colloidal stability of unary and binary suspensions in the presence and absence of PAA. Settling tests were performed on suspensions according to the method described earlier for adsorption experiments but containing 4.7 % v/v α-Al$_2$O$_3$, 3.5 % v/v m-ZrO$_2$, and 2.5 % v/v of each solid in binary systems. Relative suspension height (R.S.H.) was visually evaluated at desired times after sample preparation, from 3 min to as long as 2 wk.

Rheological measurements were made with a parallel-plate viscometer (Rheometrics, Fluids Spectrometer, Model 8400) operated in the thixotropic loop mode at 24 °C. The rheological loop consisted of four sweep rates as illustrated in Fig. 4 for 20.1 % v/v α-Al$_2$O$_3$ at pH 8.1: (1) 60 s at 50 s^{-1}, (2) 30 s to reduce shear rate to 0 s^{-1}, (3) 30 s to increase it to 50 s^{-1}, and finally (4) 30 s to reduce it again to 0 s^{-1}. Hysteresis, thixotropy, and Bingham yield values could be detected with this procedure. Detailed properties were evaluated using the steady shear mode with controlled strain and shear rates, using samples prepared as described earlier for the adsorption studies; see that section for details on initial concentrations of solid and, if present, polymer.

Both rheology and adsorption were studied, in some cases, on the same unary α-Al$_2$O$_3$ suspensions, the solid and liquid portion remaining after centrifugation and extraction of an aliquot of solution for adsorption analysis were remixed and the solid was ultrasonically redispersed at high volume fraction (ϕ), to provide 20.1 % v/v α-Al$_2$O$_3$. 3 cm^3 of this suspension were then placed in the rheometer's parallel plate chamber and the gap was set to 0.5 ± 0.2 μm. The sample was then sheared for 1 min at the highest rate to which it was to be ultimately subjected, 50 s^{-1}; this procedure standardized the degree to which samples with different properties rheologically relaxed. Measurements of shear stress and viscosity (n_i) were automatically recorded at desired sweep rates over the range of shear rate 'i', 0 to 50 s^{-1}. Figure 5 gives sample data for polymer-free suspensions.

Visual observations on the relative fluidity were also helpful in characterizing many systems, particularly those that differed greatly in rheology but were similar regarding formal composition.

Pressure Filtration

Selected suspensions were consolidated by pressure filtration to fabricate cylindrical pellets (filter cakes), from unary and binary suspensions. A

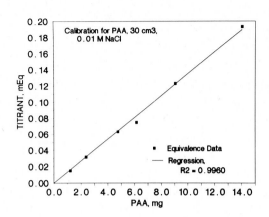

Fig. 3. Calibration data for PAA in 0.01 M NaCl, titration rate is 0.025 meq min^{-1}.

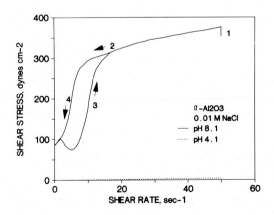

Fig. 4. Rheological loop for 20.1 % v/v α-Al$_2$O$_3$ in 0.01 M NaCl at pH 4 and 8. See text for details.

plexiglass filtering chamber was constructed for this purpose; the bottom of
the chamber was equipped with a teflon-supported 0.5-μm Nuclepore or Millipore
filter and Ar(g) was used with an applied pressure of 20 psi to force the liquid
through the filter. Typical dimensions for diameter and thickness of the
pellets are, respectively, 24 and 2 to 5 mm.

Pellets were fabricated using suspensions in which the initial concentration of solid was maintained at either 5 or 20 % v/v; ionic strength was 0.01 M
NaCl. Samples were mixed, aged at ambient room temperature for one day using
in a reciprocal shaker, and subjected to ultrasonication prior to filling the
filtration chamber. Owing to the slow filtration rate and mechanical weakness
of some samples, a standard procedure of allowing the pellet to remain in the
chamber under pressure for approximately 18 h was adopted, after which time
it was removed and air-dried, uncovered for 1 d, and finally placed for an
extended period of 18 to 24 h in a vacuum oven at 50 °C. A small piece was
broken off each pellet and thinly coated (relative to the piece's mass and
volume) with paraffin at a temperature just exceeding its melting point of
~50 °C. The density of the coated pieces was calculated using the mass of
the coated piece in air, that in water, and Archimedes' principle (ref. 14).

Electrophoresis

Microelectrophoretic mobility, μ_e, was measured using an automated
electrokinetic analyzer (Pen Kem, System 3000). Concentration of solid was
~0.001 % v/v; samples were subjected to ultrasonication prior to filling the
sample chamber. Figure 6 shows representative data for polymer-free systems.

Pycnometry

Nominally 25-cm^3 pycnometers were calibrated (ref. 15) with doubly
distilled water at 25.0 °C, using the value of 0.99707 g cm^{-3}. Density of
PAA-solutions at desired concentrations and pH-values was determined according
to established procedures (ref. 15); see Fig. 7 for sample data. Partial molal
volume was calculated from the data in this figure.

RESULTS
Unary α-Al_2O_3 Systems

General Stability Behavior. Curve A in Fig. 8 summarizes the information
contained in a series of kinetic studies and illustrates the effect of pH on
the stability of α-Al_2O_3, as reflected by its ability to remain in suspension
under conditions of 0.01 M NaCl and an aging time of 15 h. A flocculation zone
exists, whose onset is characterized by the critical flocculation pH (c.f.p.)
and which consists of destabilized and settled systems. Flocculation is most
severe when pH > 8, according to measurements of R.S.H.

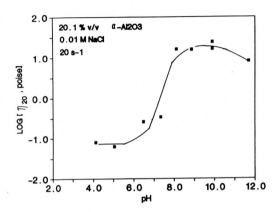

Fig. 5. Logarithm of viscosity for 20.1 % v/v α-Al₂O₃ in 0.01 M NaCl as a function of pH; shear rate is 20 s⁻¹.

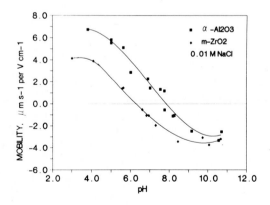

Fig. 6. Electrophoretic mobility, μₑ, vs pH for α-Al₂O₃ and m-ZrO₂ in 0.01 M NaCl.

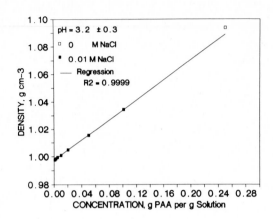

Fig. 7. Sample pycnometric data for PAA in 0.01 M NaCl at low pH.

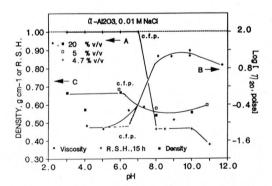

Fig. 8. Comparison of (A) relative sedimentation height (R.S.H.), (B) log η_{20}, and (C) density of pellets showing the effect of pH for systems containing 0.01 M NaCl; see text for details. Dotted lines are used for interpolation to define c.f.p. values.

Curves B and C in Fig. 8 give the corresponding profiles for suspension viscosity and density of the pellets prepared by pressure filtration. These data were obtained with systems at the same ionic strength as those for Curve A but with higher concentrations of solid, ~20 vol % for viscosity and 20 and 5 vol % for density. Transitions in colloidal stability, as reflected by c.f.p. values of 7.0, 6.3, and 6.0 shown in this figure, correlate well with each other and with the values of μ_e in Fig. 6.

The breadth of the flocculation zone evaluated by R.S.H. and log n_{20}-values extends to pH values greater than 12 and is a qualitative measure of the sensitivity of the α-Al$_2$O$_3$ suspension to pH. The zone depicted by density measurements is somewhat narrower. Slightly different breadths were obtained for different aging times and concentration of solids, with the latter parameter essentially determining the accuracy with which stable and unstable systems can be distinguished.

Effect of PAA on Stability. Figure 9 shows the general effect of poly-acrylic acid with a molecular weight of 90,000 on the stability of α-Al$_2$O$_3$, as reflected by viscosity measured in 0.01 M NaCl. If sufficient polymer is present, the suspension can be effectively stabilized, even under pH conditions that would normally induce the rapid flocculation reflected in the previous figure. A rapid decrease in viscosity was always accompanied by reduced hysteresis and thixotropic character and its onset is denoted by the critical stabilization concentration (c.s.c.). Data gathered (ref. 7) over the ranges of PAA concentration from 0 to 10 % w/w, based on solid, and pH 3 to 11 exhibit profiles similar to the one in Fig. 9. Those data and the ones in Fig. 9 are summarized by the stability diagram in Fig. 10. Different regions are evident in this figure. Regions I and II consist of stable α-Al$_2$O$_3$ suspensions with particles respectively having net positive and negative surfaces, based on microelectrophoretic data. Region III, denoted by hatching, is one of instability for which particles are rapidly flocculated and settled. The dashed curve within this region indicates p.z.c. conditions that correspond to reversal of charge from positive to negative in the presence of polymer; see Fig. 11. Finally, Region IV represents a transition between II and III. Its lower boundary is defined by both the c.s.c. data represented in the previous figure and the conditions that just provide monolayer coverage of PAA, as explained in the next section; the upper boundary corresponds to the c.f.c. conditions in Fig. 9.

This figure demonstrates that the pH at which the PAA induces a lowering of viscosity and an increase in Newtonian character depends on the additive's concentration. Similarly, the specific level of PAA that induces a marked lowering of viscosity and an increase in Newtonian character depends on pH.

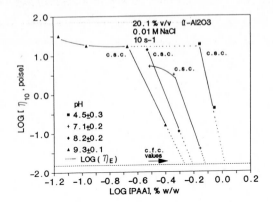

Fig. 9. Logarithm of viscosity for α-Al₂O₃ suspensions with polyacrylic acid in 0.01 M NaCl. Viscosity η_E is related (refs. 16, 17) to that of the solution, η_0, by the equation, $\eta_E = \eta_0 (1-\phi)^{-2.5}$. Extrapolation of data to log η_E used to obtain c.f.c., the critical flocculation concentration of polymer. Interpolation yields the c.s.c., the critical stabilization concentration.

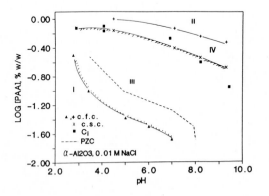

Fig. 10. Stability map for α-Al₂O₃ in the presence of PAA and 0.01 M NaCl. Squares represent minimum initial [PAA], C_i, for which $\Gamma = \Gamma_{max}$. Hatching devotes unstable side of transitions between stable and unstable conditions. See text for description of each region.

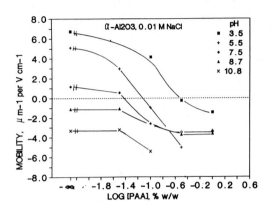

Fig. 11. Microelectrophonetic mobility, μ_e, vs logarithm of [PAA] for α-Al$_2$O$_3$ in 0.01 M NaCl under various pH conditions.

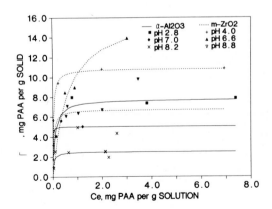

Fig. 12. Adsorption isotherms for PAA in suspensions of α-Al$_2$O$_3$ (solid lines) and m-ZrO$_2$ (dashed lines) under various pH conditions and in 0.01 M NaCl.

Adsorption. Figure 12 contains data gathered in adsorption experiments for the pH range 3 to 9 for each solid in the presence of 0.01 M NaCl; variability in pH is less than ±0.5 pH-units. The curves in this figure were determined by linear regression analysis of the data plotted as C_e/Γ vs Γ, a treatment that assumes Langmuirian behavior. (ref. 16) Indeed, most sets of data analyzed in this fashion were adequately described by the quantitative expressions obtained. Consequently, Langmuirian-type behavior seems present for this combination of α-Al_2O_3 and PAA. The saturation level of adsorption is denoted as Γ_{max}; see Fig. 10 for a summary of the pH-dependence of these values.

It is evident from this figure and the Fig. 10 that the saturation level of polymer decreases as the pH is increased. The data in this figure suggest adsorption of the high-affinity type, i.e., the low concentration regime of each curve corresponds to essentially 100 % adsorption of polymer.

Unary m-ZrO_2 Systems

Limited data were taken on the stability of this unary system in the absence and presence of PAA. However, extensive data on adsorption and electrokinetics allowed stability to be evaluated.

General Stability Behavior. Figure 13 summarizes the sedimentation data gathered on 3.5 and 5.0 % v/v m-ZrO_2 in 0.01 M NaCl over a wide pH range and respectively, after 15 h and 2 wk aging. The data in Fig. 13 gathered on 5-% systems were obtained in an earlier phase of this study and were previously reported by Baik et al. (ref. 4) Finally, Table 2 summarizes the observations and experiments on viscosity of related systems.

Well-defined flocculation is evident from the data in Fig. 13 and Table 2. It is bounded by the critical flocculation and stabilization pH values. As is the case for Al_2O_3, the c.f.p. define transitions in colloidal stability.

Effect of PAA on Stability and Electrokinetics. Observations on the effect of PAA on the sedimentation and viscosity for m-ZrO_2 suspensions parallel those described earlier for Al_2O_3. The qualitative affect of PAA on suspensions of m-ZrO_2 is also included in Table 2.

Figure 14 summarizes microelectrophoretic experiments for suspensions of m-ZrO_2 in the presence of PAA; ionic strength is 0.01 M NaCl.

Adsorption. Figure 12 also summarizes adsorption experiments for the PAA:m-ZrO_2 system in the pH range 4 to 9 and in 0.01 M NaCl. The data for m-ZrO_2 qualitatively resemble those for α-Al_2O_3. Overall (ref. 7), the saturation level of PAA on m-ZrO_2 also decreases as the pH is increased and adsorption appears to be of the high-affinity type. Contrasting this behavior is the isotherm obtained at pH 6.6. It shows that a substantial fraction of PAA remains in solution prior to the concentration at which the surface is saturated.

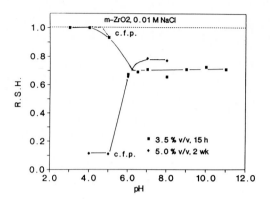

Fig. 13. Relation sedimentation height, R.S.H., for m-ZrO$_2$ suspension in 0.01 M NaCl.

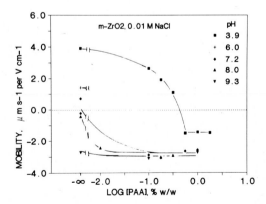

Fig. 14. Microelectrophoretic mobility, μ_e, vs logarithm of [PAA] for m-ZrO$_2$ in 0.01 M NaCl under various pH conditions.

TABLE 2

Qualitative assessment of viscosity in m-ZrO_2 suspensions.

[PAA], % w/w[a]	Log [PAA]	Viscosity[b]	Region[c]
		pH 4	
2.25	0.35	L	II
1.5	0.18	M	IV
1.0	0.00	H	III
0.75	-0.12	H	III
0	--	L	I
		pH 6	
1.65	0.22	L	II
1.1	0.041	M	IV
0.75	-0.12	H	III
0.55	-0.26	H	III
0	--	H	III
		pH 8	
1.30	-0.52	L	II
0.75	-0.12	M	IV
0.50	-0.30	H	III
0.38	-0.42	H	III
0	--	H	III
		pH 10	
0.70	-0.15	L	II
0.46	-0.34	M	IV
0.31	-0.51	H	III
0.23	-0.64	H	III
0	--	M	I or III

[a]Based on solids.
[b]L = low, M = medium, and H = high.
[c]Assignments refer to phenomena already described in Fig. 10 and the accompanying text.

Binary Systems

Figure 15 summarizes density measurements gathered on pressure-casted pellets fabricated using binary suspensions that did and did not contain PAA. The ratio of α-Al_2O_3:m-ZrO_2 is 9:1, based on volume and corresponding to a number ratio of 0.59:1. The dashed curve is included for comparison and represents data for α-Al_2O_3 in the absence of both m-ZrO_2 and PAA.

The PAA-containing systems were prepared by equilibrating each type of solid at the indicated conditions of pH and ionic strength with the concentration of polymer needed to develop monolayer coverage, Γ_{max}. Subsequently, each unary PAA/α-Al_2O_3 system was mixed with the corresponding unary PAA/m-ZrO_2 suspension. The concentration of polymer to which each solid

was exposed prior to preparation of binary suspensions is give in Fig. 16.
The data for α-Al$_2$O$_3$ in Fig. 16 are based on Figs. 10 and 12, whereas those
for m-ZrO$_2$ are derived from Fig. 12 and other data (ref. 7).

Density profiles gathered in the absence of PAA for the unary α-Al$_2$O$_3$
system and the composite pellets exhibit similar shape and pH sensitivity.
A marked shift in the profile that arises in the presence of m-ZrO$_2$, relative
to the unary α-Al$_2$O$_3$ system, is evident in Fig. 15. Stability as reflected
in density of the binary pellets in the presence of PAA clearly exhibits a
different pH sensitivity than that characterizing the polymer-free systems.

Finally, Fig. 17 summarizes microelectrophoretic data, converted to zeta
potential following guidelines of Smith (ref. 18) to account for ionic strength
and particle size, for the unary suspensions in the presence and absence of PAA.
Concentration of polymers for each solid is given by Fig. 16.

CONCLUSIONS AND CLOSING REMARKS

Based on the data presented, we conclude:

1. Adsorption of polyacrylic acid (M_w = 90,000) onto α-Al$_2$O$_3$ and m-ZrO$_2$
(Fig. 12) appears Languirian and is of the high-affinity type.

2. For pH values less than approximately 7, pH-sensitivity of the
saturation levels of adsorption appear to be principally determined by PAA
(Fig. 16); individual character of each polymer-solid combination is more
pronounced in basic media and may imply that the fundamental interaction
between polymer and solid involves Coulombic-type association of anionic
sites on the PAA backbone and cationic surface sites.

3. Stability of unary systems containing PAA (Figs. 8 and 10 and
Table 2) resembles that in polymer-free systems since it is governed by
the net electrostatic character developed upon adsorption (Figs. 11 and
14). Consequently, stability varies with the degree of coverage. Density
of solution and probably also its rheology are not significantly affected
by equilibrium concentrations of PAA (Figs. 7 and 12).

4. Sensitivity of stability to pH in binary, α-Al$_2$O$_3$:m-ZrO$_2$ systems that
contain PAA is determined by the degree of anionic character of PAA, when each
solid has previously been exposed to the concentration of polymer that provides
monolayer coverage; this sensitivity is also reflected in the corresponding
electrokinetic behavior (Fig. 17).

Finally, theoretical considerations of adsorption and its effects have not
been included herein, primarily owing to space limitations. Such treatments are
in progress (refs. 4, 7) and will be reported at a later date. Some preliminary
results of those studies are noteworthy, however. Summarily, we find (ref. 7)
that theoretical determinations of adsorption and stability in both unary and
binary suspensions must consider the electrostatic patch model of Gregory (refs.

234

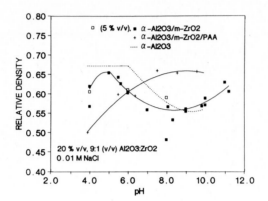

Fig. 15. Density of pressure-casted pellets as a function of pH in the presence and absence of PAA.

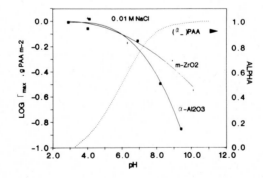

Fig. 16. Logarithm of saturation adsorption of PAA, Γ_{max}, vs pH for α-Al$_2$O$_3$ and m-ZrO$_2$. Dashed curve represents α_- for PAA as a function of pH and is taken from Fig. 2.

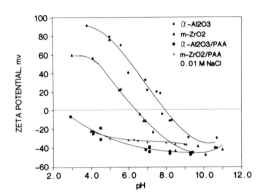

Fig. 17. Zeta potential vs pH for conditions displayed in Fig. 16.

19, 20), the pH sensitivity of PAA configuration (refs. 11, 12), and heterocoagulation phenomena that ultimately originate from the acid-base character
of the solids (refs. 3, 6) and the polymer (refs. 11, 12).

ACKNOWLEDGMENTS

This work was conducted within the Basic Energy Sciences Program by
the Division of Materials Sciences, U.S. Department of Energy, under contract
No. DE-AC05-840R21400 with Martin Marietta Energy Systems, Inc. The help of
F. Stooksbury in preparing the manuscript is gratefully acknowledged.
Discussions with S. Baik and P. F. Becher on the processing of Al_2O_3-ZrO_2
composites in the early stages of this work were helpful and are happily noted.

REFERENCES

1 (a) P. F. Becher, M. V. Swain and S. Somiya, Eds., Advanced Structural
   Ceramics, Mater. Res. Soc. Symp. Proc., Vol. 78, Pittsburgh, 1987, 306 p.;
   (b) W. M. Kriven, The transformation mechanism of spherical zirconia
   particles in alumina, Adv. Ceram., 12 (1984) 64-77; (c) J. G. P. Binner,
   R. Stevens and S. R. Tan, Adv. Ceram. 12, (1984) 428-435; (d) N. Claussen
   and M. Ruhle, Design of transformation-toughened ceramics, Adv. Ceram.,
   3 (1981) 137-163; (e) A. G. Evans, D. B. Marshall, and N. H. Burlingame,
   Transformation toughening in ceramics, Adv. Ceram., 3 (1981) 202-216;
   (f) I. J. McColm, Refractory oxide ceramics, in Ceramic Science for
   Materials Technologists, I. J. McColm, Blackie & Son, Glasgow, 1983,
   Chapter 5, pp. 235-310.

2   J. F. Lynch, Ed., Engineering Property Data on Selected Ceramics, Volume
    III, Single Oxides, Battelle Columbus Laboratories, Columbus, OH., 1981,
    Section 5.4, pp. 5.4.1-1 and 5.4.5-1.
3   R. O. James and G. A. Parks, Characterization of aqueous colloids by
    their electrical double layer and intrinsic surface chemical properties,
    Surf. Colloid Sci., 12 (1982) 119-216.
4   S. Baik, A. Bleier and P. F. Becher, Preparation of Al_2O_3-ZrO_2 composites
    by adjustment of surface chemical behavior, Mater. Res. Soc. Symp. Proc.,
    73 (1986) 791-800.
5   (a) J. Cesarano, I. A. Aksay and A. Bleier, Stability of aqueous α-Al_2O_3
    suspensions with polymethacrylic acid (PMAA) polyelectrolyte, Am. Ceram.
    Soc., in press; (b) J. Cesarano, I. A. Aksay and A. Bleier, Polyelectrolyte
    adsorption on α-Al_2O_3 and aqueous suspension stability, in Extended
    Abstracts American Ceramic Society 87th Annual Meeting, Am. Ceram. Soc.,
    Columbus, OH. (1985) Paper No. 95-B-85, p. 26.
6   (a) W. C. Hasz, Surface Reactions and Electrical Double Layer Properties of
    Ceramic Oxides in Aqueous Solution, M.S. Thesis, Massachusetts Institute of
    Technology, Cambridge, MA, December 1983, 165 p.; (b) W. C. Hasz and
    A. Bleier, Surface reactivity of silica and alumina ceramic powders, in
    Advances in Materials Characterization II, Materials Science Research,
    Vol. 19, R. L. Snyder, R. A. Condrate, Sr., and P. F. Johnson, Eds., Plenum,
    New York (1985) pp. 189-201; (c) A. Bleier, Acid-base properties of ceramic
    powders, in Advances in Materials Characterization, Materials Science
    Research, Vol. 15, D. R. Rossington, R. A. Condrate, and R. L. Snyder, Eds.,
    Plenum, New York (1983) pp. 499-514.
7   A. Bleier and C. G. Westmoreland, Unpublished data.
8   J. Westall, J. Zachary and F. Morel, MINEQL: A computer program for the
    calculation of chemical equilibrium composition of aqueous systems,
    Technical Note No. 18, Ralph Parsons Laboratory, Massachusetts Institute
    of Technology, Cambridge, MA (1976) 91 p.
9   (a) D. E. Yates, S. Levine and T. W. Healy, Site-binding model of the
    electrical double layer at the oxide/water interface, J. C. S. Faraday,
    I-70 (1974) 1807; (b) J. A. Davis, R. O. James and J. O. Leckie, Surface
    ionization and ionization at the oxide/water interface. I: Computation
    of electrical double layer properties in simple electrolytes, J. Colloid
    Interface Sci., 63 (1978) 480; (c) J. A. Davis and J. O. Leckie, Surface
    ionization and ionization at the oxide/water interface. II: Surface
    properties of amorphous iron oxyhydroxide and adsorption of metal ions,
    J. Colloid Interface Sci., 67 (1978) 90; (d) J. A. Davis and J. O. Leckie,
    Surface ionization and ionization at the oxide/water interface. III:
    Adsorption of anions, J. Colloid Interface Sci., 74 (1980) 32; (e) J. A.
    Davis and J. O. Leckie, ACS Symp. Ser., 93 (1979) 299-317.
10 Westall, J., Chemical equilibrium including adsorption on charged surfaces,
    Adv. Chem. Ser., 189 (1980) 33.
11 C. Tanford, Physical Chemistry of Macromolecules, John Wiley & Sons, New
    York, 1961, 710 p.
12 (a) R. Arnold and J. Th. G. Overbeek, The dissociation and specific
    viscosity of polymethacrylic acid, Rec. Trav. Chim., 69 (1950) 192;
    (b) J. J. Hermans and J. Th. G. Overbeek, The dimensions of charged long
    chain molecules in solutions containing electrolytes, Rec. Trav. Chim.,
    67 (1948) 761.
13 P. Molyneux, Water-Soluble Synthetic Polymers: Properties and Behavior,
    Vol. II, CRC, Boca Raton, FL, 1984, 266 p.
14 McGraw-Hill Dictionary of Scientific and Technical Terms, 3rd Ed.,
    S. P. Parker, Ed., McGraw-Hill, New York, 1984, p. 100.
15 D. P. Shoemaker, C. W. Garland, J. I. Steinfeld and J. W. Nibler,
    Experiments in Physical Chemistry, 4th Ed., McGraw-Hill, New York, 1981,
    pp. 161-179.
16 P. C. Hiemenz, Principles of Colloid and Surface Chemistry, 2nd Ed.,
    Marcel Dekker, New York, 1986, pp. 198-199.
17 R. Roscoe, Br. J. Appl. Phys., 3 (1952) 267.

Interfacial Phenomena in Biotechnology and Materials Processing,
edited by Y.A. Attia, B.M. Moudgil and S. Chander
Elsevier Science Publishers B.V., Amsterdam, 1988 — Printed in The Netherlands

SURFACE CHEMICAL AND ADSORPTION CHARACTERISTICS OF MAGNETIC FERRITE AND OXIDE PARTICLES IN NONAQUEOUS SYSTEMS

G. F. Hudson[1], S. Raghavan[1], and M. A. Mathur[2]

[1]Department of Materials Science and Engineering, University of Arizona, Tucson, AZ 85721
[2]IBM Corporation, General Products Division, Tucson, AZ 85744

ABSTRACT

Magnetic particles suitable for high-density recording are being actively studied by magnetic tape researchers. Barium ferrite and chromium dioxide are two such magnetic particles under study by several scientists. A knowledge of the acidic and basic properties of these particles is critical for the proper choice of dispersants in coating formulations. Using the technique of flow microcalorimetry, the acidic and basic sites on doped and un-doped barium ferrites and chromium dioxide have been investigated. The adsorption of 4-nitrophenol (basic site probe) and pyridine (acidic site probe) onto these particles has been measured and analyzed. An explanation of the experimental observations based on the surface chemical properties of the magnetic particles is given.

BACKGROUND

A recent trend in magnetic tape studies has been the evaluation of magnetic particles suitable for high-density recording. Barium ferrite ($BaFe_{12}O_{19}$) and chromium dioxide (CrO_2) are two such particles currently being evaluated by many researchers. To determine optimum coating formulations, it is desirable to have some knowledge of the surface properties of these particles. Such information is useful for achieving good dispersions of magnetic particles in the magnetic ink. In particular, an understanding of the acidic and basic properties of particles can help in the evaluation of new dispersants and binders.

Flow microcalorimetry (FMC) is a versatile technique that has been successfully used as a method to evaluate the adsorption energetics of different adsorbates onto a variety of materials possessing different surface chemical properties. Jones and Pope (ref. 1) investigated the adsorption of lubricating oils onto ferric oxide. Noll and co-workers (ref. 2) studied the effect of chain length on the adsorption of alcohols onto silica gel. Fowkes et al. (ref. 3) examined the adsorption of pyridine and triethylamine from cyclohexane onto Fe_3O_4 and α-Fe_2O_3. Joslin and Fowkes (ref. 4) investigated the heat of adsorption of a variety of nitrogen and oxygen bases from neutral hydrocarbons onto α-Fe_2O_3. In a detailed study, Joslin (ref. 5) reported the adsorption of pyridine onto α-Fe_2O_3 and investigated the effect of water on the system. Hudson and Raghavan (ref. 6) earlier reported experiments on the adsorption of pyridine onto chromium dioxide and γ-Fe_2O_3.

Small, plate-like, hexagonal particles of barium ferrite have shown great potential for high-density, perpendicular, recording tape (ref. 7). One of the techniques to produce barium ferrite is via a glass crystallization method (ref. 8). A mixture of BaO, B_2O_3, and Fe_2O_3 is melted and rapidly cooled by dropping it between rotating steel nip rollers to form amorphous glass flakes. These flakes are heat-treated for crystallization and then washed in acid to remove excess BaO and B_2O_3. By adding Co(II)O and Ti(IV)O_2 to the original melt (for coercivity control), it is possible to obtain a "doped" ferrite product, namely, $BaFe_{(12-2x)}Co_xTi_xO_{19}$ (ref. 9).

Chromium dioxide is typically produced by hydrothermal decomposition of CrO_3 and Cr_2O_3 at a temperature of 300 to 500°C in the presence of oxygen at extremely high pressures (ref. 10). Particles of CrO_2 thus obtained are typically needle-like, single crystals of very low porosity and submicron in size (ref. 11). As pure chromium dioxide will disproportionate in humid atmospheres to give Cr(III) and Cr(VI) products, it is normally slurried in a sodium bisulfite solution to stabilize the surface against degradation by the formation of a thin layer of CrOOH (Cr_2O_3) (ref. 12). It is possible that some of the sulfite used in the stabilization procedure is oxidized and strongly adsorbed onto the chromium dioxide particles. Hudson et al. (ref. 13) found that approximately 0.4 μmole/m² of sulfate is present on the surface of chromium dioxide identical to that used in this study.

EXPERIMENTAL MATERIALS AND METHODS

The magnetic particles evaluated in our experiments included doped barium ferrite (d-BaFe, Toda 4565), undoped barium ferrite (u-BaFe, Ferro Ottawa Corp.) and chromium dioxide (CrO_2, duPont D-500-2). The surface areas and chemical analysis of these particles are given in Table 1.

TABLE 1

Chemical and physical properties of the magnetic particles studied

	d-BaFe	u-BaFe	CrO_2
Surface area, m²/g	43.6	21.5	26.0
Chemical analysis	50.79% Fe 11.71% Ba 3.81% Co 3.29% Ti	60.03% Fe 11.99% Ba <0.02% Co <0.01% Ti	0.4 μmole/m² of sulfate present

The surface areas were determined by nitrogen gas adsorption and the chemical composition was determined by inductively coupled plasma analysis. In addition, x-ray photoelectron spectroscopy (XPS) analysis of the doped barium ferrite sample showed that titanium was uniformly distributed throughout the sample, that is, there was no preferential concentration of Ti at the surface. Because of the overlap of Ba and Co XPS peaks, cobalt was not detectable but was also assumed to be uniformly distributed.

Reagent-grade hexane (10 ppm water) and high-performance liquid chromatography (HPLC) grade tetrahydrofuran (THF, 15 ppm water) were used as solvents for this investigation. The solvents were dried over a 3 Å molecular sieve and bathed in argon gas to control the moisture content. The moisture content was determined by Karl-Fisher titration. Reagent-grade pyridine (96 ppm water), used as the acidic probe, was also dried in the same fashion and bathed in argon gas. The basic molecular probe consisted of 4-nitrophenol.

The heat of adsorption was measured using a Microscal Mark 3V FMC equipped with syringe micropumps and HPLC changeover valves. This instrument has been well described elsewhere (ref. 14). A Waters Associates R401 differential refractometer was connected in series to measure solute concentrations in the effluent stream from the FMC. Data from the FMC was processed using a Spectraphysics recording integrator and a strip chart recorder was used to record data from the refractometer.

The heat of adsorption was determined by first dry-loading an optimized weight of sample into the sample chamber; a fresh sample bed was prepared for each experimental run. The sample bed was initially equilibrated in solvent flowing at 3.3 ml/hr. Following this, the flow rate was doubled for approximately 20 minutes to ensure uniform packing of the sample bed. The flow rate was then returned to 3.3 ml/hr and the bed was allowed to equilibrate again before measurements were made. After switching to the solute-containing solvent, we measured both the heat of adsorption and amount of solute adsorbed, as described previously by the authors (ref. 6). The heat of adsorption is reported as cal/g, molar heat of adsorption as kcal/mole, and adsorption density as μmole/m^2.

Electrophoretic mobility measurements were made using a Zeta-Meter electrophoresis unit equipped with a continuous-flow cell. The electrophoresis cell was constructed of glass and PTFE and had an inlet port and an outlet port to allow circulation of a dilute (0.01%) suspension of particles. Using this cell, it was possible to cycle the pH of the suspension and study aging effects, if any. Potassium chloride was used to control the ionic strength. Reported mobility values were at 10^{-3} $\underline{M}$ KCl.

RESULTS

The bulk of the experimental work concentrated on the differences between doped barium ferrite (d-BaFe) and undoped barium ferrite (u-BaFe). A limited number of experiments were performed with the chromium dioxide sample, mainly for comparison with earlier work (ref. 6).

Initial experiments were performed to determine the molar heat of adsorption and adsorption density of pyridine from hexane solutions onto the BaFe samples. Pyridine is used as a molecular probe for acidic surface sites. Earlier work by the authors (ref. 6) had shown that monolayer coverage of pyridine on chromium dioxide, as predicted by the Langmuir equation, was approximately 6.3 μmole/m^2 with a molar heat of adsorption of 10 kcal/mole. Adsorption isotherms and molar heat of adsorption of pyridine on d-BaFe and u-BaFe are seen in Figs. 1 and 2, respectively.

240

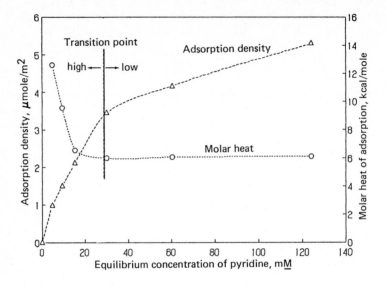

Fig.1 Adsorption of pyridine from hexane onto d-BaFe

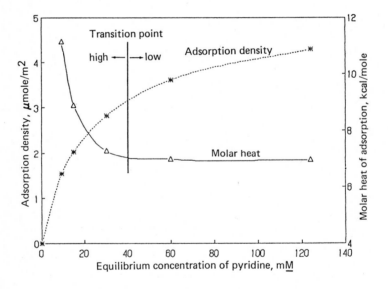

Fig.2 Adsorption of pyridine from hexane onto u-BaFe

Although adsorption, unlike the heat values, was not observed to level off at high pyridine concentrations, at 124 millimolar (mM) pyridine the adsorption densities were 5.3 μmole/m² and 4.3 μmole/m² on d-BaFe and u-BaFe, respectively. The molar heat of adsorption stabilized above pyridine concentrations of 30 mM for both samples. In that region, the molar heat of adsorption of pyridine of d-BaFe was approximately 6.0 kcal/mole, while for u-BaFe it was approximately 7.0 kcal/mole. These results indicate that the surface of d-BaFe is slightly more basic than the surface of u-BaFe, and that chromium dioxide is the least basic (most acidic) (see Table 2) of the three particles.

TABLE 2

Molar heat of adsorption and adsorption density of 35.2 m$\underline{M}$ 4-nitrophenol in hexane with 10 vol. % THF

	Molar Heat of Adsorption kcal/mole	Adsorption Density μmole/m²
CrO$_2$	0.46 ± 0.06	1.50 ± 0.1
d-BaFe	3.72 ± 0.40	2.53 ± 0.06
u-BaFe	1.86 ± 0.26	2.44 ± 0.06

Having examined the acidic nature of the surfaces, we then performed experiments using 4-nitrophenol as a basic site probe. The solvent for this test was hexane with 5 vol. %THF added to ensure solubility of the 4-nitrophenol. The adsorption isotherm for d-BaFe is seen in Fig. 3. Monolayer coverage appears to have been achieved near 60 m$\underline{M}$ 4-nitrophenol at an adsorption density of 5.9 μmole/m². The corresponding molar heat of adsorption at this concentration was approximately 2.1 kcal/mole.

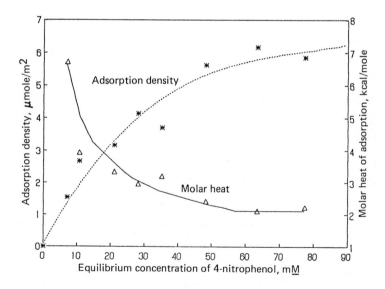

Fig.3 Adsorption of 4-nitrophenol from hexane containing 5 vol. % THF onto d-BaFe

The relative error for heat values using this solvent system was typically ± 10 to 20%, larger than the ± 10% or less reproducibility obtained using straight hexane. Until this could be further elucidated, we decided to compare the three particles at one concentration of 4-nitrophenol and to increase the volume of THF to 10% in hexane to ensure total solubility of the 4-nitrophenol. These results are summarized in Table 2. The data in Table 2 confirms test results using pyridine, namely, that the order of increasing surface acidity is d-BaFe, u-BaFe, followed by chromium dioxide.

Because a distinct difference in the surface chemical properties of the three particles was apparent from the initial experiments, electrophoretic measurements were made with the ferrite samples in an aqueous 10^{-3} $\underline{M}$ potassium chloride solution as a function of pH to gain further information on the acidic and basic character of the samples. Results, seen in Fig. 4, show that the point of zero charge (PZC) of the d-BaFe sample occurs at a slightly higher pH than that of the u-BaFe sample. This is another indication that the surface of d-BaFe is more basic than that of u-BaFe. The curve for chromium dioxide is from a previous study by the authors (ref. 13) and indicates that it is the most acidic of the three particles.

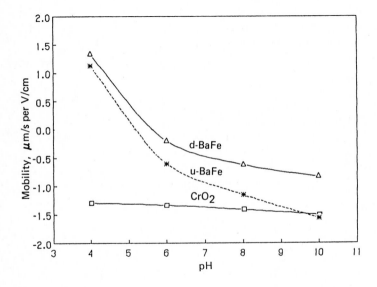

Fig.4 Electrophoretic mobility of d-BaFe, u-BaFe, and chromium dioxide in aqueous solution (10^{-3} M KCl)

DISCUSSION

The pyridine adsorption data was fit to both the Langmuir and Freundlich isotherms (ref. 15). Values for R^2 for each isotherm and predicted values for monolayer coverage (from Langmuir isotherm) are listed in Table 3. (Data for the chromium dioxide was taken from (ref. 6).)

TABLE 3

R^2 values for the fit of experimental data to Langmuir and Freundlich isotherms

Sample	Langmuir	Freundlich
Pyridine		
d-BaFe	0.9883	0.8621
u-BaFe	0.9943	0.9648
CrO_2	0.9143	0.9951
4-Nitrophenol		
d-BaFe	0.9161	0.9284
Monolayer capacity predicted by Langmuir, $\mu mole/m^2$		
d-BaFe	6.44 (pyridine), 8.44 (4-nitrophenol)	
u-BaFe	5.03 (pyridine)	
CrO_2	6.53 (pyridine)	

Using the monolayer capacities predicted by the Langmuir equation, the calculated areas per adsorbed pyridine molecule are 26, 33, and 25 $Å^2$ on d-BaFe, u-BaFe, and CrO_2, respectively, and 20 $Å^2$ for 4-nitrophenol on d-BaFe. From the monolayer capacities of pyridine and 4-nitrophenol on d-BaFe, the total number of acidic and basic sites on d-BaFe can be calculated to be 0.9×10^{15} sites/cm², which is remarkably close to the theoretical limit of 10^{15}. From these calculations, the ratio of basic sites to acidic sites on d-BaFe is approximately 1.84:1.

From Figs. 1 and 2, it is easily seen that while the molar heat of adsorption of pyridine on the ferrite samples levels off to some finite value, the values for adsorption density do not. (In spite of this, the Langmuir equation fits the data quite well!) This leads to the conclusion that the acidic sites on the ferrite surface are not homogeneous. By assuming that the high-energy acidic sites are completely neutralized at the point where the molar heat of adsorption levels off (indicated by the vertical line in Figs. 1 and 2), the ratio of high- and low-energy sites may be calculated. Results of these calculations are listed in Table 4. Note that the decrease in the percentage of high-energy acidic sites caused by doping is nearly identical to the total dopant level of Co and Ti of 7.1% in the d-BaFe sample. This indicates that the substitution of Co and Ti for Fe in the ferrite lattice leads to an increase of the basicity of the ferrite surface.

TABLE 4

Distribution of high- and low-energy acidic sites on d-BaFe and u-BaFe

	d-BaFe	u-BaFe
% high	52.8	60.0
% low	47.2	40.0

The increase in the basicity as a result of doping is also evidenced by the electrophoretic mobility measurements, which show that d-BaFe has a higher PZC than u-BaFe. According to the mechanism proposed by Roy and Fuerstenau (ref. 16), the incorporation of Ti into Fe sites of the ferrite lattice should make the ferrite an n-type semiconductor. This in turn would cause a shift in the PZC to a lower pH. Similarly, the incorporation of Co into the Fe sites of the ferrite lattice should make the ferrite a p-type semiconductor and cause a shift in the PZC to a higher pH. Experimentally, we found that on doping with approximately equal amounts of Co and Ti, the PZC was shifted to a slightly higher pH value. This indicates that Co doping has a much stronger effect than Ti doping. The reason for this can only be speculated. The crystal structure of barium ferrite is shown in Fig. 5 (ref. 17). In doped samples, cobalt and titanium replace iron in the octahedral sites, with a large fraction of the cobalt going into the spinel layer (ref. 18). If preferential cleavage occurs along the dotted lines in Fig. 5, then cobalt will have a more dominant effect on the surface chemistry than titanium.

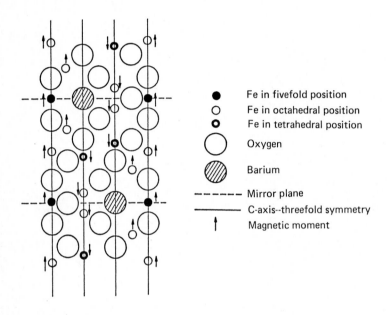

Fig.5 Unit cell of barium ferrite, taken from (ref. 17).

As mentioned earlier, experiments performed using 4-nitrophenol in hexane-THF mixtures produced results with large standard deviations. Besides, the adsorption density of 4-nitrophenol seemed to depend on the THF content of the solvent and decreased with an increase in THF content. Because THF is a Lewis base, it should not interfere with the measurement of basic sites using 4-nitrophenol but it might interact with 4-nitrophenol. One method to decouple the effect of THF would be to study the adsorption of 4-nitrophenol at different THF contents and extrapolate to near zero % THF. Ideally, it would be best to study the adsorption of 4-nitrophenol with a more neutral solvent than THF. Unfortunately, very limited solubility of 4-nitrophenol in neutral solvents such as hexane limits solvent choices. Thus, the technique used to obtain data on the nature of basic sites on particles could use some refinements.

CONCLUSIONS

Using pyridine and 4-nitrophenol as molecular probes, we found that doped barium ferrite is slightly more basic than undoped barium ferrite. Of the two dopants, Co and Ti, present in the doped ferrite investigated, cobalt appears to exert more influence on the surface chemical properties than Ti.

ACKNOWLEDGMENTS

Special thanks are extended to Drs. R. K. Agnihotri and R. L. Bradshaw of IBM Tucson for their advice, help, and interest in this work.

REFERENCES

1   W.G. Jones and M.I. Pope, The influence of water content on the adsorption of lubricating oil base stocks onto ferric oxide, studied by flow-microcalorimetry, Thermochimica Acta, 101 (196) (1986) 155-167.
2   L.A. Noll, G. Woodbury, Jr., and T.E. Burchfield, Dependence of adsorption of cosurfactant on chain length, Colloids and Surfaces, 9 (1984) 349-354.
3   F.M. Fowkes, C. Sun, and S.T. Joslin, Enhancing polymer adhesion to iron surfaces by acid base interactions, in: L. Leidheiser, Jr. (Ed.), Corrosion Control by Organic Coatings, National Association of Corrosion Engineers, Houston, TX, 1981, pp.1-3.
4   S.T. Joslin and F.M. Fowkes, Surface acidity of ferric oxides studies by flow microcalorimetry, Ind. Eng. Chem. Prod. Res. Dev., 24 (3) (1985) 369-375.
5   S.T. Joslin, Characterization of the surface sites on alpha-ferric oxide using flow microcalorimetry, PhD Dissertation, Lehigh University, 1984.
6   G.F. Hudson and S.Raghavan, A flow microcalorimetric investigation of the acidity of chromium dioxide and γFe_2O_3, Colloids and Surfaces, in press.
7   T. Ido, O. Kubo, and H. Yokoyama, Coercivity for Ba-ferrite superfine particles, IEEE Trans. Magn., MAG-22 (5) (1986) 704-706.
8   B.T. Shirk and W.R. Buessem, Magnetic properties of barium ferrite formed by crystallization of a glass, J. Am. Cer. Soc., 53 (4) (1970) 192-196.
9   T. Fujiwara, Barium ferrite for perpendicular recording, IEEE Trans. Magn., MAG-21 (5) (1985) 1480-1485.
10 B.L. Chamberland, The chemical & physical properties of CrO_2 & tetravalent chromium oxide derivatives, CRC Critical Reviews in Solid State and Materials Sciences (Nov. 1977) 1-31.
11 Y. Imaoka, K. Takada, T. Hamabata, and F. Maruta, Advance in magnetic recording media—from maghemite and chromium dioxide to cobalt adsorbed gamma ferric oxide, Ferrites, D. Reidel Pub. Co., Boston, 1982, pp.516-520.
12 DuPont Technical Bulletin, Chromium Dioxide Magnetic Particles Product Information.
13 G.F. Hudson, I. Ali, and S. Raghavan, Electrokinetic characteristics of CrO_2 and Cr_2O_3, in: Proc. Int. Symposium on Surfactants in Solution, August 1986, New Delhi, India.
14 G. Steinberg, What you can do with surface calorimetry, CHEMTECH, 11 (1981) 730-737.
15 D.O. Hayward and B.M.W. Trapnell, Chemisorption, Butterworth & Co., London, 1964, pp.160-175.
16 P. Roy and D.W. Fuerstenau, The effect of doping on the paint of zero charge on alumina, Surface Science, 30 (1972) pp.487-490.
17 K.J. Standley, Oxide Magnetic Materials, Clarendon Press, Oxford, 1972, pp.149-157.
18 J. Smit, F.K. Lotgering, and U. Enz, Anisotropy properties of hexagonal ferrimagnetic oxides, J. Appl. Phys., 31 (1960) 137S-141S.

Interfacial Phenomena in Biotechnology and Materials Processing,
edited by Y.A. Attia, B.M. Moudgil and S. Chander
Elsevier Science Publishers B.V., Amsterdam, 1988 — Printed in The Netherlands

DISPERSION STABILITY OF ALUMINA IN THE PRESENCE OF JEFF AMINES

A. Srinivasa Rao
Department of Ceramics
Center for Ceramics Research
Rutgers University
Piscataway, NJ 08855-0909

SUMMARY

In order to understand the suspension stability of concentrated alumina slips in presence of three surface active agents, electrophoresis, rheology and sedimentation experiments were carried out on one commercial wet classified alumina powder and three surface active agents that exhibit the characteristics of cationic and non ionic additives. The results suggest that the alumina powder is associated with some acidic surface impurity and all three additives adsorb onto the negatively charged alumina surface. The flocculation behavior of the alumina particles in water and in additive solution with smallest polymer chain length appear to depend upon the electrophoretic mobility of the particles. The stability of alumina in additive solution with longer polymer chain length remains independent of the particle surface charge.

INTRODUCTION

It has been suggested that fine powders tend to agglomerate in dry powder state [1]. Since the agglomerate structure is not uniform, the microstructural uniformity of the sintered body varies; as a result the reliability of the ceramic processing will be poor. Dry powder agglomerates tend to deagglomerate in the dispersion phase provided their surface forces are well optimized. Surface electrical charges of oxides in liquids can be controlled either by regulating the pH of the suspension or by the addition of suitable surface active agent [2-3]. The present investigation was undertaken in order to understand the stability of some commercial alumina powders in presence of few cationic surface active agents. In this paper we present some of our results obtained following electrophoresis, rheology and sedimentation experiments.

EXPERIMENTAL

One commercial alumina powder (Alcoa A-16) and three commercial cationic surface active agents (Jeff Amine M-302, Jeff Amine M-310 and Jeff Amine M-320 (manufactured by Jeffereson Chemicals) that were supplied by the Mansanto Electronic Materials Company and Gulf Research and Development Company, were used in this investigation. Although the alumina powder was produced from the same conventional method, the powder was subjected to wet classification procedure prior to the shipment to our laboratory. The additives Jeff Amine

M-302, M-310 and M-320 are ethoxylated amines, however they differ in their chemical composition. The chemical structure of the additives is given as:

$$R - OCH_2 - \underset{\underset{CH_3}{|}}{CH} - OCH_2 - \underset{\underset{CH_3}{|}}{CH} - N \begin{cases} (\,CH_2\ CH_2O\,)\ H \\ (\,CH_2\ CH_2O\,)\ H \end{cases}$$

While Jeff Amine M-302 contains 2 ethylene oxide chains, Jeff Amines M-310 and M-320 contain 10 and 20 chains respectively. It was suggested by the manufac-urer that these additives exhibit the characteristics of both cationic and an-ionic surface active agents depending upon their chemical activity of the ad-sorbing sites and also upon their concentration in suspension. The chemical analysis of the additives as was supplied by the manufacturer is given in Table 1[4].

	Jeff Amine		
	M-302	M-310	M-320
Moles Ethylene Oxide	2.0	10.0	20.0
Amine Content (meq/g)	2.71	1.35	0.75
Hydroxyl Number (mg KOH/g)	293.0	147.0	91.0
Specific Gravity at 20°C	0.9453	1.0251	1.0645
pH of 1% Solution	9.75	9.7	9.65
Surface tension, dyne cm^{-1} of 1% Solution		32.2	33.9
HLB* no.	6.0	15.0	19.0

* Hydrophile-lipophile balance, determined experimentally.

Table 1. Physical and chemical properties of additive solutions as was suggested by the manufacturer.

Dispersions containing 10-60%(wt) of alumina were prepared by mixing pre-determined amounts of alumina and additive solutions using an ultrasonic probe for five minutes. The pH of these slips was adjusted by adding either 1 mol dm^{-3} HCl acid or NaOH followed by thorough mixing. All suspensions used in electrophoresis experiments were prepared in 10^{-3} mol dm^{-3} KNO$_3$ solution in order to maintain the electrical double layer thickness fairly constant in the pH range 3–10.

The particle size distribution of the alumina powder was determined using Coulter Counter and the surface of the powder was analyzed using FTIR spectroscopy. The electrophoretic mobility of the alumina powder was determined in 40%(wt) suspensions using the Mass Transport Analyzer whose design is based on the Hittrof principle [5,6]. The rheology of the slips was investigated using Brookfield Digital viscometer at low shear rates ($9-90s^{-1}$) and the sedimentation experiments were carried out in 100 ml graduated cylinders by following the rate of fall of the settling particle interface over the first two hour period.

RESULTS

The particle size distribution of the alumina powder measured using Coulter Counter is given in Table 2. These results suggest that the alumina powder used in this investigation is poly dispersed and the median diameter of the particles is 11 microns. Figure 1 shows the FTIR spectrum of the 11 micron wet classified alumina powder surface and also the spectrum of pure spectroscopic grade alumina powder. These results indicate that the surface of alumina used in this investigation is different to that of pure alumina. However, the additional peaks observed at wave length 3600, 2900, 1600 and 1200 cm^{-1} appear to closely resemble that of sodium citrate. It has been suggested that during powder processing, such as the grinding operation, either sodium citrate or citric acid is added [7]. It is possible that such an addition of sodium citrate, and the incomplete removal of the additive from the surface of alumina powders after powder processing may explain the observed surface contamination of the alumina powders investigated here. Since the supplier of this powder (Mansanto Electronic Materials Company) were interested in the flocculation behavior of powders in as received state, the present study was limited to the as received powders only.

The electrophoretic mobility of both 11 micron alumina powders in 10^{-3} mol dm^{-3} KNO_3 and in additive and 10^{-3} mol dm^{-3} KNO_3 solutions is shown in Figure 2. These results suggest that the alumina powder behaves electrophoretically negatively charged in the pH range 3–11. An addition of the additive to the slips decreases the net negative charge of the dispersed alumina particles perhaps due to electrostatic charge neutralization. In addition, the results also suggest that above a critical additive concentration, the electrophoretic mobility of the alumina particles remain nearly independent of the additive concentration. Figure 3 shows the viscosity of 50%(wt) 11 micron wet classified alumina slips in water and additive solutions. These results suggest that the viscosity of the alumina slips at a given pH initially appears to increase with an increase in the additive concentration. However, at higher concentrations this trend is reversed. The sedimentation rate (calculated from the rate of fall of the settling particle interface plots) of alumina particles from 50%(wt) slips prepared in water or additive solution is given in Figure 4. These results suggest that the sedimentation rate of the alumina powders in water and additive solutions tend to remain constant in the pH range 3-8. In the pH range pH 8-10.3, the sedimentation rate increases sharply, and above pH 10.3 this trend is reversed. both additives Jeff Amines M-310 and M-320 appear to be very effective in stabilizing the system from flocculation. Additive Jeff Amine M-302 although at low concentrations promotes the particle agglomeration due to flocculation, it tends to stabilize the suspensions at high additive concentrations.

Particle Diameter (micron)	Cumulative Percent
0 - < 2	0
> 2 - < 4	2
> 4 - < 6	4
> 6 - < 8	12
> 8 - < 10	19
> 10 - < 12	43
> 12 - < 14	10
> 14 - < 16	4
> 16 - < 18	2
> 18 - < 20	1
> 20 - < 22	1
> 22 - < 24	1
> 24 - < 30	1
≥ 30	0
Mean Diameter (micron)	10.9

Table 2. Particle size distribution of 11 micron wet classified alumina powder as was determined using Coulter Counter.

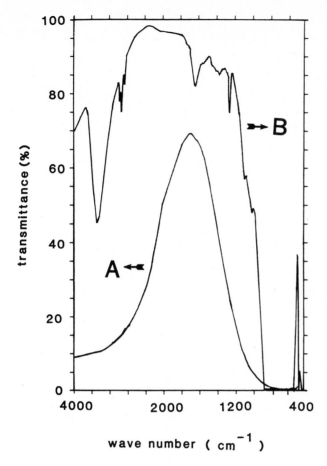

Figure 1. FTIR spectrum of (A) pure spectroscopic grade and (B) as-received 11 micron wet classified powders dispersed in potassium bromide powder compact.

DISCUSSION

An increase in the sedimentation rate and the slip viscosity due to an increase in flocculation of dispersed particles is well established [8]. It is also known that in oxide suspensions, the decrease in the ionic dissociation of surface hydroxyl gorups increases the degree of flocculation of the slip. For practical ceramic processing either flocculated slips or slips that are well stabilized at either too acidic or too basic conditions are not acceptable. Therefore, it is important to understand the stability of the slips in the pH range which is more acceptable for practical processing. For pure oxides such as alumina, this has been well optimized [9-11]. However, the process of optimization of as received powders that were subjected to different surface treatments is very difficult. If the powders were not properly cleaned after surface

252

treatment, small traces of the additives used in the chemical processing of the
powders, such as sodium citrate in our samples, may remain adsorbed on the surface of the powders.   Therefore, the optimization of surface properties of such
commercial ceramic oxide becomes more complex.   If the suspensions also contain
surface active agents, the additives may introduce different surface chemical
changes.   For example, the additive may dissolve the adsorbed surface impurity
and change the properties of the suspending fluid, and/or the desorbed surface
impurity may form a complex with the additive and readsorb onto the surface of
the dispersed solids.   It would be very difficult to fully relate the observed
rheology and sedimentation results and interpret the behavior in terms of the
electrokinetic behavior of the dispersed alumina particles.

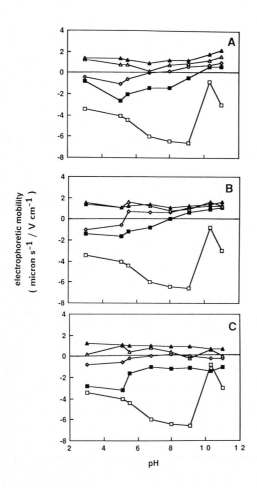

Figure 2.   Electrophoretic mobility of 11 micron wet classified alumina in 40%
(wt) suspensions prepared in (□) 0, (■) 0.2, (◇) 0.5, (▲) 1.0 and (▲) 2.0%(wt)
additive, and 10^{-3} mol dm^{-3} KNO_3 solutions and at different pH.   Additive solution (A) Jeff Amine M-302, (B) Jeff Amine M-310 and (C) Jeff Amine M-320.

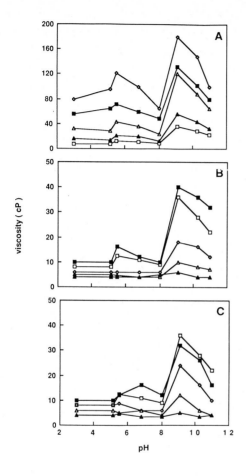

Figure 3. Viscosity plotted as a function of pH of 50%(wt) 11 micron wet
classified alumina slips prepared in (□) 0, (■) 0.2, (◇) 0.5, (▲) 1.0 and
(▲) 2.0%(wt) (A) Jeff Amine M-302, (B) Jeff Amine M-310 and (C) Jeff Amine M-320
solution.

FTIR results shown in Figure 1 and the electrophoresis results shown in
Figure 2 indicate that 11 micron wet classified alumina powder is associated
with acidic impurity and all three Jeff Amines annul the contribution due to the
acidic part of the surface. Although one can argue that the charge reversal of
alumina may result due to the adsorption of cations on the negative surface of
the alumina powders, it is difficult to determine why the viscosity of dispersions containing nearly uncharged particles is lower than the slips containing
charged particles (Figure 4). It is possible that the surface charge on the
alumina particle, in and by itself, does not play an important role in the
process of stabilization of the alumina slips in Jeff Amines.

254

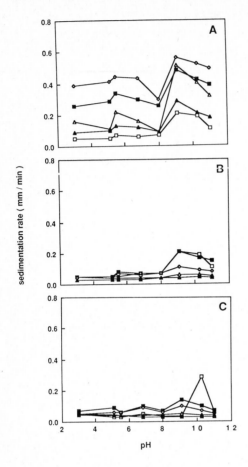

Figure 4. The sedimentation rate of 11 micron wet classified alumina from
50%(wt) suspensions prepared in (□) 0, (■) 0.2, (◇) 0.5, (▲) 1.0 and (▲) 2.0%(wt)
(A) Jeff Amine M-302, (B) Jeff Amine M-310 and (C) Jeff Amine M-320 solutions
and at different pH.

Since the flocculation of the particles in suspension is determined not
only by the surface charge but also by the steric effect of the adsorbed polymer
in solution, additional and useful information can be obtained from the estimation of the diameter of the alumina particle during settling. For example, the
shape of the plots of the alumina particle diameter versus the zeta potential
or the electrophoretic mobility indicates the state of the suspension, i.e.,
whether the dispersion is stable (diameter of the flocc ≈ diameter of the primary
particle) or flocculated (diameter of the flocc ≫ diameter of the primary particle). The sedimentation data represented in Figure 4 was analyzed using modified Steniour equation [12-13] and the mean diameter of the settling alumina

particles in suspension was estimated. Figure 5 shows the normalized flocc size (represented as the ratio between the mean diameter of the flocc that was estimated from sedimentation experiment to the mean diameter of the primary particle as was determined using Coulter Counter) as a function of the electro- phoretic mobility of alumina particles in concentrated dispersions. These results suggest that the 11 micron wet classified alumina powders in suspensions of water and Jeff Amine M-302 flocculate when the net surface charge is nearly zero and remain as primary particles when the particle charge is high. Alumina particles on the other extreme remain as primary particles in Jeff Amine M-320 solution, although the net surface charge is zero. The stability of alumina slips in Jeff Amine M-310 at low concentrations appears to depend upon the sur- face charge of the particles and remain independent of particle charge at high additive concentrations.

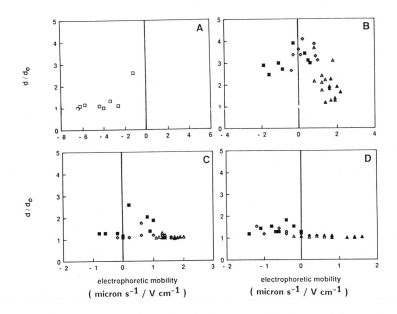

Figure 5. d/d_o as a function of electrophoretic mobility of 11 micron wet classified alumina particles in (A) water, (B,C,D) (■) 0.2 (◇) 0.5, (▲) 1.0 and (▲) 2.0%(wt) additive solutions. Additive solution (B) Jeff Amine M-310 (C) Jeff Amine M-310 and (D) Jeff Amine M-320.

CONCLUSION

From the present investigation the following conclusions can be derived:

1. The alumina powder used in this investigation is associated with some sur- face impurity whole FTIR spectrum resemble that of sodium citrate.

2. All three additives are adsorbed onto the negatively charged surface of 11

micron wet classified alumina powder and very effectively reverse the surface charge of the alumina particles.

3. The slip viscosity of alumina powders increases with an increase in the concentration of additives at low concentrations. However, this trend is reversed at high concentrations.

4. The sedimentation rate of alumina powder in additive solutions appears to remain independent of pH in the range 3-8 and the sedimentation rate decreases with an increase in the additive concentration.

5. The flocculation of 11 micron wet classified alumina powder in water and Jeff Amine M-302 decreases with an increase in the electrophoretic mobility of the dispersed particles. However, in additive Jeff Amine M-320, the alumina particles remain independent of the particle surface charge. The flocculation behavior of alumina in Jeff Amine M-310 solution at low concentrations appears to depend upon the surface charge. At high additive concentrations, the surface charge has very little effect on alumina particle agglomeration in suspension.

ACKNOWLEDGEMENT

The author would like to acknowledge the encouragement and discussions with the late Professor G. D. Parfitt. The author would also like to thank Dr. D. Marcus of the Gulf Research and Development Center, Pittsburgh for providing the additives Jeff Amines, M-302, M310 and M-320 and the Mansanto Electronic Materials Company, St. Peters, MO, for supplying wet classified alumina powders.

REFERENCES

1   P. A. Hartley, Cohesive Strength of Powder Agglomerates Containing Sub-
    micron Particles, Ph.D. Thesis, Carnegie-Mellon Univeristy, Pittsburgh,
    PA (1985).
2   S. Baik, A. Blier and P. F. Becher, Properties of $Al_2O_3-ZrO_2$ composites by
    surface chemical behavior, to be published in MRS Proceedings.
3   E. M. DeLiso, W. R. Cannon and A. S. Rao, Dilute aqueous dispersions of
    zirconia and alumina powders, Mater. Res. Soc. Symp. Series, 60, 43 (1986).
4   Advanced Technical Data, Jeffereson Chemical Company, Bellaire, TX (1978).
5   R. P. Long and S. Ross, An improved mass transport cell for measuring the
    electrophoretic mobilities, J. Coll. Sci., 20, 438 (1965).
6   S. D. James, The mass transport method for determining electrophoretic
    mobilities, J. Coll. Sci., 63, 577 (1978).
7   M. Sacks and T. Tseng, Role of sodium citrate in aqueous milling of alu-
    mina, J. Amer. Ceram. Soc., 66 [4], 242 (1983).
8   G. D. Parfitt, Dispersion of Powders in Liquids,   Applied Science Pub-
    lishers, Inc., Englewood, NJ 1981.
9   G. A. Parks, The isoelectric points of solid oxides, solid hydroxides and
    aqueous hydroxy complex systems, Chem. Reviews, 65, 177 (1965).
10 G. A. Parks, Aqueous surface chemistry of oxides and complex oxide min-
    erals, Adv. in Chem. Ser., 67, 121 (1967).
11 M. Robinson, J. A. Pask and D. W. Fuerstenau, Charge of alumina and
    magnesia in aqueous media, J. Amer. Ceram. Soc., 47, 61 (1964).
12 J. L. Bhatty, L. Davis and D. Dollimore, The use of hindered settling data

to evaluate particle size or flocc size, and the effect of particle liquid
association on such sizes, Surface Technology, $\underline{15}$, 323 (1982).

13 D. Dollimore and K. Karimian, Sedimentation of suspensions: Factors
affecting the hindered settling of alumina in a variety of liquids,
Surface Technology, $\underline{17}$, 239 (1982).

PART 3

INTERFACIAL PHENOMENA IN MINERALS PROCESSING

Interfacial Phenomena in Biotechnology and Materials Processing,
edited by Y.A. Attia, B.M. Moudgil and S. Chander
Elsevier Science Publishers B.V., Amsterdam, 1988 — Printed in The Netherlands

THE INTERACTION OF COPPER SILICATE AND COPPER HYDROXIDE SURFACES WITH AQUEOUS OCTYL HYDROXAMATE

R. HERRERA-URBINA and D.W. FUERSTENAU

Department of Materials Science and Mineral Engineering,
University of California, Berkeley, CA 94720

SUMMARY

The interaction between octyl hydroxamate and copper silicate was found to be highly pH dependent; octyl hydroxamate uptake by the solid increases as the solution pH decreases, and maximum contact angles occur when the solution is slightly acidic. Since copper ion dissolution from the mineral lattice increases significantly at low pH values, depletion of aqueous octyl hydroxamate under these conditions might be due to copper hydroxamate formation in solution. The zeta potential measurements suggest that both the copper hydroxamate complex and the hydroxamate anions adsorb specifically at the copper silicate/solution interface. Infrared spectra indicate the formation of copper octyl hydroxamate on the surface of hydroxamate-treated copper silicate.

INTRODUCTION

Interfacial phenomena at solid/liquid interfaces may drastically change upon the adsorption of organic chelating agents. Because of their ability to form highly stable compounds with specific chemical species, these reagents have found application in a variety of processes that are based on modification of the characteristics of the solid/liquid interfacial region. Both the electrical charge and wetting behavior of solid surfaces containing ions that react with chelating agents may be affected by the formation of surface chelates. The transformation of hydrophilic solid surfaces into hydrophobic ones is of utmost importance in a number of technological disciplines. In the separation of minerals by froth flotation, for example, the success of the process depends on selectively making a desired mineral hydrophobic while maintaining the gangue hydrophilic. This is achieved upon the addition of chemical reagents (collectors) that adsorb at the solid/liquid interface and displace water molecules from the surface of high surface energy solids.

Hydroxamic acids have long been identified as chelating agents, and they are known to have a strong affinity for heavy metal cations. Depending on their hydrocarbon chain length, they may also form insoluble hydrophobic metal hydroxamates. This property of hydroxamate compounds makes them powerful dewetting agents, a characteristic that has been fully exploited in mineral flotation. Sodium and potassium octyl hydroxamate salts have been extensively tested as flotation collectors for minerals containing metal ions that are known to form insoluble hydrophobic metal octyl hydroxamates (1). Among these metal cations, copper (II) exhibits a strong tendency to react with hydroxamate compounds. Although the ability of octyl hydroxamate to effectively render the surface of copper silicate hydrophobic was recognized as early as 1965 (2), the interaction between this chelating agent and copper oxide minerals is not fully understood. Furthermore, the effect of such process variables as pH, time of agitation and sample preparation has not been investigated.

The interaction between octyl hydroxamate and copper silicate and copper hydroxide surfaces is also of importance in corrosion control studies involving copper. Under corrosive environments, surface products on copper may involve oxides, hydroxides, sulfates, carbonates, etc. Prevention of corrosion processes can be achieved upon the formation of protective insoluble films on the substrate. Because of its high affinity to chelate copper, octyl hydroxamate has also been shown to be a good corrosion inhibitor for copper in neutral chloride solutions (3). This paper presents results of experimental techniques that were used to investigate the interaction of octyl hydroxamate with both copper silicate and copper hydroxide. This interaction was then related to the wetting behavior of copper silicate in the presence of potassium octyl hydroxamate.

EXPERIMENTAL MATERIALS AND METHODS

The copper silicate used in this investigation was a high-purity chrysocolla sample from Miami, Az. It was purchased from Ward's Natural Science Establishment, Inc., Rochester, New York. The as-received material was crushed, then ground in a porcelain mortar, and finally sieved to produce a minus 200 mesh size fraction that was used for both electrokinetic measurements and hydroxamate uptake determinations. The specific surface area of this fine powder, as determined by B.E.T. CO_2 adsorption methods was found to be 135.9 m^2/g.

Copper hydroxide was obtained as a precipitate after the pH of a 200 g/l $Cu(NO_3)_2 \cdot 3H_2O$ solution was raised to 8.5 by the dropwise addition of a 1.0 M KOH solution. This precipitate was separated from the liquid by filtration, then rinsed with triply distilled water whose pH had been previously adjusted to 8. The washing procedure was repeated several times until the specific conductance of the filtrate remained constant at about 230 micromhos/cm. After drying at room temperature, the solid was ground to obtain a minus 400 mesh size fraction that was used for zeta potential measurements.

The potassium salt of octyl hydroxamic acid was prepared and purified in our laboratories following the procedure described in the literature (2). The elemental analysis of this compound was found to be 53.3% carbon, 9.4% hydrogen, 6.3% nitrogen and 14.9% potassium with the remainder, 16.1%, being oxygen.

Aqueous solutions of KOH and KNO_3 were used for pH adjustment, while KNO_3 solutions were added to maintain the ionic strength constant. These chemicals were reagent grade materials.

With the exception of contact angle measurements, the first step of the experimental techniques used throughout this investigation involved conditioning of the solid particles in aqueous indifferent electrolyte solutions in the absence and the presence of octyl hydroxamate. Unless otherwise indicated, agitation of the suspension was carried out for twenty minutes at constant pH with a plastic-coated three bladed marine propeller rotating at 1100 rpm. The working temperature was $21 \pm 1^{o}C$, and the ionic strength controlled by the addition of 0.002 M KNO_3. The solution pH was always adjusted to the desired value before adding the solids.

The electrokinetic behavior of both copper silicate and copper hydroxide particles suspended in aqueous solutions was assessed through zeta potential measurements using a Zeta-Meter 3.0 apparatus manufactured by Zeta Meter, Inc., New York. After conditioning of a 0.1 g/1 suspension, an aliquot was transferred into the cell and the zeta potential of 10 to 15 particles was measured. The zeta potential values reported are the average of these measurements. A molybdenum anode and a platinum cathode were used for zeta potential measurements in the absence of octyl hydroxamate; however, since this chelating agent forms molybdenum complexes, a platinum anode had to be used when handling suspensions containing hydroxamate.

The uptake of hydroxamate by copper silicate was calculated by determinng the difference between added hydroxamate and its concentration in solution after conditioning of a 2 g/1 suspension for 72 hours. During this time, the suspensions were agitated on a tumbler rotating at 48 rpm. Aqueous hydroxamate was determined from spectrophotometric analysis of the supernatant using the well-known ferric hydroxamate method (4).

Contact angle measurements were made using a NRL Contact Angle Goniometer manufactured by Rame-Hart, Inc., Mountain Lakes, New Jersey, on a polished section of copper silicate that had been previously contacted with octyl hydroxamate solutions. The specimen was conditioned in these solutions for 20 minutes before the air bubble was placed against its surface under pressure for 2 minutes. The bubble was then freed and the contact angle was measured after a period of ten minutes of bubble-sample contact.

RESULTS AND DISCUSSION

Electrokinetic Measurements

When measuring the zeta potential of silicates, great care should be exercised in the preparation of the sample. The reason for this stems from the fact that various cations in the silicate structure tend to be leached out at low pH values, hydrolyze and readsorb on the silicate surface. Most cations hydrolyze in aqueous solutions because of their strong tendancy to form bonds with oxygen atoms and because of the self-ionization of water molecules (5).

The zeta potential of copper silicate is plotted in Figure 1 as a function of the conditioning pH, which was kept constant to avoid unreliable results due to mineral dissolution. Because of the semisoluble nature of this mineral, especially at low pH values, the effect that copper dissolved from the lattice has on the surface electrical characteristics of copper silicate upon raising the pH was also investigated. For these experiments the copper silicate suspension was first conditioned at pH 4 for 20 minutes. Then, the pH was adjusted to the desired value and the suspension conditioned for 20 more minutes at constant pH before the zeta potentials were measured. This material is designated as partially leached.

Since aqueous copper hydrolyzes and precipitates as copper hydroxide when the pH is raised to a value that depends on the total metal concentration, the interaction between both the aqueous species and the solid phase with

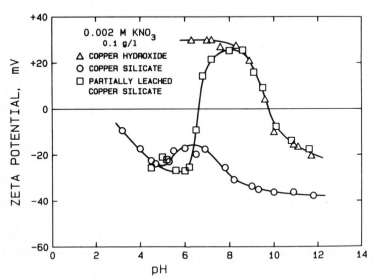

Fig. 1 - The zeta potential of copper silicate, partially leached copper silicate and copper hydroxide in 0.002 M aqueous KNO_3 at various pH values.

partially leached copper silicate will determine the electrical character-istics of its surface. Therefore, copper hydroxide was precipitated from solution and its zeta potential was measured at different pH values. The zeta potential of both natural and partially leached copper silicate as well as that of copper hydroxide is also presented in Fig. 1 as a function of pH. The electrokinetic behavior of both unleached and partially leached copper silicate and copper hydroxide shows a strong pH dependence. Unleached copper silicate is negatively charged in the pH range 3 to 12. Below pH 5 the electric charge at the copper silicate/aqueous solution interface becomes more negative as the hydroxide ion concentration in solution increases. Then a discontinuity is observed and the zeta potential becomes more positive between pH 5 to 8. At pH values above 8, the mineral surface again becomes more negatively charged. Similar zeta potential results have been reported by Bowdish and Plouf (6). Preferential dissolution of copper from the solid lattice below pH 5 leaves a negatively charged silica skeleton that electrically resembles quartz. At pH values above 5, hydrolysis of copper ions results in the formation of copper hydroxo complexes and precipitation of copper hydroxide. Metal hydroxo complex species are extremely surface-active and their adsorption in the Stern layer causes the zeta potential of copper silicate to become less negative in the pH range 5 to 8. In alkaline solutions the negative surface charge on copper silicate is mainly due to preferential adsorption of hydroxide ions.

In addition to the pH, the extent of copper dissolution also affects significantly the electrokinetic behavior of copper silicate. This is clearly shown when the zeta potential of copper silicate partially leached at pH 4 for 20 minutes is measured at different pH values. In these experiments, after the leaching step the pH was readjusted to the desired value and the suspen-sion conditioned for 20 more minutes at constant pH before measuring the zeta potential. Under these conditions, the zeta potential changes from negative to positive at around pH 6.5 and again to negative at pH 9.7. Various investi-gators have also reported similar zeta potential reversals for leached copper silicate (7-9), and for copper silicate in cupric salt solutions (7- 10).

The zeta potentials of copper hydroxide were measured at different pH values in the pH range 6 to 12, and they were found to coincide with those of partially leached copper silicate above pH 8.5. Outside of this pH range, copper hydroxide dissolves completely after 20 minutes of agitation. The isoelectric point of precipitated copper hydroxide was found to occur at pH 9.7. Parks (11) has reported a pH value of 9.4 ± 0.4 for the isoelectric point of this material. Yoon and Salman (12), however, have reported that the point of zero charge of copper hydroxide occurs at pH 7.7.

Reversal of the zeta potential of leached copper silicate, from negative to positive values coincides with the onset of $Cu(OH)_2$ precipitation as given by the solution chemistry of the Cu^{2+}-H_2O-CO_2 system shown in Fig. 2. This system has been chosen because under normal working conditions (that is, unless particular precautions are taken) carbon dioxide is always present in aqueous solutions and its hydrolysis products (anions) will compete with hydroxide ions to complex metal cations. In the case of copper, its metal hydroxide is less soluble than copper carbonate and a carbonate phase does not precipitate in acidic or weakly basic air-saturated solutions. At high pH, however, Cu (II) has a strong tendency to form the soluble carbonate complex, $Cu(CO_3)_2^{2-}$, which causes an increase in the solubility of the hydroxide precipitate.

As shown in Fig. 2, when the total copper concentration is 0.0001 M, only 1.7% of the total metal is present as the first hydroxo complex at pH 6.5. Charge reversal occurs only after about 60% of the total copper precipitates as copper hydroxide. At pH 8.5 essentially all of the copper is present as $Cu(OH)_2$, and it is above this pH value that the surface of leached copper silicate behaves as though it were $Cu(OH)_2$. These zeta potential results suggest that in addition to copper, hydrogen and hydroxide ions, the positively charged hydroxo complex of copper is also a potential determining ion for copper hydroxide.

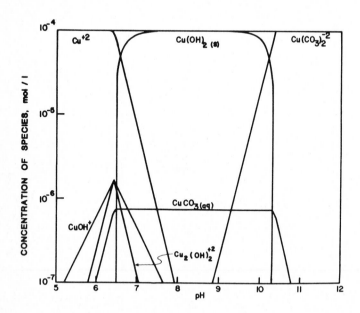

Fig. 2 - Species distribution diagram of the Cu^{2+}-H_2O-CO_2 system for 0.0001 M total copper concentration.

The zeta potential of copper silicate and copper hydroxide in octyl hydroxamate solutions of varying pH was also measured. These measurements are presented in Fig. 3 along with the zeta potential of copper octyl hydroxamate, which was precipitated from a solution containing copper nitrate and octyl hydroxamic acid. The electrical nature of the copper silicate/solution interface undergoes no apparent change upon the addition of 0.0004 M octyl hydroxamate, despite the fact that a chelation reaction takes place between surface copper sites and octyl hydroxamate. In the case of copper hydroxide, however, its zeta potential changes considerably, and it corresponds to that of copper hydroxamate at slightly basic pH values. The isoelectric point of copper hydroxamate occurs at pH 5, while that of copper hydroxide in 0.0004 M octyl hydroxamate occurs at pH 7. The shift in the isoelectric point of copper hydroxide to more acidic pH values upon the addition of octyl hydroxamate is similar to that reported for ferric oxide (13) and manganese dioxide (14) under similar conditions. The ability of the hydroxamate anion to adsorb onto negatively charged copper silicate/hydroxide surfaces indicates its specificity for reacting with surface copper sites to form a copper hydroxamate chelate.

During chelation involving hydroxamic acids, the acidic group (-OH) loses a proton and becomes an anionic donor. The carbonyl oxygen donates a pair of

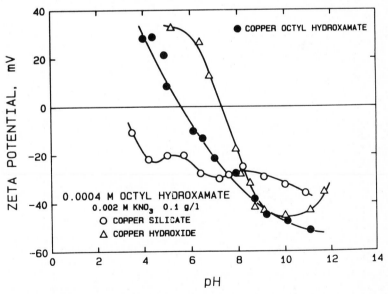

Fig. 3 - The zeta potential of copper silicate and copper hydroxide in 0.0004 M aqueous octyl hydroxamate solutions of varying pH, and the zeta potential of precipitated copper octyl hydroxamate as a function of pH at 0.002 M KNO_3.

268

electrons to the metal ion and the five-membered ring is closed. Whether surface copper sites or the hydroxo complexes of copper react with this anionic chelating reagent will depend on such parameters as pH, degree of solid dissolution, mode of sample preparation, etc. A general reaction for chelation of surface copper by hydroxamate may be the following:

$$
\begin{array}{c}
| \\
- Cu - OH + \\
|
\end{array}
\quad
\begin{array}{c}
O = C - R \\
| \\
HO - N - H
\end{array}
\qquad
\begin{array}{c}
| \\
- Cu \\
|
\end{array}
\begin{array}{c}
O = C - R \\
| \\
O - N - H
\end{array}
\quad + H_2O
$$

Figure 4 presents the zeta potential of partially leached copper silicate in 0.0001 M octyl hydroxamate solutions at different pH values. In this system, two isoelectric points are also observed: one at pH 3.3 and the other at pH 5.8. The first zeta potential reversal may be related to the adsorption of the positively charged copper hydroxamate complex onto copper silicate. The leaching step dissolves enough copper from the solid so that the concentration of the copper chelate in solution become significant. This is clearly seen from the species distribution diagram (Fig. 5) for the system Cu^{2+}-octyl hydroxamate-H_2O system. Comparing Figures 4 and 5, it can be seen that the solid is positively charged only when the charged copper hydroxamate

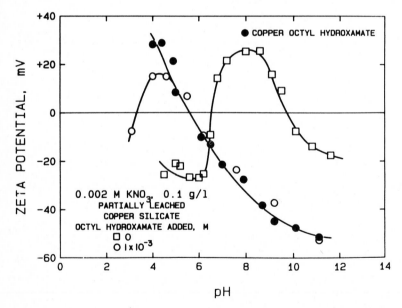

Fig. 4 - The zeta potential of partially leached copper silicate in the absence and the presence of 0.001 M octyl hydroxamate solutions at 0.002 M KNO_3 and various pH values and the zeta potential of copper octyl hydroxamate as a function of pH.

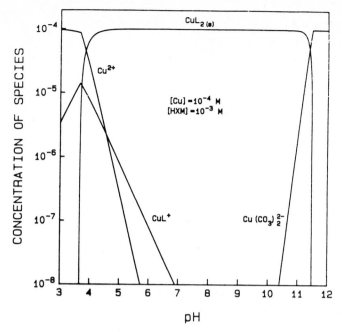

Fig. 5 - Species distribution diagram of the Cu^{2+} - octyl hydroxa-mate-H_2O system for 0.0001 M total copper and 0.001 M total octyl hydroxamate concentrations.

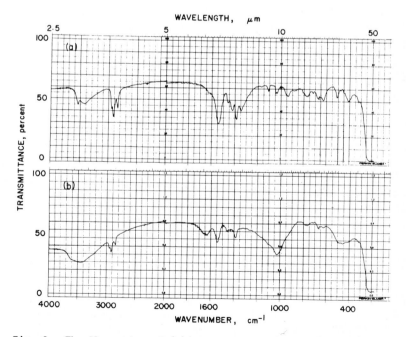

Fig. 6 - The IR spectra of (a) copper hydroxide and (b) copper silicate contacted with octyl hydroxamate solutions.

complex is present in solution in significant concentrations. Fig. 5 also shows that copper hydroxamate precipitates in solution above pH 3.7. Therefore, in the copper silicate-hydroxamate system copper hydroxamate may also precipitate on the solid surface. In fact, in slightly alkaline solutions, the electrokinetic behavior of partially leached copper silicate contacted with octyl hydroxamate resembles that of copper hydroxamate. These zeta potential measurements clearly suggest that both the copper hydroxamate complex and the hydroxamate anion specifically adsorb at the surface of partially leached copper silicate. For comparison, Fig. 4 also shows the zeta potential of partially leached copper silicate and of copper hydroxamate.

The presence of copper hydroxamate chelates on copper containing solids treated with octyl hydroxamate solutions has been identified by a number of reseachers through infrared spectroscopy (3,10,17,18). Similar results were obtained in this research work, as illustrated in Fig. 6 where the IR spectra of hydroxamate-treated copper hydroxide and copper silicate powders are presented. The most relevant information from these spectroscopic data is the band at 1540 cm^{-1}. This band corresponds to the characteristic C=O stretching vibration of the hydroxamic acid group that appears at 1660 cm^{-1} in the free acid (19). The displacement of the C=O stretching frequency indicates C=O...M coordination, where M is the metal cation. The greater the displacement of the band to longer wavelengths, the stronger is the C=O...M bond (20).

Hydroxamate Uptake and Wettability Studies

Kinetic studies on the uptake of octyl hydroxamate by copper silicate indicate that this system does not reach equilibrium even after 5 days (21). With three days of agitation at the natural pH of the system, namely pH 9.3, however, almost 90 percent of the reaction is complete. Hence, the uptake of aqueous hydroxamate by copper silicate was determined at different pH values after 72 hours of agitation of the suspension. Hydroxamate uptake by copper silicate was found to be significantly affected by pH. As can be seen from the hydroxamate uptake determinations at different pH values presented in Fig. 7, at a fixed residual hydroxamate concentration the amount of aqueous octyl hydroxamate abstracted by copper silicate increases as the pH decreases.

Since the hydroxamate ligand forms stable chelate complexes with copper ions, its adsorption on surface copper sites is expected to take place mainly through chemical mechanisms. Because of the strong effect of pH on the dissolution of copper silicate, however, its interaction with octyl hydroxamate at different pH values seems to involve more complex phenomena. In the case of slightly soluble solids, their interaction with reagents that react chemically with lattice ions may proceed by way of three different mechanisms (22):

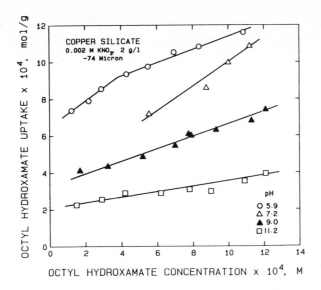

Fig. 7 - The uptake of octyl hydroxamate by copper silicate at various final
pH values after 72 hours of suspension agitation as a function of the
residual hydroxamate concentration.

Chemisorption. Interaction of the ligand with surface sites without
movement of species from their lattice positions; this type of interaction is
limited to a monolayer of adsorbed ligand species.

Surface Reaction. Interaction of the ligand with ions at the interface.
This phenomenon involves movement of species from their lattice positions and
it may result in the formation of monolayers of reaction products.

Bulk Precipitation. Interaction of aqueous ions with the ligand away from
the solid surface out into the bulk solution. Under these conditions, lattice
species will precipitate in the bulk as surfactant salts. These precipitates
may coat the surface of the solid through heterocoagulation.

Octyl hydroxamate uptake by copper silicate through surface reaction and
even bulk precipitation will certainly predominate in weakly acidic conditions
because the dissolution of copper from the solid lattice is significant (even
in the pesence of hydroxamate) under these conditions. Because of its small
size and high diffusion coefficient, the hydrogen ion will penetrate the
mineral lattice, and metal cations will be released into the bulk. The rate
of copper silicate dissolution, therefore, will significantly affect the
uptake of hydroxamate from solution.

The interaction between copper silicate and octyl hydroxamate in weakly
alkaline solutions, however, seems to involve mainly chemisorption. Since the
concentration of hydrogen ions is small under these conditions, the rate of
dissolution is substantially reduced and the hydroxamate uptake also decreases

Table 1 - Contact angle of copper silicate in octyl hydroxamate solutions of varying pH.

1×10^{-4} M Hydroxamate		4×10^{-4} M Hydroxamate	
pH	Contact angle, degrees	pH	Contact angle, degrees
3.8	zero	3.1	zero
4.4	96	3.8	115
4.7	74	4.9	93
5.1	67	6.1	82
6.3	74	7.6	70
7.0	51	8.9	71
8.2	zero	10.0	zero
9.3	zero	11.0	zero
10.2	zero		

perhaps due to the formation of a monolayer as opposed to multilayer formation or even bulk pecipitation in acidic solutions. Chelation of both surface and bulk copper by octyl hydroxamate is controlled by the thermodynamic stability constants of copper octyl hydroxamate complexes, and it is affected by the hydrogen ion concentration. In strongly acidic solutions, the hydroxamic acid molecule is more stable than copper hydroxamates because hydrogen competes with copper for the ligand. In basic solutions under these conditions, hydroxo and carbonate complexes of copper are more stable than copper hydroxamates because the hydroxide and carbonate ions compete with the hydroxamate anion to complex aqueous copper.

The effect of pH on surface copper chelation by octyl hydroxamate is clearly delineated by assessing the wetting behavior of copper silicate in octyl hydroxamate solutions of varying pH through contact angle measurements. These measurements are summarized in Table 1 for two different hydroxamate concentrations. The pH range in which an air bubble makes contact with the surface of copper silicate is broader as the hydroxamate concentration increases. In both cases, however, maximum contact angles occur under slightly acidic conditions. Clearly, chemical reaction of octyl hydroxamate with surface or interfacial copper will result in the formation of a hydrophobic copper hydroxamate film, which gives finite contact angles.

CONCLUSIONS

The surfaces of both copper silicate and copper hydroxide interact with aqueous hydroxamate by means of a chemical mechanism that results in the formation of insoluble hydrophobic copper hydroxamate chelates. This chemical interaction between copper compounds and octyl hydroxamate, however, is highly pH dependent because this parameter affects both the solubility of the solids and the stability of metal octyl hydroxamates. Depending on the pH, octyl hydroxamate may chelate surface and interfacial copper, as well as bulk copper ions. Interaction in the bulk is not desirable in surface-based processes because it consumes the chelating agent without modifying the electrical nature or wetting behavior of the solid/liquid interface. IR spectra were used to identify copper octyl hydroxamate on the surface of both copper silicate and copper hydroxide samples contacted with octyl hydroxamate. Zeta potential measurements suggest that the hydroxamate anion specifically adsorbs at the solid/liquid interface in these systems. The ability of octyl hydroxamate to render the surface of copper silicate hydrophobic was assessed through contact angle measurements. The maximum contact angle was obtained in slightly acidic octyl hydroxamate solutions.

ACKNOWLEDGEMENT

Financial support for this research work was provided by the National Science Foundation.

REFERENCES

1   D.W. Fuerstenau and Pradip, Mineral Flotation with Hydroxamate Collectors, in Reagents in the Mineral Industry, The Institution of Mining and Metallurgy, 1984, 161-168.
2   H.D. Peterson, M.C. Fuerstenau, R.S. Richard and J.D. Miller, Chrysocolla Flotation by the Formation of Insoluble Surface Chelates, Trans. AIME, 232 (1965) 388-392.
3   T. Notoya and T. Ishikawa, Corrosion Inhibition of Copper with Potassium Octylhydroxamate, Bull. Fac. of Eng., Hokkaido University, 98 (1980) 13-19.
4   J.B. Neilands, Struct. Bonding, Vol. 1, 1966, p. 62.
5   J. Kragten, Atlas of Metal-Ligand Equilibria in Aqueous Solutions, John Wiley & Sons, New York, 1978.
6   F.W. Bowdish and T.M. Plouf, Effect of pH on the Zeta Potential of Chrysocolla, Trans. AIME, 254 (1973) 66-67.
7   J.W. Scott and G.W. Poling, Chrysocolla Flotation, Canadian Met. Quarterly, 12 (1973) 1-8.
8   G. Gonzales, The Recovery of Chrysocolla with Different Long-Chain Surface Active Agents as Flotation Collectors, J. Appl. Chem. Biotechnol., 28 (1978) 31-39.
9   G. Gonzales and H. Soto, The Effect of Thermal Treatment on the Flotation of Chrysocolla, Int. J. Miner. Process., 5 (1978) 153-162.
10 B.R. Palmer, B.G. Gutierrez and M.C. Fuerstenau, Mechanisms Involved in the Flotation of Oxides and Silicates with Anionic Collectors: Part I, Trans. AIME, 258 (1975) 257-260.
11 G. Parks, The Isoelectric Points of Solid Oxides, Solid Hydroxides, and Aqueous Hydroxo Complex Systems, Chem. Reviews, 65 (1965) 177-198.
12 R.H. Yoon and T. Salman, Zero Point of Charge of Cupric Hydroxide and Surface Area Determination by Dye Adsorption, Can. Met. Quarterly, 10 (1971) 171-177.
13 S. Raghavan and D.W. Fuerstenau, The Adsorption of Aqueous Octyl Hydroxamate on Ferric Oxide, J. Coll. Interf. Sci., 50 (1975) 319-330.

14 R. Natarajan and D.W. Fuerstenau, Adsorption and Flotation Behavior of
    Manganese Dioxide in the Presence of Octylhydroxamate, Int. J. Miner.
    Process., 11 (1983) 139-153.
15 D.W. Fuerstenau, R. Herrera-Urbina and A.S. Ibrado, unpublished results.
16 D.W. Fuerstenau and J.S. Hanson, unpublished results.
17 L.Evrard and J. De Cuyper, Flotation of Copper-Cobalt Oxide Ore with
    Alkylhydroxamates, in XIth Int. Miner. Process. Congress, Cagliari,
    Italy, 1975, 655-669.
18 J. Lenormand, T. Salman and R.H. Yoon, Hydroxamate Flotation of Malachite,
    Can. Met. Quarterly, 18 (1979) 125-129.
19 A. Soler, M.C. Bonmati and J. Martinez Gomez, Estudio en el Infrarojo de
    los Acidos Hidroxamicos, I. Estado solido y disolucion, Anales de Quimica,
    Series C, 78 (1982) 149-156.
20 L. Bellamy and R.F. Brauch, The Infrared Spectra of Chelate Compounds,
    J. Chem. Soc., 12 (1954) 4491-4494.
21 D.W. Fuerstenau, R. Herrera-Urbina and J. Laskowski, Surface Properties
    of Flotation Behavior of Chrysocolla in the Presence of Potassium
    Octylhydroxamate, in Anales II Congreso Latinoamericano de Flotacion
    1985, Concepcion, Chile, Vol. II, 1985, pp. Ju 1.1 - Ju 1.21.
22 S. Chander and D.W. Fuerstenau, Electrochemical Reaction Control of Contact
    Angles on Copper and Synthetic Chalcocite in Aqueous Potassium Diethyl
    Dithiophosphate Solutions, Int. J. Miner. Process., 2 (1975) 333-352.

Interfacial Phenomena in Biotechnology and Materials Processing,
edited by Y.A. Attia, B.M. Moudgil and S. Chander
Elsevier Science Publishers B.V., Amsterdam, 1988 — Printed in The Netherlands

ADSORPTION OF METALLIC IONS AND SURFACTANTS ONTO THE COAL SURFACE

Qiang Yu,[1] Weibai Hu[1] and Monxiong Guo[2]

[1]Department of Metallurgy and Metallurgical Engineering, University of Utah,
Salt Lake City, Utah 84112

[2]Beijing Graduate School, China Institute of Mining, Beijing, China

SUMMARY

The adsorption of Ca^{2+} and Fe^{3+} ions on the coal surface and its effect on
the adsorption of sufactants and flotability of coal has been studied by FTIR
and zeta-potential measurement. Ca^{2+} and Fe^{3+} ion specifically adsorb on the
coal surface and change its point of zero charge. The adsorption of Fe^{+3} ion
is very strong and in the form of π-metal complex combined the aquated Fe^{3+} ion
with the polynuclear aromatic rings of the organic molecule of coal. It affects the adsorption of surfactants (OT, CTAB, and FCA) and increases the hydrophilicity on the coal surface. The adsorption of Ca^{2+} is weaker and almost
in the form of physical and/or ionic exchange adsorption with functional groups
of coal. The coadsorption and overlapping adsorption of these ions and surfactants on the coal surface also occur.

INTRODUCTION

Surface characteristics and flotation behavior are effected by metal ions
and surfactants. Klassen (ref. 1) proposed that the adsorption of metal ions
on the coal surface could change the characteristics of the hydrate film on the
coal surface and the behavior of coal flotation. Wen (ref. 2) and Sun et al.
(ref. 3) found that metal ions can adsorb on the coal surface as ion exchanging
adsorption with the functional groups of coal surface or electrostatic adsorption, which changes the electrostatic property, stability, and movability of
the hydrate film on the coal surface. Thus the coal flotation can be improved.
Glanville and Wightman (ref. 4) reported that adding a certain amount of metal
ions to the anionic surfactant OT solution enhances the immersion of coal powder in the solution. However, metal ions have little effect on coal powder in
nonionic surfactant (Triton X-100) solution. This paper will present the study
of the adsorption of Ca^{2+} and Fe^{3+} ion on the coal surface and its effect on
the adsorption of surfactants (OT, CTAB and FCA) and flotability of coal.

EXPERIMENTAL

Vacuum flotation

The flotability of Datong fine coal (-28 mesh) was determined by vacuum
flotation. For each test about 50 ml of salt solution was put into a 100-ml

beaker. 3.0 grams of the coal was added to the solution. Then the solution was constantly stirred for 4 minutes to condition the slurry. The slurry prepared was transferred into a vacuum flotation cell with a funnel. Then 75 ml of this salt solution was added to the cell slowly to maintain a suitable level of the slurry. Vacuum flotation was done for 4 minutes at -740 mm Hg. After flotation, the concentrate and tailing were dried, and yield was determined.

Surfactants

Three kinds of commercial surfactants were chosen for this study, all of them are soluble in water: nonionic surfactant FCA ($R-(OC_2H_5)_{5-7}-OH$), anionic surfactant OT ($R-COO-CH_2-CH(SO_3Na)-OOC-R$), and cationic surfactant CATB ($R-CH_2-N(CH_3)_3Cl$).

Zeta-potential measurement

125 ml of salt solution was made and about 1 gram of -200 mesh Datong coal was added to the solution. Then the solution was constantly stirred for about 10 minutes to place the coal particles in suspension. The samples prepared were used the following day in order to allow for the adsorption of metal ions on the coal surface. Zeta-potential measurements were made by using a DPM-1 microelectrophoresis meter. An average of 10 to 20 readings were taken for every sample to find the zeta-potential.

FTIR examination

200 ml was prepared of 0.1 M of salt solution, and about 2 grams of oxidized and demineralized Datong coal (-200 mesh and <0.15 % ash) and 0.5 gram of surfactant was added to the solution for each sample. Then the solution was stirred for about 5 minutes. The samples prepared were used the following day for the adsorption of metal ions and surfactants on the coal surface. The samples were filtered and dried in a vacuum dryer at 70°C. The KBr pelleting method was used for preparing the sample for FTIR examination. In order to prevent the effect of moisture on the results of FTIR examination, the pelleting procedure was done under an infrared light, and the KBr pellets prepared were placed in a dryer for 7 days. The examinations were done by using Nic 170 SX FTIR.

RESULTS AND DISCUSSION
Vacuum flotation

The results of vacuum flotation of Datong fine coal in different salt solutions are shown in Figure 1. The yield of flotation decreases with the increase of salt concentration in the range of study. Ca^{2+}, Mg^{2+} and Na^+ have little effect on coal flotation while Al^{3+} and Fe^{3+} affect it to some extent.

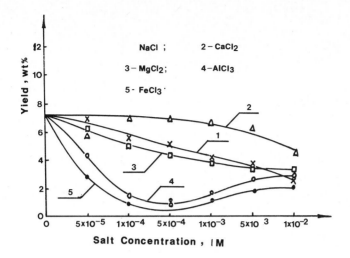

Fig. 1. The vacuum flotation of Datong fine coal in different salt solutions.

Among them the yield of vacuum flotation decreases from 7% to 5% in 0 to 1×10^{-2} M $CaCl_2$ solution, but it decreases significantly from 7% to 0.5% in 0 to 1×10^{-2} M $FeCl_3$ solution.

Zeta-potential measurement

Figures 2, 3 and 4 show the variation in zeta-potential of the coal surface in the different salt and surfactant solutions. Fe^{3+} ions adsorb on the coal surface, and it changes the coal-surface charge positively, and the zeta potential rises to over zero, as shown in Figure 2. It proves that Fe^{3+} ions adsorb strongly on the coal surface, which is hardly affected by pH. The adsorption of Ca^{2+} on the coal surface can also change its zeta-potential and PZC. It seems that the specific adsorption or ion exhange adsorption between Ca^{2+} ions and the coal surface has occurred. The adsorption is greatly effected by the pH of the solution, since there is a larger variation of the zeta-potential of the coal surface with pH. Its adsorption is weaker than that of Fe^{3+} on coal surface.

The zeta-potential of the coal surface also changes with the salt solution concentration while adding surfactant into the salt solution. Since Ca^{2+} ions adsorb on the coal surface weakly, they have little effect on the adsorption of surfactants. Because of specific adsorption of Fe^{3+} ion, it affects obversely on the adsorption of surfactants on the coal surface. The adsorption of Fe^{3+} ions can change the coal surface charge positively (> +10 mv) whether there are surfactants in the solution or not. The zeta-potential of the coal surface varies slowly with the concentration of salt solution in the range studied.

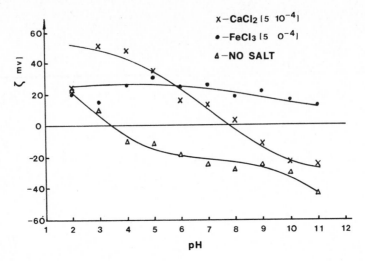

Fig. 2. The relationship between the zeta-potential of the Datong coal surface and pH in different salt solutions.

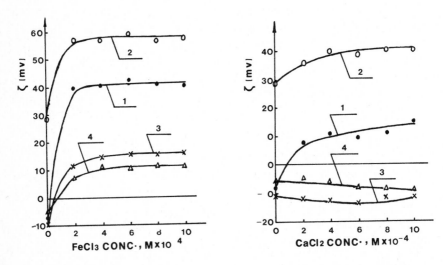

Fig. 3. The effect of $FeCl_3$ on the zeta-potential of Datong coal surface in the different surfactant solutions. 1 - none, 2 - CTAB, 3 - OT, 4 - FCA.

Fig. 4. The effect of $CaCl_2$ on the Datong coal surface in the different surfactant solutions. 1 - none, 2 - CTAB, 3 - OT, 4 - FCA.

The reason may be that the saturated adsorption of Fe^{3+} ions on the coal surface is reached even at low Fe^{3+} ion concentration ($<2 \times 10^{-2}$ M).

The results of zeta-potential measurement in figures 3 and 4 proved that Ca^{2+} and Fe^{3+} ions have a small effect on the adsorption of nonionic surfactant FCA on the coal surface, and the zeta-potential changes gradually to close to the PZC. While there are Ca^{2+} and Fe^{3+} ions in cationic surfactant CATB solution, the zeta-potential will increase with the concentration of Ca^{2+} and Fe^{3+} ions in the solution because the coadsorption of Ca^{2+} and Fe^{3+} ions makes the coal surface positively charged. The adsorption of Ca^{2+} and Fe^{3+} ions also affects the zeta-potential of the coal surface adsorbed by anionic surfactant OT. The zeta-potential is between that by FCA and CTAB or 5 mv more than FCA (see figures 3 and 4).

FTIR examination

Semiquantitative FTIR examination was used for the study of the interaction among metal ions, surfactants, and coal surface. Figures 5 through 8 show the infrared spectrum of these samples studied. The data of attenuational coefficient K of FTIR examination for all samples are given in table 1.

Adsorption of Ca^{2+} and Fe^{3+} ions on the coal surface

Table 1 shows that almost no change takes place in infrared sectrum after adsorption of Ca^{2+} ion on the coal surface. The Kal and Karo values did not change while K_{OH} and K_{1700} decreased, and K_{1600} increased. It proves that $-(COO)_2Ca$ (carboxyl) salt has been formed in the oxidized Datong coal sample

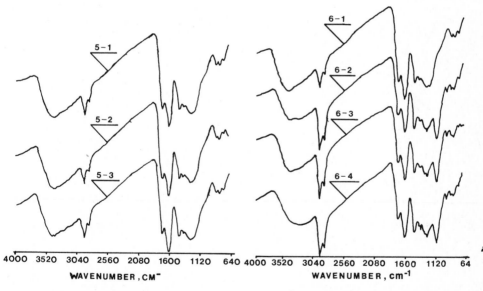

Fig. 5. The FTIR of oxidized Datong coal after adsorption of metal ions in different salt solutions.
5-1 oxidized Datong coal
5-2 oxidized Datong coal+FeCl$_3$
5-3 oxidized Datong coal+CaCl$_2$

Fig. 6. The FTIR of oxidized Datong coal after adsorption of surfactant FCA in different salt solutions.
6-1 oxidized Datong coal
6-2 oxidized Datong coal+FCA
6-3 oxidized Datong coal+FeCl$_3$+FCA
6-4 oxidized Datong coal+CaCl$_2$+FCA

280

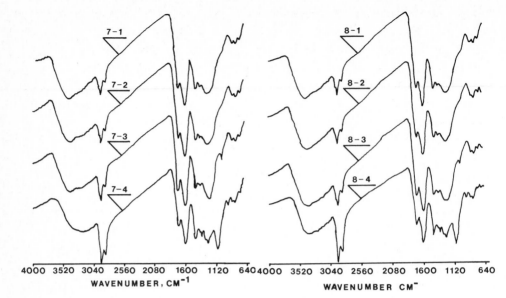

Fig. 7. The FTIR of oxidized Datong coal after adsorption of surfactants FCA and CATB in CaCl$_2$ solution.
7-1 oxidized Datong coal
7-2 oxidized Datong coal+CaCl$_2$
7-3 oxidized Datong coal+CaCl$_2$+CATB
7-4 oxidized Datong coal+CaCl$_2$+FCA

Fig. 8. The FITR of oxidized Datong coal after adsorption of surfactants FCA and CATB in FeCl$_3$ solution.
8-1 oxidized Datong coal
8-2 oxidized Datong coal+FeCl$_3$
8-3 oxidized Datong coal+FeCl$_3$+CATB
8-4 oxidized Datong coal+FeCl$_3$+FCA

TABLE 1.
Attenuational coefficient K of FTIR examination for Datong coal samples.

Sample Name	Weight (mg)	K_{OH}	Kal	K_{1700}	K_{1600}	Karo
oxid. Datong coal	1.4875	0.3494	0.3624	0.4129	0.5494	0.0538
oxid. coal+OT	1.4770	0.2227	0.4558	0.3126	0.4878	0.0519
oxid. coal+FCA	1.4175	0.2005	0.6067	0.2801	0.4106	0.0394
oxid. coal+FeCl	1.5050	0.3683	0.3543	0.3847	0.5801	0.0392
oxid. coal+FeCl+OT	1.5225	0.3367	0.3787	0.3684	0.5271	0.0326
oxid. coal+FeCl+FCA	1.3580	0.2066	0.5361	0.2425	0.3603	0.0242
oxid. coal+CaCl	1.4875	0.3182	0.3710	0.3882	0.6343	0.0549
oxid. coal+CaCl+OT	1.3580	0.2424	0.3570	0.3373	0.4688	0.0450
oxid. coal+CaCl+FCA	1.5050	0.2176	0.5475	0.2480	0.3718	0.0400

adsorbed by Ca^{2+}. This means ionic exchange adsorption has occurred between Ca^{2+} and the oxygen-functional gropups of the coal surface. Ca^{2+} adsorbs mainly on the hydrophilic zone of the coal surface, but adsorbs physically on the hydrophobic zone or polynuclear aromatic rings of coal. These adsorptions are both electrostatic ionic exchanging adsorption and reversible, it depends on the components of the solution and its pH. Because this adsorption can not change the properties of hydrate film on the coal surface, it has almost no affect on the coal floatability.

After Fe^{3+} adsorbs on the coal surface, the K_{OH} of the infrared spectrum increases. It shows that there are more water molecules (H-O-H) in the sample, and that makes the strength of infrared spectrum of extensional retraction vibration of the hydroxy (-OH) in the sample or K_{OH} increase (containing those water coadsorbed with Fe^{3+} on the coal surface). It may mainly be that $(Fe(OH_2)_3)^{3+}$ or $((H_2O)_4Fe(OH)_2Fe(OH_2)_4)^{4+}$ aquated ions adsorbs on the coal surface. The decrease at K_{1700} show that there are $-(COO)_nFe^{+(3-n)}$ salts or complex formed with oxygen-functional groups on the coal surface. It is similar to the adsorption of Ca^{2+} ion on the coal surface.

On the other hand, Table 1 and Figure 5 also show that the adsorption of Fe^{3+} ions on the coal surface makes Karo of infrared spectrum decrease and change at the 900 - 700 cm^{-1} region. It proves that the adsorption hinders the out-off-plane deformation vibration of the aromatic hydrogen of coal. It means that contact or specific adsorption occurs between $(Fe(OH_2)_3)^{3+}$ or $((H_2O)_4Fe(OH)_2Fe(OH_2)_4)^{4+}$ aquated ions and polynuclear aromatic rings of the coal molecule.

Since the structure of bonding electron layer of Fe^{3+} may take on the form of $(3d^2, 4S, 4P^3)$ mixing orbits bonding with coordinate groups (water molecules and polynuclear aromatic rings of coal) to form inner sphere complexes, or in the form of $(4S, 4P^3, 4d^2)$ mixing orbits bonding with coordinate groups to form outer sphere complexes. Therefore in water molecule or hydroxy ion, as one of coordinate group, the 2P orbit of oxygen can supply a pair of indispensible electrons to empty mixing orbit to connect the complex. The electrons at the aromatic polynuclear of coal is movable and polarizable. It can also supply a couple of electrons to empty mixing orbit to join the complex. At the same time, the electrons at 4d orbit of Fe^{3+} ion can transfer to coordinate groups to form feedback bond to make the complex more stable. Such Fe^{3+} ions, in the center, can adsorb on the coal surface in the form of π-complex combined H_2O or OH and aromatic polynuclear of coal.

The adsorption of Fe^{3+} ion on the coal surface is considered as three forms: (1) electrostatic adsorption, (2) ionic exchange adsorption and (3) π-complex adsorption. The first and second adsorptions occur mainly on the hydrophilic zone of the coal surface and the third one at hydrophobic zone. So, Fe^3 ion adsorbs strongly on the whole coal surface. It makes zeta-potential of the coal surface increase, and a thicker hydrate film formed on the coal surface. Because of π-complex formed, the hydrophilicity of coal surface increases, the flotability of coal decreases and the yield of vacuum flotation decreases.

Affect of metal ions on the adsorption of surfactants on coal surface

Table 1 and Figures 6, 7 and 8 show that Ca^{2+} and Fe^{3+} ions have harmful effect on the adsorption of surfactants on the surface of oxidized Datong coal, particularly on the surfactant OT. The reduction of Kal proved that the adsorption of surfactants on the coal surface decreases. The main reasons are as follows:

(i) Interaction between metal ions and surfactants. An unsoluble deposit was formed after surfactant OT was added in to the $CaCl_2$ or $FeCl_3$ solution. It was an organic metal complex produced by the reaction between anionic surfactant OT and Ca^{2+} and Fe^{3+} ions or by salt-out action. There are a lot of metal ions in coal flotation slurry. It would consume a large amount of the surfactant OT. The surfactant concentration in the slurry drops. Thus the adsorption of surfactant OT on the coal surface will be reduced. Ca^{2+} and Fe^{3+} ions have not reaction with FCA, therefore, nonionic surfactant FCA is best as a promoter of coal flotation.

(ii) Competitive adsorption between Ca^{2+} ion and FCA. We proved that the surfactant FCA can specifically adsorb on the whole coal surface in the form of P-type hydrogen bond with -COOH or -OH groups of the coal surface or in the form of π-type hydrogen bond with aromatic hydrogen of coal (5). Becuase Ca^{2+} adsorbs weakly but FCA adsorbs strongly on the coal surface, the competitive adsorption of Ca^{2+} ion and FCA on the coal surface in the soltuion results that Ca^{2+} ion has almost no effect on the adsorption of FCA. Table 1 and Figures 6, 7, and 8 show that Ca^{2+} ion has no effect on P-type hydrogen bond and π-type hydrogen bond adsorption of FCA, and also no effect on the adsorption amount of FCA on the coal surface.

(iii) Competitive adsorption and coadsorption between the Fe^{3+} ion and FCA. Aquated Fe^{3+} ion and FCA both can specifically adsorb on the coal surface. A competitive adsorption between Fe^{3+} ion and FCA as well as coadsorption of Fe^{3+} ion and FCA on the coal surface will take place in their solution. It leads to the decreasing adsorption of FCA on the coal surface because of the reduction of Kal of FTIR shown in Table 1 and Figure 8. The change of attenuation coefficient of extention-retraction vibration of the hydroxyl group (K_{OH}) in FTIR shown in Table 1 is as follows:

$$K_{OH}(\text{oxid. coal} + \text{FCA}) = K_{OH}(\text{oxid. coal} + FeCl_3 + \text{FCA}) < K_{OH}(\text{oxid coal})$$

$$< K_{OH}(\text{oxid. coal} + FeCl_3)$$

It shows the coadsorption or the overlapping adsorption of the aquated Fe^{3+} ion and FCA on the coal surface.Because the -O- bond of polar group of FCA can remove the water molecules of aquated Fe^{3+} ion combined with the polynuclear aromatic rings of coal, so FCA may adsorb on Fe^{3+} ions instead of the water molecules as a new supplement to form a three element complex (coal-metallic-surfactant complex). The hydrocarbon bonding of nonpolar groups of FCA among themselves or between it and coal surface may assist adsorption of FCA on aquated Fe^{3+} ions. Thus, it will reduce the water molecules in aquated Fe^{3+} ions adsorbed on the coal surface or reduce K_{OH}, that means to reduce greatly the effect of aquated Fe^{3+} ions adsorbed on the coal surface on the flotability of coal particles.

In addition since aquated Fe^{3+} ions can specifically adsorb on polynuclear aromatic rings of coal and FCA can also act with aromatic hydrogen of coal. The former is three-elements complex adsorption and the latter is π-type hydrogen bond adsorption. Both of them strong action with aromatic ring of coal. They all hinder the out-off-plane deformation vibration of the aromatic hydrogen and make Karo decrease further (See Table 1).

CONCLUSIONS

1. Ca^{2+}, Mg^{2+} and Na^+ ions have little effect on the flotation of coal, but Fe^{3+} and Al^{3+} ions do have effect on it.

2. Ca^{2+} and Fe^{3+} ions have a harmful effect on the adsorption of surfactant OT on Datong oxidized coal. The adsorption of surfactant FCA on the coal surface is not effected by metal ions. It changes the zeta-potential of the coal surface close to the PZC whether the solution contains metal ions or not.

3. Fe^{3+} ion specifically adsorbs on the whole coal surface, besides of electrostatic and ion exchange adsorption, and mainly in the form of π-metal complex combined the aquated Fe^{3+} ion with the polynuclear aromatic rings of coal. It increases the zeta-potential of coal surface and reduces the flotability of coal. The Fe^{3+} ion has great effect on the adsorption of surfactants on the coal surface. The Ca^{2+} adsorbs weakly on the coal surface in the form of physical adsorption and ion exchange adsorption. It has little effect on the adsorption of surfactants.

4. The surfactant FCA can specifically adsorb on the whole coal surface. It can remove water molecules of aquated Fe^{3+} ions combined with the polynuclear aromatic rings of coal, and adsorb on Fe^{3+} ions to form a new three-element complex (coal-metallic ion-surfactant complex). Thus, it reduces the affect of aquated Fe^{3+} ions on coal flotability.

284

REFERENCES

1 V. N. Klassen, Coal Flotation, Mining Industry Publisher , 1957 (in Russian).
2 G. W. Wen and S. C. Sun, An Electrokinetic Study on the Oil Flotation of Oxidied Coal, Separation Science Technology, 16 (10), 1980.
3 J. A. L. Campbell and S. C. Sun, Bituminous Coal Electrokinetics, Transactions Vol. 247, Society of Mining Engineers, AIME, June 1970, p.111.
4 Glanville and Wightman, Action of Wetting Agent on Coal Dust, Fuel, Vol 68, November, 1979, p. 819.
5 W. B. Hu, Q. Yu and M. X. Kuo, New Reagent Design for Coal Flotation, Second Annual Pittsburgh Coal Conference Proceedings, 1985, p. 25.

Interfacial Phenomena in Biotechnology and Materials Processing,
edited by Y.A. Attia, B.M. Moudgil and S. Chander
Elsevier Science Publishers B.V., Amsterdam, 1988 — Printed in The Netherlands

ADSORPTION OF OLEATE ON DOLOMITE AND APATITE

BRIJ M. MOUDGIL, T.V. VASUDEVAN, D. INCE AND M. MAY

Mineral Resources Research Center, Dept. of Materials Science and Engineering,
University of Florida, Gainesville, Florida 32611

SUMMARY
   Separation of dolomite from apatite using fatty acid as the collector in
the two-stage conditioning process was attributed to the amount of the collector adsorbed as well as changes in the nature of the adsorbed species upon reconditioning. In this study it was determined that surface precipitation is
the predominant mechanism by which oleate adsorbs both on apatite and dolomite.
Also, no significant desorption occurred upon reconditioning at pH 4 or below.
FTIR studies revealed that changes in adsorbed species after second stage
conditioning is the major reason for observing the desired flotation
selectivity.

INTRODUCTION

   The removal of dolomite impurities from apatite is essential, before sub-

sequent chemical processing of the phosphate concentrate for the production of

phosphoric acid. A two-stage conditioning technique developed by Moudgil and

Chanchani (ref. 1), was successfully employed to reduce the dolomite content

of phosphate concentrate to the desirable levels. The success of the above

method was reported to be based on differences in both the amount and nature

of the adsorbed collector (oleate) species on the two minerals. A study of

adsorption of oleate on apatite and dolomite is therefore essential to

evaluate the effect of major parameters in order to optimize the

adsorption/flotation conditions.

   In thermodynamic analysis of flotation of apatite with oleate,

Ananthapadmanabhan and Somasundaran (ref. 2) determined that surface precip-

itation of the dissolved calcium-oleate complex was the major reason for the

induced hydrophobicity and the resulting flotation. In a subsequent study of

adsorption of oleate on apatite Moudgil et al., (ref. 3) confirmed that in the

region of collector concentration employed for flotation, surface precip-

itation of calcium oleate complex was the predominant mechanism of adsorption.

However, no such study has been reported for the adsorption of oleate on

dolomite. In this paper, the effect of pH on both the adsorption kinetics and

nature of the adsorbed species on the two minerals, will be evaluated so as to

elucidate the mechanism by which selectivity is achieved using the two-stage

conditioning process.

EXPERIMENTAL

Materials

(i) <u>Dolomite</u>. Hand picked dolomite was obtained from International Minerals and Chemicals Corporation. The sample was crushed, ground, deslimed to remove fines and dried. The dried material was sized and the -65 +100 mesh fraction was used for adsorption studies. The MgO content of the sample was found to be 18.9%. The surface area, as determined by BET was 6.0 m^2/g.

(ii) <u>Apatite</u>. High grade phosphate product obtained from Agrico Chemical Company was acid scrubbed, sized to -65 +100 mesh fraction and deslimed. The P_2O_5 content of the sample was 35.3% and the BET surface area was found to be 11.5 m^2/g.

(iii) <u>Oleic acid</u>. Carbon-14 tagged oleic acid was obtained from ICN pharmaceuticals in nitrogen sealed ampules of 25 µci in 0.1 ml benzene. Non-radioactive gold label grade (>99% pure) material was purchased from Aldrich Chemical Company. A mixture of ^{14}C - labelled and cold oleic acid was used for adsorption experiments.

(iv) <u>Other Reagents</u>. Acidic and basic pH modifiers and salt (KNO_3) used to adjust the ionic strength were obtained from Fisher Scientific Company.

Methods

(i) <u>Analysis</u>. The analysis of radioactive oleic acid was carried out using a Beckman Model LS1800 liquid scintillation counter.

(ii) <u>Preparation potassium oleate solution</u>. The saponification of gold label oleic acid was carried out by dissolving a known amount of sample in deaerated, triple distilled water at pH 11.0. The radioactive tracer was prepared using the same procedure, with ^{14}C - labelled oleic acid. The ionic strength of the solution in both cases was maintained at 3×10^{-2} N KNO_3.

(iii) <u>Equilibrium adsorption tests</u>. Suspensions of 1.25 and 2.5 wt% solids were prepared by weighing 0.2 and 0.4 g of dolomite, respectively, into a 20 ml scintillation vial to which 6 ml of 0.055 N KNO_3 and 5 ml of KOH were added. The initial pH of the suspensions were so adjusted, that the final pH after aging for 24 hrs, was about 10. A known amount of potassium oleate solution was then added and the final volume made up to 16 ml using KNO_3 solution at pH 10. The adsorption was carried out to equilibrium by tumbling the vials at 10.5 rpm for 1 hr. After equilibration, the supernatant was decanted and the solids were rinsed with 3×10^{-2} N KNO_3 at pH 10. The amount adsorbed was determined using direct measurement of oleate concentration on solids. The amount of precipitate formed, if any, was evaluated by measuring the oleate content of the supernatant before and after centrifuging it for 4 hrs at 15,000 rpm.

(iv) Adsorption studies under flotation (two-stage conditioning) conditions. One weight percent suspensions of apatite or dolomite were prepared in the same manner, as described previously. The samples were aged for 2 hrs at pH 10, conditioned for 2.5 minutes with 1.87×10^{-4} kmol/m^3 potassium oleate and then reconditioned at pH 4 for an additional 2.5 minutes. This conditioning schedule was followed to simulate the flotation test procedure. The supernatant was decanted, solids rinsed and adsorption on solids measured as described earlier. Adsorption measurements under single stage conditioning at pH 10 was also carried out for comparison purposes.

(v) Fourier-Transform Infrared Spectroscopic studies. A 1 wt% suspension of apatite/dolomite of -325 mesh size fraction, was aged for 20 min. followed by conditioning with 5×10^{-3} kmol/m^3 oleate solution for 1 hr. at a desired pH level. At the end of the conditioning period, the solids were separated from the supernatant and rinsed with distilled water of the same pH as during conditioning, and dried at 50°C for 24 hrs and stored in a vacuum desiccator until used.

The diffused reflectance IR spectra was obtained using a Bomem DA3.10 FTIR spectrometer equipped with a 25 cm path length Michelson interferometer fitted to a KBr beam splitter.

RESULTS AND DISCUSSION

An examination of the previous investigations on adsorption of oleate on apatite and dolomite (refs. 1-3) indicated that the selectivity observed in the separation of the two minerals may be due to differences in one or more of the following factors: i) mechanism of adsorption, molecular adsorption or surface precipitation; ii) adsorption kinetics, which determines the amount adsorbed; and iii) nature of the adsorbed oleate species. Both equilibrium and kinetic studies were conducted to evaluate the possible differences in mechanisms and amount of adsorption, while FTIR study was carried out to determine the nature of adsorbed species.

Equilibrium Adsorption Study

In an analysis of flotation of salt-type minerals with hydrolyzable surfactants, Ananthapadmanabhan and Somasundaran (ref. 2) determined that the surface precipitation of metal-surfactant complex is the predominant reason for the observed flotation of the minerals. A more detailed study (ref. 3) on the adsorption of oleate on apatite indicated that the adsorption isotherm can be divided into four major regions. The first region, in which the concentration of oleate was below the solubility limit of calcium oleate precipitation, chemisorption was found to be the mechanism of adsorption. In the concentration range (region II) which is slightly above the solubility limit, both molecular adsorption and surface precipitation of calcium oleate were detected. With further increase in the oleate concentration (region III) surface precipitation of calcium oleate was found to be the predominant adsorption mechanism. At very high concentrations of oleate (region IV), bulk precipitation was dominant and the adsorption isotherm reached a plateau value. The concentration of oleate at which flotation was carried out fell in the region III of adsorption isotherm and thus it was concluded that surface precipitation was the major reason for the flotation of apatite. In order to understand the differences, if any, between the adsorption mechanisms on the two minerals equilibrium adsorption study of oleate on dolomite was carried out in the present study. The adsorption isotherm was evaluated only in regions II, III and IV, as these were the regions of interest in flotation. The results obtained (Figure 1) indicated that, as in the case of apatite,

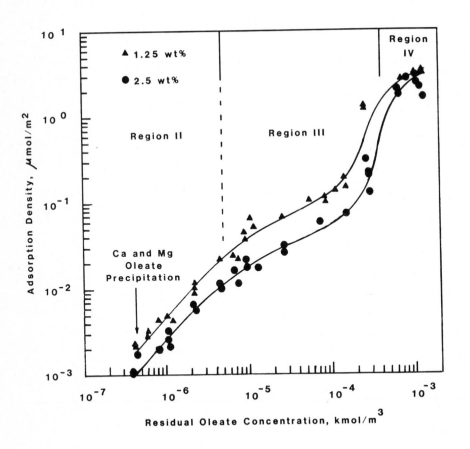

Fig. 1. Adsorption of oleate on dolomite at two solids loadings.

adsorption density on dolomite was inversely proportional to pulp density of solids in region III, which was identified to be due to surface precipitation of metal-oleate complex. The adsorption behavior in region IV was found to be similar to that of apatite, the isotherms approaching a plateau value. It can, therefore, be concluded that in the region of oleate concentration employed in flotation, surface precipitation of metal-oleate complex is the predominant adsorption mechanism in the case of both apatite and dolomite.

Adsorption Study under Flotation Conditions

   The possible differences in the magnitude of adsorption densities on the two minerals was evaluated by carrying out adsorption studies under conditions similar to those employed in flotation under both single and two-stage conditioning methods. A similar study conducted previously (ref. 1) indicated that the adsorption density was higher on dolomite (1.55 $\mu mol/m^2$) than on apatite (0.11 $\mu mol/m^2$) and that there was no change in adsorption densities after second stage conditioning at lower pH (4). It was argued that since the adsorption remained unchanged upon reconditioning at pH 4 or below, the observed selectivity in two-stage conditioning was due to the differences in the amount of the collector adsorbed and the nature of the adsorbed species in the two cases. It should be noted, however, that in the above study adsorption was measured using a solution depletion technique and it was shown later in two different studies on the adsorption of oleate on hematite (ref. 4) and on apatite and dolomite (ref. 5) that solution depletion method at low pH ($<$7) may result in erroneous results. This was due to coating of the container by oleic acid, which separates out in the form of emulsion at lower pH value, thus resulting in misleading conclusions. It was, therefore, decided to repeat the adsorption tests conducted by Moudgil and Chanchani (ref. 1), using direct measurement of oleate adsorption on the minerals. The results obtained are summarized in Table 1. It can be seen from these data that adsorption density on dolomite was higher than apatite and it essentially remained unchanged upon reconditioning. It should, however, be noted that the magnitude of the adsorption density was lower than in the previous study. This probably is due to different apatite and dolomite samples employed in the present study. Next, it was decided to examine the nature of the adsorbed species on the two minerals.

TABLE 1

Adsorption of oleate on apatite and dolomite

Pulp density: 1 wt%

		Single Stage Conditioning	Two-Stage Conditioning
Conditioning pH		10	10
Conditioning time		2.5	2.5
Reconditioning pH		10	4
Reconditioning time		2.5	2.5
Adsorption Density ($\mu Mol/m^2$)	Apatite	0.23	0.23
	Dolomite	0.37	0.43

FTIR Spectrometric Study

     To identify the nature of adsorbed species by IR spectroscopy, Moudgil and Chanchani (ref. 1) observed that both oleate and oleic acid were present on the surface of apatite at pH 10, while only oleic was found at lower pH value. In the case of dolomite, however, the nature of the adsorbed species could not be identified due to masking of the surface spectra of the relevant species by that of the dolomite itself. In the present study, the instrument used was equipped such that differences in FTIR spectra between the untreated and oleate treated samples of both apatite and dolomite could be obtained. The results obtained are shown in Figures 2 and 3. These spectra indicated that as observed in the earlier study (ref. 6) both oleate and oleic acid were present on apatite at pH 10. A similar result was indicated for dolomite. After reconditioning at lower pH (pH 4 or below) oleic acid was detected to be the predominant specie with some oleate present on both minerals. It is known that oleate is a more efficient collector for flotation than oleic acid. Considering the trend towards greater adsorption density on dolomite at pH 10 it is possible that relatively more surfactant remains in the form of oleate on dolomite than on apatite after reconditioning at lower pH. This would lead to better flotation response of dolomite after reconditioning and selective removal of dolomite from apatite during the two-stage conditioning process.

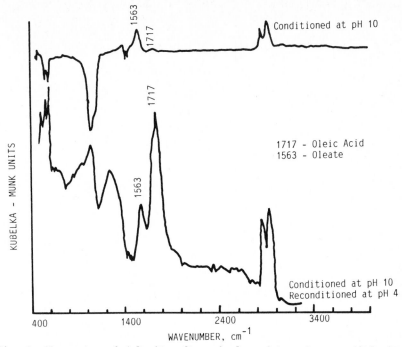

Fig. 2. IR spectra of dolomite after single and two-stage conditioning.

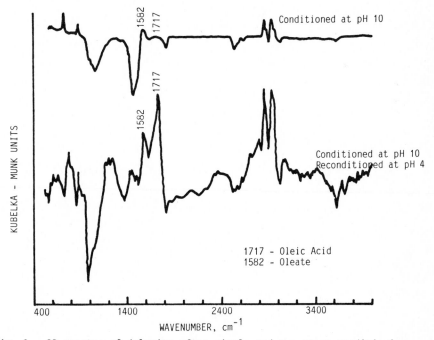

Fig. 3. IR spectra of dolomite after single and two-stage conditioning.

CONCLUSIONS

Previous studies have indicated that the selectivity observed in the separation of apatite and dolomite by two-stage conditioning process could be due to differences in mechanism of adsorption, adsorption kinetics or the nature of the species adsorbed on the two minerals. Equilibrium adsorption studies conducted on the two minerals revealed that surface precipitation of metal-oleate complex was the predominant mechanism of adsorption on both apatite and dolomite and thus could not account for the differences in their flotation behavior. Adsorption study, conducted under flotation conditions, showed a trend towards higher adsorption density on dolomite using both single and two-stage conditioning methods. The nature of the adsorbed species observed on apatite and dolomite were found to be similar after two-stage conditioning. It was determined that both oleate and some oleic acid were present on the two minerals after single-stage conditioning at alkaline pH. After reconditioning, oleic acid was predominant on both minerals with relatively more oleate on dolomite. Considering that oleic acid is a poor collector as compared to oleate, the selectivity obtained during the two-stage conditioning process, therefore, appears to be due to the amount and the change in the nature of the adsorbed species on the two minerals.

Acknowledgements

Financial support of this work was provided by the Occidental Chemical Co., International Minerals and Chemical Corp. - Florida, Gardinier, Inc., NSF-PYI Award (MSM #8352125) and the Florida Institute of Phosphate Research (Grant #85-02-067). Any opinions, findings, and conclusions or recommendations expressed in this work are those of the authors and do not necessarily reflect the views of the Florida Institute of Phosphate Research.

REFERENCES

1   B. M. Moudgil and R. Chanchani, "Flotation of apatite and dolomite using sodium oleate as the collector," Min. Metall. Process., 2(1) (1985) 13-19.

2   K. P. Ananthapadmanabhan and P. Somasundaran, "Surface precipitation of inorganics and surfactants and its role in adsorption and flotation," Colloids and Surfaces, 13(2/3) (1985) 151-167.

3   B. M. Moudgil, T. V. Vasudevan, and J. Blaakmeer, "Adsorption of oleate on apatite," Min. Metall. Process., 4(1) (1987) 50-54.

4   L. J. Morgan, K. P. Ananthapadmanabhan, and P. Somasundaran, "Oleate adsorption on hematite: problems and methods," Int. J. of Min. Process., 18 (1986) 139-152.

5   D. Ince, "Effect of sodium chloride on the selective flotation of dolomite from apatite," Ph.D. Dissertation, University of Florida, 1987.

6   R. Chanchani, "Selective flotation of dolomite from apatite using sodium oleate as the collector," Ph.D. Dissertation, University of Florida, 1984.

Interfacial Phenomena in Biotechnology and Materials Processing,
edited by Y.A. Attia, B.M. Moudgil and S. Chander
Elsevier Science Publishers B.V., Amsterdam, 1988 — Printed in The Netherlands

APPLICATION OF FLUOROSURFACTANTS IN SELECTIVE FLOTATION OF COAL

R.W. Lai and M.L. Gray

Pittsburgh Energy Technology Center, U.S. Department of Energy, Pittsburgh, PA
15236

SUMMARY
    Four fluorosurfactants with different molecular structures were examined
to compare their effectiveness as surface modifiers of coal and pyrite. The
effects of fluorosurfactants on the surface of coal and pyrite were characterized by measuring the contact angles, the spreading coefficient, and the
flotation response.
    Among the fluorosurfactants evaluated, the anionic fluorosurfactant is
most effective in enhancing the surface hydrophobicity of the coal, followed
by the amphoteric, cationic, and nonionic fluorosurfactants. The mechanism of
fluorosurfactant adsorption and resulting surface properties of coal were
interpreted in terms of the differences in molecular structure.
    Meanwhile, the effect of fluorosurfactant on the contact angle of pyrite
was investigated, and the conditions needed to maximize the hydrophobicity
difference between coal and pyrite were evaluated. Based on the differences
in surface hydrophobicity of coal and pyrite, a series of preliminary
flotation tests was conducted to confirm the feasibility of selective flotation of coal from pyrite.

INTRODUCTION

    In this study, the surface wettability modification of coal through the

adsorption of fluorosurfactants in an aqueous phase was investigated, and

preliminary tests on the application of fluorosurfactant on the selective

flotation of coal from pyrite were conducted. The fluorosurfactants were

chosen over the hydrocarbon-type surfactants because of their unique structure

and surface activity.

    Four types of fluorosurfactants with different ionic functional groups

were chosen for this study to elucidate the mechanism of surface adsorption

and to study the effect of various surfactants on the surface properties and

flotation response of the coal. The four types of surfactants are anionic,

amphoteric, cationic, and nonionic Zonyl (a Dupont trademark)

fluorosurfactants.

    The major differences in properties between the Zonyl fluorosurfactants

and the nonfluorinated, or hydrocarbon-type, surfactants are attributed to the

fluorocarbon chain, $F(CF_2CF_2)_x...$, in the fluorosurfactant molecules. The

fluorocarbon portions of these molecules give the surfactants an extreme

tendency to orient at interfaces. This tendency results from the small affinity of fluorocarbon molecules for water and the low interaction between fluorocarbon chains. Consequently, fluorosurfactants lower the surface tensions of aqueous solutions more than hydrocarbon surfactants do. When adsorbed on a solid surface, the fluorosurfactants tend to orient to produce a fluorocarbon-like surface. This effect is an advantage when the surface must become hydrophobic.

THE FLUOROSURFACTANTS

The fluorosurfactants chosen for this study are anionic, nonionic, cationic, and amphoteric types having the chemical structure and physical properties listed in Tables 1 and 2.

TABLE 1. Chemical Structures of Zonyl Fluorosurfactants

$R_f = F(CF_2CF_2)_{3-8}$

Zonyl FSA	$R_fCH_2CH_2SCH_2CH_2CO_2^-Li^+$
Zonyl FSN	$R_fCH_2CH_2O(CH_2CH_2O)_xH$
Zonyl FSC	$R_fCH_2CH_2SCH_2CH_2N^+(CH_3)_3CH_3SO_4^-$
Zonyl FSK	$R_fCH_2CH(OCOCH_3)CH_2N^+(CH_3)_2CH_2CO_2^-$

TABLE 2. Typical Physical Properties

Ionic Type	Zonyl FSA Anionic	Zonyl FSN Nonionic	Zonyl FSC Cationic	Zonyl FSK Amphoteric
Density @ 25°C	1.17	1.06	1.16	1.25
Aqueous Surface Tension, dynes/cm				
0.01% solids	22	24	21	21
0.10% solids	18	23	19	19

EXPERIMENTS

To understand the surface wettability modification of coal in the presence of the aforementioned fluorosurfactants, the following physical parameters were measured in the laboratory: surface tension, contact angle, and flotation response.

Additionally, the spreading coefficient, which involves the surface energy terms of solid, gas, and liquid interphases, was measured indirectly.

Surface Tension

A Fisher Surface Tensiomat Model 21 was used to measure the surface tension by the DuNoy ring method. This method involves the determination of the force required to detach a ring of platinum-iridium wire from the surface of the liquid. In the absence of any surfactants, the surface tension of water at 24°C was 73 dynes/cm². In the presence of 100-600 ppm of the various ionic types of fluorosurfactants, the surface tension was drastically reduced to less than 30 dynes/cm². The results of typical surface tension measurements using the anionic Zonyl FSA are depicted in Figure 1.

Contact Angle

Variation of the hydrophobicity or hydrophilicity of coal surface in the presence of various fluorosurfactants was measured using the contact angle as an indicator. Contact angles on a Pittsburgh seam coal and a coal-pyrite were measured with a Gartner Scientific Goniometer using the captive bubble method (1,2).

Zeta Potential

The zeta potential of coal was measured with a Lazer Zee, Model 501, from Penkem, Inc. Sulfuric acid and sodium hydroxide were used for pH adjustment.

Flotation

In flotation tests, a 100-mL Pyrex glass cell with a fritted bottom for gas bubbling was used to float the coal. In each test, a 2-gram powdered coal (100 x 200 mesh) sample was first conditioned for one minute, then underwent a three-minute flotation. The froth and the residue were collected and weighed, and the flotation recovery was calculated.

RESULTS AND DISCUSSIONS
Surface Tension and Contact Angle Measurements

Fluorosurfactants used in this study can readily reduce the surface tension of aqueous solution from 73 dynes/cm to as low as 20 dynes/cm at ambient temperature. A typical plot of the effect of fluorosurfactant concentration on the surface tension, where the surface tension was reduced to 20 dynes/cm at the anionic fluorosurfactant concentration of 100 ppm, is shown in Figure 1. The other three fluorosurfactants studied exhibit the same

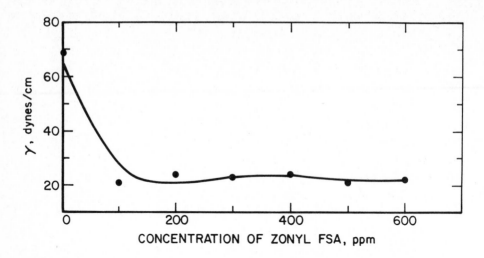

FIGURE 1. EFFECTS OF ANIONIC FLUOROSURFACTANT ON SURFACE TENSION IN DISTILLED/DEIONIZED WATER

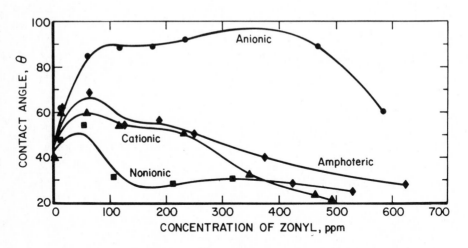

FIGURE 2. EFFECTS OF FLUOROSURFACTANT ON CONTACT ANGLE MEASURED ON PITTSBURGH SEAM COAL.

capability of reducing the aqueous surface tension to the 20 dynes/cm range. (These figures are not included in the paper.)

The effect of fluorosurfactant concentration on the contact angle of the coal is shown in Figure 2. The contact angles of the coal in the absence of the fluorosurfactants were around 38-42 degrees. In the presence of the anionic fluorosurfactant Zonyl FSA, the contact angle increased markedly, reaching a range of 80-90 degrees at an FSA concentration of 100 to 500 ppm.

The amphoteric Zonyl FSK and cationic Zonyl FSC showed only a moderate effect in increasing the contact angle, while the nonionic Zonyl FSN does not exhibit a noticeable effect in increasing the contact angle (Figure 2). In the figure, there is a common trend of decreasing contact angles as the concentration of these fluorosurfactants increases over 50 ppm. This is due to the side effect of increase in the alkalinity of the solutions.

For fluorosurfactants FSK, FSC, and FSN, the contact angles peaked at around 50 ppm concentration, then decreased to a point (<20 degree) where the air bubble would not adhere to the surface of the coal (see Figure 2).

The dependence of the contact angle on the pH is illustrated in Figure 3. All four fluorosurfactants at 100 ppm give the maximum contact angle around the neutral pH.

Judging from the relative magnitude of the contact angle in Figure 3, the effectiveness of enhancing the hydrophobicity of the coal, and hence the effectiveness of the fluorosurfactant adsorption on coal, can be estimated as follows:

Nonionic FSN < Cationic FSC < Amphoteric FSK < Anionic FSA

The mechanism of fluorosurfactant adsorption and the relative effectiveness of adsorption on coal can be inferred from contact angle measurements presented in Figure 3 and zeta potential measurements shown in Figure 4.

Three types of adsorption mechanisms are considered to be involved in the coal/fluorosurfactant system: a pure electrostatic attraction, hydrogen bonding, and an acid (carboxylic) to acid (carboxylic) dimerization.

For a nonionic fluorosurfactant, only the hydrogen bonding is considered to be operative during the adsorption. The ether-oxygen on the nonionic fluorosurfactant is responsible for the interaction with the hydroxyl group on the coal surface at a neutral pH, where the surface is near the point of zero charge and has a relatively high concentration of OH groups available for hydrogen bonding (3).

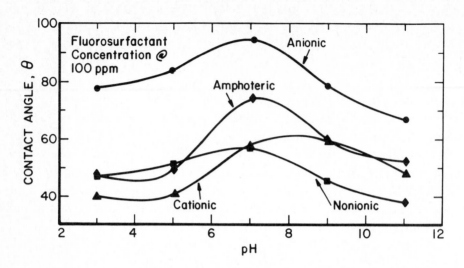

FIGURE 3.   EFFECTS OF FLUOROSURFACTANT ON CONTACT
ANGLE MEASURED ON PITTSBURGH SEAM COAL

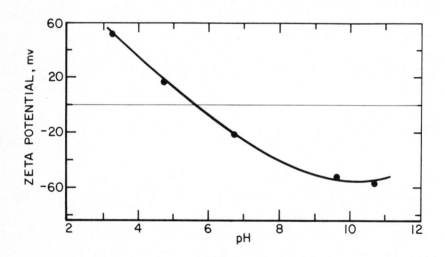

FIGURE 4.   ZETA POTENTIAL OF 100 X 200 MESH
PITTSBURGH SEAM COAL.

The zeta potential of the coal as a function of pH is shown in Figure 4. The point of zero charge, or point of zero mobility, is slightly shifted to the acid side owing to the effect of mineral matter and to a minor oxidation of the coal.

For a cationic fluorosurfactant, the contact angles in the acid range are smaller than those in the alkaline range. This is due to the negative adsorption caused by the charge repulsion between the positive charge on the coal surface and the positive charge on the cation of the cationic fluoro-surfactant in the acid pH medium. The cationic fluorosurfactant gives a slight increase in contact angle in the alkaline pH, indicating that the electrostatic attraction is operative in this pH range. However, their effectiveness is not significant as compared with that of the acid-to-acid dimerization attraction between the acid group on the coal surface (3) and the acid group on the fluorosurfactant. The adsorption through the acid-to-acid dimerization of the type (4)

$$
\text{COAL-C} \overset{\text{OH}\cdots\text{O}}{\underset{\text{O}\cdots\text{OH}}{\big\langle}} \text{C-SURFACTANT}
$$

is considered to be much stronger than the electrostatic or the hydrogen-bonding type adsorption for the coal and fluorosurfactant system. This acid-acid dimerization adsorption is mainly credited for the increase in the contact angles for anionic and amphoteric fluorosurfactant.

For an anionic fluorosurfactant, the electrostatic attraction and the acid-acid dimerization adsorption are operative in the acidic pH range; the hydrogen bonding and the acid-acid dimerization attraction are operative at neutral pH. The decrease in contact angle at the alkaline pH is presumably due to an increase in the repulsive force between the negative charge of the ionized carboxylate ion on the surfactant and the negative charge of the carboxylate ion on the coal surface.

The Spreading Coefficient

Spreading coefficient is a measurement of the spreading wetting of coal between the air and aqueous phases (5). This coefficient is defined as

$$
\text{Spreading Coefficient} = \gamma_{LA} \, (\cos \theta - 1)
$$

where γ_{LA} is the surface tension of the solution, and θ is the contact angle measured across the aqueous phase.

The effect of fluorosurfactants on the spreading coefficient on coal is shown in Figure 5. These spreading coefficients show a trend similar to that of the contact angle measurement in Figure 2. Clearly, the anionic fluoro-surfactant is most effective in removing the coal from the aqueous phase into the air phase, indicating a favorable situation for froth flotation where coal particles must transfer from the aqueous phase to the air phase.

Contact Angle of Pyrite

The contact angle measurements of pyrite, along with those of the coal for comparison, as a function of the anionic FSA fluorosurfactant concentration are shown in Figure 6. The anionic fluorosurfactant enhances equally the contact angle of the coal and pyrite, implying that the separation of coal and pyrite by flotation is not feasible without a special treatment of the pyrite.

To obtain a separation of coal from pyrite, either the coal or the pyrite must be selectively depressed through a chemical treatment. To achieve this, the depression of pyrite with sodium sulfide was tried out in the laboratory. The effectiveness of sodium sulfide in the selective depression of pyrite as measured by the contact angle is illustrated in Figure 7. In the figure, a substantial difference in the contact angle of the coal and pyrite was obtained. The decrease in the contact angle of the coal at a high sodium sulfide concentration is due to an increase in the pH associated with the addition of sodium sulfide.

Flotation

In a model study, a coal-pyrite powder (minus 100 mesh) was used in the tests to evaluate the effect of pH on its floatability in the presence of sodium sulfide. The pyrite is floatable near the neutral pH, but not floatable at a high pH or at a low pH (Figure 8). Strategically, the flota-tion of coal from pyrite at an alkaline pH is not favorable because the floatability of the coal will suffer (see Figure 7). Consequently, for efficient rejection of pyrite using the reagents studied in this research, coal flotation should be conducted at an acidic pH of less than pH 4, as implied in Figure 8.

A set of preliminary flotation tests was conducted to illustrate the feasibility of separating coal from pyrite using the fluorosurfactant. The results of a series of tests on a middle Kittanning coal from Butler County, Pa., are shown in Table 3. The results indicate that at a 38-42% weight recovery of coal, the rejection of total sulfur exceeded 80%. Because of this

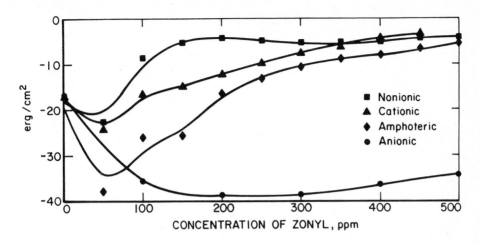

FIGURE 5. EFFECTS OF FLUOROSURFACTANT ON SPREADING COEFFICIENT OF PITTSBURGH SEAM COAL.

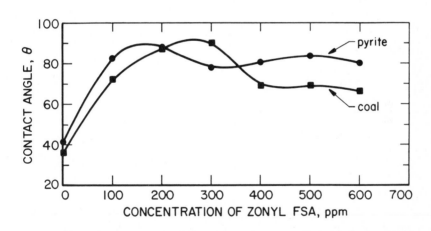

FIGURE 6. EFFECTS OF ANIONIC FLUOROSURFACTANT ON THE CONTACT ANGLE OF COAL AND PYRITE.

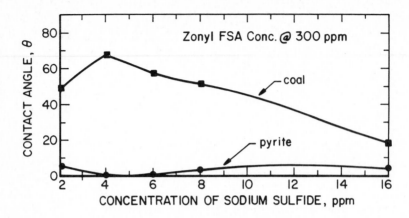

FIGURE 7. EFFECTS OF SODIUM SULFIDE ON THE CONTACT ANGLE OF COAL AND PYRITE.

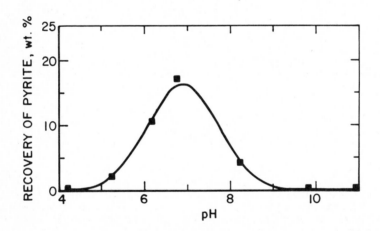

FIGURE 8. EFFECT OF PH ON PYRITE FLOTATION IN THE PRESENCE OF SODIUM SULFIDE AND ZONYL FSA.

encouraging result, more tests are planned to optimize the test conditions to improve the ultimate coal recovery and pyrite rejection in the laboratory.

TABLE 3. Flotation of Middle Kittanning Coal with Anionic Fluorosurfactant

Test	FSA ppm	pH	Product	Wt. %	S, %	% Dist. of Sulfur
1	300	1.76	Froth	39.9	1.93	14.3
			Tails	60.1	7.70	85.7
2	300	1.98	Froth	37.8	1.77	12.7
			Tails	62.2	7.37	87.3
3	600	2.03	Froth	41.9	2.31	18.4
			Tails	58.1	7.41	81.6
			Feed	100.0	5.32	

ACKNOWLEDGMENT

The authors thank R.A. Patton and P. Steffan for the experimental work.

REFERENCES

1　Gaudin, A.M., Flotation, 2nd ed., McGraw-Hill, New York, 1957, p. 148.
2　Rosenbaum, J.M., and D.W. Fuerstenau, On the Variation of Contact Angles with Coal Rank, Int. J. Min. Processing, 12, 1984, pp. 313-316.
3　Fuerstenau, D.W., M.C. Williams, and R.H. Urbina, The Physical and Chemical Properties of Coal and Their Role in Coal Beneficiation, Proceedings Int. Conf. on Coal Science, Oct. 28-31, 1985, Sydney, N.S.W. Pergamon Press, pp. 517-520.
4　Lai, R.W., W.W. Wen, and J.M. Okoh, Effect of Humic Substances on the Flotation Response of Coal, submitted for publication in Coal Preparation: A Multinational Journal.
5　Rosen, M.J., Surfactants and Interfacial Phenomena, John Wiley & Sons, 1978.

DISCLAIMER

Interfacial Phenomena in Biotechnology and Materials Processing,
edited by Y.A. Attia, B.M. Moudgil and S. Chander
Elsevier Science Publishers B.V., Amsterdam, 1988 — Printed in The Netherlands

FLOTATION OF OXIDIZED CHALCOPYRITE WITH HYDROXAMATE COLLECTORS

K.K. DAS AND PRADIP

Tata Research Development and Design Centre, 1, Mangaldas Road, Pune 411 001.
India.

SUMMARY

The surface oxidation of sulphide minerals is known to reduce their
flotation recovery with conventional thiol reagents. Freshly ground pure
chalcopyrite mineral samples when oxidised with oxidants such as glacial
acetic acid, ammonia and hydrogen peroxide, exhibit the expected
deterioration in its flotation response to xanthate collector. Potassium
octyl hydroxamate, on the other hand, has been demonstrated to float the
oxidised chalcopyrite very well. Hallimond tube test results indicate
two peaks in the flotation recovery namely at pHs 6 and 9. Based on our
results, a chemisorption mechanism has been proposed. The adsorption of
hydroxamate is believed to occur at pH 6 through the formation of copper
hydroxy complex at the surface and at pH 9 through the co-adsorption of
hydroxamate anions and neutral hydroxamic acid molecules. The electrokinetic
and surface tension results support the proposed mechanism.

INTRODUCTION

Sulphide ores containing chalcopyrite ($CuFeS_2$) are the main source of
copper in the world. These ores are beneficiated using thiol type flotation
collectors. A significant portion of world copper ore reserves are, however,
also in the form of oxide and mixed oxide/sulfide ore deposits. The flotation
recovery of oxide copper minerals from such ores, in particular, from mixed
sulfide-oxide ores is a difficult problem since oxide copper minerals do not
respond to conventional sulfide mineral flotation (ref. 1).

Although sulphidization treatment prior to flotation has been the most
commonly used method to make oxide minerals amenable to xanthate flotation, it
suffers from a basic drawback, that is, the sulphidizing agents like sodium
sulfide and hydrosulfide at higher dosages are also very effective depressants
for sulfide minerals. A close control of dosage as well as Eh conditions in
the pulp during flotation is, therefore, critical for the success of this
technique. An alternative method of overcoming this problem has been proposed
where more specific collectors consisting of copper chelating functional groups

are employed either alone or in conjunction with xanthates (refs. 2,8). Alkyl hydroxamates (refs. 3,4) have been reported to be promising collectors for oxide copper minerals (refs. 5-8).

Sulphide minerals are metastable and hence prone to oxidation and weathering (ref. 9). A process similar to what has been happening over geological periods, can also occur at the surface of chalcopyrite under certain conditions, giving rise to an oxidized surface similar in terms of flotation response, to other oxide copper minerals. For instance, at Malanjkhand copper ore plant in India, significant amounts of chalcopyrite in the ore is believed to be lost in tails due to such oxidation otherwise known as 'tarnishing'. A study has been undertaken in our laboratory to simulate this 'tarnishing' of chalcopyrite surface by deliberate oxidation of the mineral using oxidants such as hydrogen peroxide.

This paper presents the results on the flotation behaviour of oxidized chalcopyrite with conventional xanthate collectors and compares it with those obtained using a chelating type reagent like K-octyl hydroxamate. The superiority of hydroxamate collectors for such a system has been demonstrated to be due to the specific chelation of copper cations on the surface with adsorbing hydroxamate species. An adsorption mechanism has been proposed and substantiated with appropriate electrokinetic and surface tension data.

EXPERIMENTAL MATERIALS AND METHODS

Materials

Chalcopyrite samples were obtained from Wards, New York and Alminrock, India. Almost 99% pure chalcopyrite from Wards, had a characteristic bright golden yellow colour. The sample from Alminrock, India was cleaned by magnetic separation and repeated desliming. The purified chalcopyrite sample, however, remained dull, perhaps due to already tarnished surface. Malachite sample obtained through Wards, New York was 99% pure. Quartz sample used in the study was of white crystalline variety and was crushed, cleaned with dilute hydrochloric acid and washed with distilled water to get rid of iron contamination before use. All inorganic reagents used in this work were of analytical grade. Double distilled water of conductivity less than 1 micromhos/cm was used throughout the experiments.

The feed for flotation experiments was prepared by stage wise grinding of minerals in a planetary mill using agate bowls and balls. Minus 400 mesh fraction was finely ground and used for electrokinetic experiments.

Potassium ethyl xanthate was obtained from M/s Fluka Chemicals and further purified by standard method of repeated dissolution in acetone and precipitation by petroleum ether (ref. 10).

Potassium octyl hydroxamate was synthesized and characterized in the laboratory as per the standard method (refs. 3,5,11).

Methods

Flotation experiments were conducted in a modified Hallimond tube set up. 1 gm solid (- 48 + 65 mesh size), in 100 ml solution was conditioned separately in a shaker for 90 minutes and floated for 60 seconds at a nitrogen flow rate of 60 cc/minute. The pH of the solution was measured before and after the experiment.

Electrophoretic mobilities were measured at a particle concentration of 0.01% solids in a Zetameter 3.0 system. The conditioning time was kept constant at 24 hours. The specific conductance values reported here were also measured in the same setup. The conductivity of 10 ml of water and 10 ml of K-octyl hydroxamate of required concentration when added to a 30 ml of preadjusted pH water was measured. The difference in the specific conductance due to presence of hydroxamate was thus determined as a function of pH.

Surface tension measurements were carried out by standard Wilhelmy plate method.

RESULTS & DISCUSSION

Flotation Experiments

Wards chalcopyrite sample showed a very high natural flotability. In order to select an appropriate oxidant for our work, a number of oxidants were tried. Fig. 1 summarizes the flotation results obtained after oxidation with glacial acetic acid, ammonia, nitric acid and hydrogen peroxide. A treatment with 10% hydrogen peroxide solution was found adequate to completely suppress the natural flotability of the chalcopyrite sample and therefore, hydrogen peroxide oxidant was used in subsequent experiments.

Figure 2 shows the natural flotability of another chalcopyrite sample from Alminrock, India. As expected, after oxidation with H_2O_2, it does not exhibit any natural flotability. As illustrated in the figure, even as much as 100 ppm of K-ethyl xanthate is not sufficient to float this oxidized sample whereas unoxidized sample floats very well with xanthate. The sample, when subjected to hydroxamate flotation, exhibits high

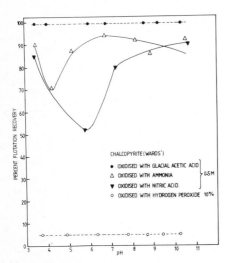

Fig. 1 . Effect of different oxidants
on the natural flotability of Wards
chalcopyrite

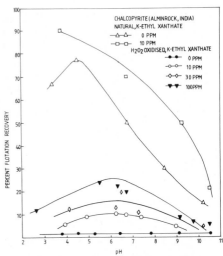

Fig.2. Flotation of chalcopyrite
(Alminrock) with K-ethyl xanthate
both natural and H_2O_2- oxidised.

recoveries at pH 6 and pH 9. The superiority of hydroxamate collector for
oxidized chalcopyrite has been illustrated in Fig. 3. With increasing
concentrations of hydroxamate the recovery is higher as expected (Fig. 4).
Above 36 ppm, almost 100% recoveries were obtained in the pH range of
6-10. It is interesting to note that at lower dosage of hydroxamate the
peak at pH9 is more prominent but this difference disappears at higher
dosages. The characteristic two peaks at pH6 and pH9 are also observed
for Wards chalcopyrite, even though a much lower concentration (only 15 ppm)
of hydroxamate is required to achieve 100% recoveries in this case (Fig. 5).

There thus exists a difference in the surface characteristics of
oxidized chalcopyrite samples from these two sources namely Wards and
Alminrock, India. Fig. 6 illustrates the degree of oxidation for the two
samples as compared to malachite $CuCO_3 . Cu(OH)_2$, a completely oxidized
copper mineral, in terms of their response to hydroxamate flotation at
pH9. Malachite is the least responsive and Wards the most, suggesting
that the surface of Alminrock sample oxidized under similar conditions has
undergone oxidation to a higher degree.

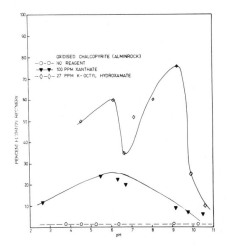

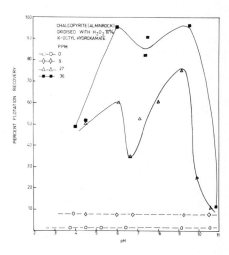

Fig. 3.   Flotation of H_2O_2 oxidised chalcopyrite with K-octyl hydroxamate and K-Ethyl xanthate collectors

Fig. 4.   Effect of collector dosage and pH on the flotation recovery of H_2O_2 oxidised chalcopyrite (Alminrock)

Electrophoretic mobilities of the chalcopyrite samples used in this work have been compared with those of cupric oxide and malachite in Fig. 7. The point of zero charge (PZC) for each of these samples were determined by measuring mobilities as a function of pH at three different ionic strengths-the other two curves are not shown in the figure for the sake of clarity. PZC values for chalcopyrite range anywhere between pH 2 to 3.5 (refs. 12,13). We have determined the PZCs of chalcopyrite samples from Wards and Alminrock to occur at pH 4.5 and 5.0 respectively. The PZC of malachite was observed to be pH 8.0, very close to the value of 7.9 reported by Lenormand etal (ref. 7). PZC of cupric oxide was even higher at pH 8.7. The mobility curve for native sulfur shown in Fig. 7 has been taken from literature (ref. 14).

Sulphide minerals tend to have PZCs below pH 4, but with increasing degree of oxidation, the PZC shifts towards alkaline pHs perhaps due to the formation of corresponding oxide, hydroxide and carbonate species on the surface (ref. 15). The electrokinetic data presented in Fig. 7 indicates that the Alminrock chalcopyrite sample (having higher PZC) is more oxidized

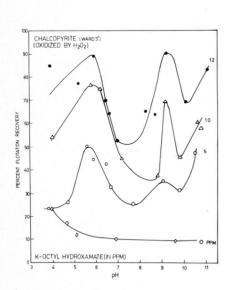

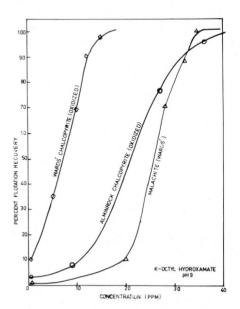

Fig. 5. Effect of collector dosage and pH on the flotation recovery of H_2O_2 oxidised chalcopyrite (Wards)

Fig. 6. Flotation recovery as a function of hydroxamate dosage at pH9 for oxidised chalcopyrite samples and malachite.

than that of Wards sample. This hypothesis is also consistent with their observed natural flotability in Hallimond tube. Furthermore, under similar conditions of oxidation, this difference in the extent of oxidation for the two samples is further accentuated as illustrated in Fig. 6, by their response to hydroxamate collectors.

Mechanism of Collector Interaction

Having thus established the efficacy of hydroxamate collectors for flotation of oxidized chalcopyrite otherwise not amenable to conventional xanthate flotation, some work was carried out to establish the mechanism of hydroxamate interaction with the oxidized chalcopyrite surface.

It is well known in literature that hydroxamate reagents form stable chelate complexes with copper and iron metal ions, the main constituents

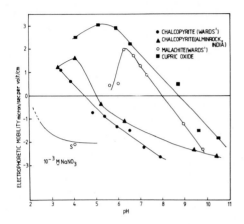

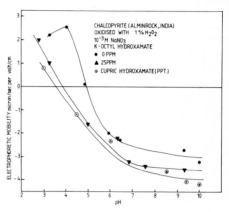

Fig. 7. Electrophoretic mobility as function of pH for chalcopyrite (Wards), chalcopyrite (Alminrock), malachite (Wards) and cupric oxide

Fig. 8. Effect of hydroxamate adsorption on the electrokinetics of oxidised chalcopyrite (Alminrock)

of chalcopyrite (refs. 4-6, 11). IR studies carried out on hydroxamate adsorption clearly confirm the formation of copper hydroxamate and Fe-hydroxamate at mineral/water interfaces (refs. 5,6,11). As illustrated in Fig. 8, electrokinetic studies also support the chemisorption mechanism for oxidized chalcopyrite/hydroxamate system. The characteristic shift in the isoelectric point (IEP) of the surface expected as a result of hydroxamate chemisorption is clearly evident in this case (refs. 16,17). Similar electrokinetic behaviour has been observed for Fe_2O_3/hydroxamate (ref. 11), MnO_2/hydroxamate (ref. 18) and malachite/hydroxamate (ref. 19) systems. The IEP of the surface progressively moves towards the PZC of the corresponding metal hydroxamate. Copper hydroxamate formed at the chalcopyrite surface imparts hydrophobicity to the mineral.

There are two distinct flotation peaks observed in this work, namely at pH6 and pH9. As discussed in the following paragraphs, these two peaks should be expected for all copper minerals when floated with hydroxamate collectors. The peak at pH9 is a characteristic of hydroxamate collectors and it has been observed for a number of mineral systems (refs. 3,4). Fuerstenau and Pradip (ref. 3) and Pradip (ref. 4) have recently reviewed the published literature on hydroxamate interaction with mineral surfaces.

The flotation results obtained in this work strongly suggest that alkyl hydroxamates chemisorb at mineral/water interfaces through two distinctly different phenomena namely, the hydrolysis of lattice cations in the mineral and the ionization of hydroxamate collector. The adsorption of hydroxamate at the oxidized chalcopyrite/water interface occurs through the following two mechanisms :

(i) The peak at pH9 is due to the adsorption of associative ion-molecular complexes of hydroxamate, very similar to those observed for oleate (ref. 22) and amine collectors (ref. 23). It has been observed that the adsorption and flotation peaks for such a system coincide with the pH at which the concentration of associative complexes is maximum, that is, at pHs closer to the pK of the collector when both the ionic as well as the neutral species are present.

The pK of K-octyl hydroxamate is reported to be around pH9 (ref. 2). It is therefore not surprising that associative complexes, $(RCONHOH)$ $RCONHO^-$ predominate at pH9. The surface tension of aqueous hydroxamate solution, measured as a function of pH also exhibits the characteristics minimum at pH9, substantiating the claim that the surface activity of these complexes is maximum at this pH (Fig. 9). Another supporting evidence for this association is presented in Fig. 10. The conductivity measurements show a peak at pH9, the characteristics of hydroxamate collectors. This peak at pH9 is observed irrespective of the mineral under consideration, provided there are metal ion sites available at the surface for chelation with hydroxamates.

(ii) Chemisorption and hence flotation peak at pH6, can be attributed to the presence of $CuOH^+$ complexes at this pH. Fuerstenau and co-workers (refs. 20,21) have clearly demonstrated that the adsorption of anionic collectors on minerals occurs through their metal hydroxy complexes. It is therefore observed that the pH of maximum adsorption and flotation for copper minerals coincides with the pH at which one would predict the maximum concentration of $CuOH^+$, based on the metal hydrolysis data available in the literature. The adsorption occurs as follows :

$$CuOH^+ + RCONHOH = RCONHOCu^+ + H_2O$$

$$CuOH^+ + RCONHO^- = RCONHOCu^+ + OH^-$$

or alternatively,

$$CuOH^+ + 2RCONHOH = (RCONHO)_2Cu + H_2O + H^+$$

$$CuOH^+ + 2RCONHO^- = (RCONHO)_2Cu + OH^-$$

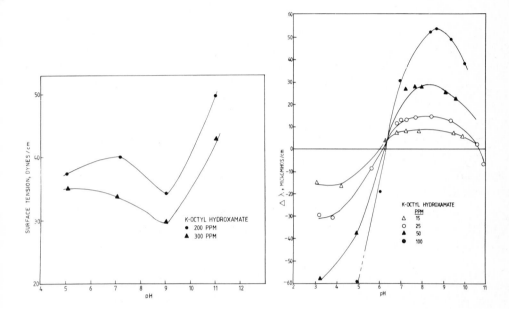

Fig. 9.   Surface tension as a function
of pH for aqueous K-Octyl hydroxamate
solutions

Fig. 10.   Effect of concentration and
pH on the specific conductance of
aqueous hydroxamate solutions.

It is important to note that metal hydroxy complexes assist
adsorption in the first layer.   At higher dosages the subsequent multilayer
adsorption may occur through hydrophobic bonding between the chains or through
hydrogen bonding between the functional groups (refs. 3, 11).   The sharp
increase in the flotation recoveries of oxidized chalcopyrite around pH6, at
higher concentrations of hydroxamate is indicative of such interactions.

In order to further confirm the mechanism of adsorption through the
formation of corresponding metal hydroxy complexes, pure quartz was
floated with and without activation, using soluble copper and iron salts
respectively.   The results are shown in Figs. 11 & 12.   Unactivated quartz
does not respond to hydroxamate flotation.   But with prior conditioning in
copper nitrate solution, the quartz floats very well, in particular, at
pH6 and pH9, the two characteristic peaks of oxidized chalcopyrite system.

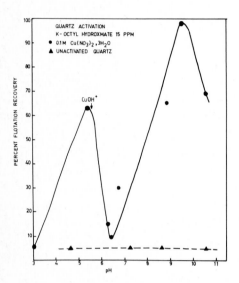

Fig. 11. Hydroxamate flotation of quartz with and without activation by Cu^{2+} ions.

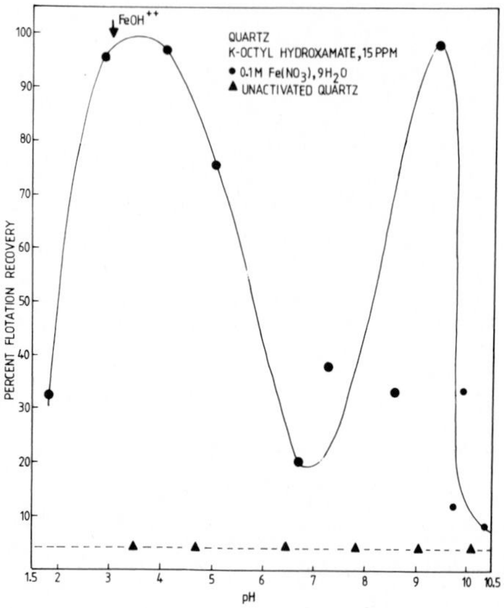

Fig. 12. Hydroxamate flotation of quartz with and without activation by Fe^{3+} ions.

For Fe^{+++}, as expected the peak shifts to pH3, the pH at which $FeOH^{++}$ predominates in the bulk.

It is to be noted here that the flotation peak at pH9, characteristic of the adsorption of associative ion-molecular complexes of hydroxamate is also observed in case of activated quartz. The presence of chelating metal ions at the surface is, therefore, essential for adsorption of even highly surface active ion-molecular complexes. These results are also consistent with the IR data on hydroxamate adsorption which indicates the formation of stable metal chelates at pH9. The ion-molecular association clearly helps adsorption but the basic interaction is still through the formation of covalent or co-ordinate bonding with the metal sites available on the surface.

In contrast to various other mineral systems where it is difficult to distinguish between the two mechanisms discussed above, the copper mineral/ hydroxamate system is of special importance. The pH of maximum concentration of $CuOH^+$ occurs at 5.5, which is clearly distinguishable from pH9, the pK of hydroxamate collectors. The presence of two distinct peaks at pH6 and pH9, therefore clearly establishes the validity of these two modes of hydroxamate adsorption which may in some cases be indistinguishable.

CONCLUSIONS

Chalcopyrite, after deliberate oxidation with aqueous hydrogen peroxide solution, does not respond to xanthate flotation. Potassium octyl hydroxamate, a chelating type reagent, however, has been found to be an effective collector for oxidized chalcopyrite. Flotation test results indicate two distinct peaks in recovery namely at pH6 and 9. Based on our results a chemisorption mechanism is proposed. Electrokinetic data also supports chemisorption mechanism. The adsorption of hydroxamate occurs at pH6 through the formation of copper hydroxy complexes at the surface and at pH9, through the co-adsorption of hydroxamate anions and neutral hydroxamic acid species. The surface tension measurements of aqueous hydroxamate solutions exhibit a minimum around pH9, which further confirms the presence of strongly surface active ion-molecular complexes in solution. Conductivity measurements also support the phenomena of ion-molecular association around pH9.

ACKNOWLEDGEMENTS

The authors wish to thank the Department of Science and Technology (DST), India for support of this research, under an Indo-U.S. STI programme. Thanks are due to Dr. R.A. Kulkarni and Mr. N. Natarajan of National Chemical Laboratory for their assistance in surface tension measurements. The authors are grateful to Dr. E.C. Subbarao, Director, Tata Research Development & Design Centre, for his constant encouragement and keen interest throughout the course of this work. Discussions with Dr. N.K. Khosla are gratefully acknowledged.

REFERENCES

1   D.W. Fuerstenau and S. Raghavan, Surface Chemical properties of oxide copper minerals in : P. Somasundaran (Ed.), Advances in Mineral Processing, Arbiter Symposium proceedings, AIME publication, Chap. 23, 1986, pp 395-407.
2   D.R. Nagraj and P. Somasundaran, Chelating agents in flotation: oximes-copper minerals system, Mining Engg., Sept. 1981, pp 1351-1357.
3   D.W. Fuerstenau and Pradip, Mineral flotation with hydroxamate collectors, in : M.J. Jones and R. Oblatt (Eds.), Proc. Reagents in Mineral Industry, IMM pub., 1984, pp. 161-68.
4   Pradip, Surface chemistry and applications of alkyl hydroxamate collectors in mineral flotation, Trans. IIM, 40(4), 1987, pp 287-304.

5   H.D. Peterson, M.C. Fuerstenau, R.S. Pickard and J.D. Miller, Crysocolla flotation by the formation of insoluble surface chelates, Trans. AIME, 232, 1965, pp. 388-92 and U.S. Patent 3,438, 494 (1969)

6   L. Evrard and J. De Cuyper, Flotation of copper-cobalt oxide ores with alkyl hydroxamates, in : Proc. 11th International Mineral Processing Congress, Cagliari, Italy, 1975, pp 655-69.

7   J. Lenormond, T. Salman and R.H. Yoon, Hydroxamate flotation of malachite, Canadian Metallurgical Quarterly, 18, 1979, pp 125-129.

8   K. Dekum, Chen Jingquing and Z. Weizhi, Application of hydroxamic acid and hydroxamic-xanthate collector system in metal ore flotation, in : M.J. Jones and R. Oblatt (Ed.), Proc. Reagents in Minerals Industry, IMM pub., 1984, pp 169-172.

9   A.N. Buckley and R. Woods, An x-ray photoelectron spectroscopic study of the oxidation of chalcopyrite, Australian J. Chem., 37, 1984, pp. 2403-13.

10   S.R. Rao, Xanthates and related compounds, Marcel Dekker pub., New York, 1971.

11   S. Raghavan and D.W. Fuerstenau, The adsorption of aqueous octyl hydroxamate on ferric oxide, J. Colloid and Interface Sci., 50(2), 1975, pp 319-330.

12   J.G. Groppo Jr. and R.H. Yoon, Selective flotation of sulfide ores using alkyl pyridinium collectors, in : B. Yarar and D.J. Spottiswood (Ed.) Proc. Interfacial Phenomena in Mineral Processing, Engineering Foundation Conference, New Hampshire, 1981, pp. 271-285.

13   D. McGlashen, A. Rovig and D. Podobnic, Assessment of interfacial reactions of chalcopyrite, Trans. AIME, 244, 1969, p. 446.

14   M.S. Moignard, D.R. Nixon and T.W. Healy, Electrokinetic properties of the zinc and nickel sulfide-water interfaces, Proc. Australasian Inst. Mining and Metallurgy, 1976.

15   T.W. Healy and M.S. Moignard, A review of electro-kinetic studies of metal sulphides in : FLOTATION- A.M. Gaudin memorial volume, no. 1, (Ed.) M.C. Fuerstenau, Chap. 9, pp 275-297, 1976.

16   Pradip, On the interpretation of electrokinetic behaviour of chemisorbing surfactant systems, "Trans. IMM, in press".

17   K. Osseo-Asare and D.W. Fuerstenau, Sulfonate adsorption and wetting behaviour at solid-water interfaces, Croatica Chemica Acta 45, 1973, pp 149-61.

18   R. Natarajan and D.W. Fuerstenau, Adsorption and flotation behaviour of manganese dioxide in the presence of octyl hydroxamate, Int. J. Mineral Processing, 11, 1983 pp 139-53.

19   Pradip and K.K. Das, unpublished results.

20   M.C. Fuerstenau, R.W. Harper and J.D. Miller, Hydroxamate vs. fatty acid flotation of iron oxide, Trans AIME, 247, 1970, 69-73.

21   B.R. Palmer, G. Gutierrez and M.C. Fuerstenau, Mechanism involved in the flotation of oxides and silicates with anionic collectors, parts 1 & 2, Trans. AIME, 258, 1975, 257-60.

22   P. Somasundaran and K.P. Anantpadmanabhan, Solution Chemistry of flotation, in : P. Somasundaran (Ed.) Advances in Mineral Processing, Arbiter symposium proceedings, AIME, Chap. 8, 1986, pp. 137-153.

23   D.W. Fuerstenau, Correlation of contact angles, adsorption density, zeta potentials and flotation rate, Trans. AIME, 208, 1957, pp. 1365-7.

Interfacial Phenomena in Biotechnology and Materials Processing,
edited by Y.A. Attia, B.M. Moudgil and S. Chander
Elsevier Science Publishers B.V., Amsterdam, 1988 — Printed in The Netherlands

EFFECTS OF PROCESS PARAMETERS ON THE SELECTIVE FLOCCULATION CLEANING OF UPPER
FREEPORT COAL

Y.A. ATTIA and K. DRISCOLL

The Ohio State University, Department of Metallurgical Engineering, Division
of Mining Engineering, 116 W. 19th Ave., Columbus, Ohio 43210 USA

SUMMARY
    The purpose of this study is to investigate the effects of process
parameters and to define conditions for "optimum" performance of the selective
flocculation process. The parameters investigated were: velocity gradient,
polymer dispersion time, rate of polymer addition, polymer stock solution
concentration, slurry pH, slurry solids content, and floc conditioning time.
Using "optimized" parameters with a pre-cleaned coal, the ash content of the
coal was reduced from 14.8% to 5.6% in a single-step, and to 2.7% when a
two-step process was employed. This research demonstrates that super-clean
coal can be produced using a two-step selective flocculation process, when
used in conjunction with a pre-cleaning technique such as gravity separation.

INTRODUCTION

    Selective flocculation is one of the newest and most promising techniques
for fine particle separation. The process uses differences in surface
chemical properties of the mixed particles to effect their separation. In
this process, a polymeric flocculant selectively absorbs on the material to be
flocculated, leaving all other material suspended in the slurry. The
flocculated particles are then separated from the slurry by differential
gravitational settling, or some other more appropriate technique.

    The process has been applied at the laboratory, pilot plant and commercial
scale on several systems, which include copper ores (ref 1.), cassiterite
(ref. 2), coal (refs. 3 and 4), potash (ref. 5), and iron ore (ref. 6). A
recent review by Yu and Attia (ref. 7) presented several applications of
selective flocculation in the mineral industry.

    Selective flocculation is a feasible process for super-cleaning (less than
3% ash and 1% sulfur) of ultrafine coal slurries for coal-water fuels (ref.
8). However, to achieve this level of cleaning, the process has to be repeated
several times. Each time, more ash is rejected from the coal. Some of the
ash minerals are typically entraped within the flocs, or entrained in the
flocculated fraction during separation. In conclusion, if the process is well
designed and the conditions for optimum selective flocculation and floc
separation were identified, super-cleaning of coal slurries might be readily

achieved with a minimum amount of effort and cost. Hence the objectives of
this study were to investigate the effects of several important process parameters on the performance of selective flocculation and delineate the conditions for "optimum" performance of this process. The parameter studies were
the velocity gradient during polymer addition, polymer dispersion time, rate
of flocculant addition, polymer stock solution concentration, slurry pH,
slurry solids content, floc conditioning time, and flocculant dosage.

EXPERIMENTAL TECHNIQUES
Chemicals and Minerals

(i) Chemicals. The flocculant used throughout the research was FR-7, a
totally hydrophobic flocculant obtained from Calgon Corporation, Pittsburgh,
Pennsylvania. The polymer is an oil-in-water emulsion, and therefore is water
dispersible. It has a molecular weight of less than one million. A
dispersant was used to maintain a stable coal suspension. The dispersant used
was a 500 mg/l concentration of reagent grade sodium metaphosphate (SMP),
supplied by Fisher Scientific. Sodium hydroxide and hydrochloric acid
solutions were used to adjust the slurry pH.

(ii) Minerals. Upper Freeport coal samples were obtained from the
O'Donnell Mine, No. 3, Armstrong County, Pennsylvania, operated by the
Rochester and Pitt Coal Company, Indiana, Pennsylvania, The coal slurry was
prepared from one kilogram of coal ground with two kilograms of distilled
water containing 3000 mg/l of SMP. The mixture had a pH of approximately
10.0. The high SMP concentration was used during grinding so that when the
slurry was diluted to five percent solids, the SMP concentration would be 500
mg/l. An attrition mill, supplied by the Fort Pitt Mine Machine Co.,
Pittsburgh, Pennsylvania was used to grind the coal to minus 500 mesh (25
microns).

EXPERIMENTAL APPARATUS

Mixing equipment. The mixing studies were performed in a baffled 400 ml
beaker (shown schematically in Fig. 1). The beaker diameter was 7.45 cm,
yielding a volume of 325 cubic centimeters. The impeller diameter was 2.48
cm; the baffle width was 0.75 cm; the impeller diameter was 0.75 cm; the
impeller blade width was 0.5 cm, and the blade length was 0.6 cm. These sizes
were determined using the relative dimensions as shown in Fig. 1.

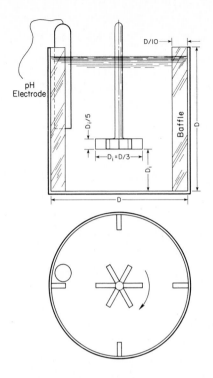

Fig. 1. Schematic Diagram of the Backflow Mixer.

The mixing device used was a T-line mixer, Model 102, manufactured by Tallboy Engineering Co., and supplied by Sepor Inc., Wilmington, California. The mixer included a rheostat by which the power supplied to the motor could be varied. The rheostat setting (0 to 100 percent maximum voltage) was calibrated to impeller speed.

PROCEDURES

(i) Selective Flocculation - Baseline Conditions. Flocculation tests on a raw Upper Freeport coal slurry were conducted using arbitrary initial conditions. The slurry was agitated for five minutes at a shear rate of 425 sec^{-1}, and the pH was adjusted to 7.0 using sodium hydroxide and hydrochloric acid solutions. A 10 mg/l dose of the polymer FR-7 was then added to the slurry from a stock solution at a concentration of 0.1% by weight (1000 mg/l). The required polymer dosage was added over a one minute interval at a shear rate of 425 sec^{-1}, and then allowed to disperse for an additional minute at the same shear rate. The floc suspension was conditioned at low shear (170 sec^{-1}) for three minutes. The flocculated slurry was then allowed to stand

for five minutes to allow the flocs to settle. After settling, the flocs were
separated from the dispersed fraction by decantation. The flocculated and
dispersed fractions were dried, weighed, and their ash contents determined by
ASTM recommended procedures. The standard baseline operating conditions for
the selective flocculation experiments were based on prior research not
presented here, and the work done by Attia and Yu (ref. 4). An additional
criterion was imposed on the baseline conditions: the parameter was set near
the middle of the range over which it would later be varied. The following
table lists the standard conditions for the various parameters investigated in
this study.

TABLE 1.
Standard Conditions for Selective Flocculation Experiments.

VARIABLE	VALUE
Slurry pH	7.0
Flocculant Addition Time	1 min.
Polymer Dispersion Time	1 min.
Floc Conditioning Time	3 min.
Shear Rate During Polymer Addition	425 sec^{-1}
Solids Content of Suspension	5 wt%
Polymer Dosage	0.2 mg/g
Polymer Stock Solution Concentration	0.1 wt%

(ii) Parametric Studies. Eight parameters were studied in the course of
the investigation; slurry pH, suspension solids content, polymer dispersion
time, stock solution concentration, floc conditioning time, shear rate during
polymer addition, polymer dosage, and rate of polymer addition. Slurry pH was
studied by varying the pH during flocculation from 5.0 to 11.0. The work by

TABLE 2.
Range of Variation of the Mixing Parameters.

VARIABLE	RANGE OF VARIATION
Slurry pH	5 to 10
Flocculant addition time	10 to 120 sec.
Flocculant dispersion time	0 to 120 sec.
Floc conditioning time	0 to 5 min.
Shear rate during polymer addition	170 to 890 sec^{-1}
Solids content of suspension	1 to 15 wt%
Polymer dosage	0.02 to 0.4 mg/g
Polymer stock solution concentration	0.01 to 1.0 wt%

Attia and Yu (ref. 4) had indicated that a neutral to slightly basic pH (7.0 to 8.5) was preferred for the selective flocculation of coal with FR-7. This investigation concentrated on this range. Table 2 lists the range of variation for the eight process parameters investigated.

RESULTS AND DISCUSSION

This section details the experimental results for the eight process parameters. Where possible, the apparent optimum value for the parameter was identified. There were several cases however, where two optima were observed; one that would maximize recovery[1], and one that would maximize upgrading[2]. It should be remembered that the ultimate goal of this study was to produce a super-clean coal in a minimum number of cleaning stages. Therefore, the optimum value presented was the one that would maximize the upgrading, while still maintaining a high coal recovery.

Effect of pH

Fig. 2. shows a plot of the process performance as a function of slurry pH. It can be seen that the maximum recovery and upgrading occurred in the range of pH 7 to 8, and that both upgrading and recovery decreased as the pH was raised or lowered. The low upgrading at low pH was caused by several possible factors. First, the shale minerals were approaching their point of zero charge (pzc = 3 to 5) where the minerals have a less negative surface charge, which increased the possibility of coagulation. In addition, the hydrophobic flocculant probably began to absorb on the shale minerals as well as the coal, that is the flocculant might have no longer been selective. Additionally, the dispersant, SMP, was not completely ionized in the acidic pH range, thus the suspension may not have been as well dispersed as at higher pH values. The reduced recovery may have been a result of the high hydrogen ion concentration causing a change in the polymer structure, e.g., coiling. In the acidic solution, the polymer would be more tightly coiled since the hydrophilic portion of the polymer (an acrylic acid derivative) was not as strongly ionized. This would reduce the effective length of the polymer, thus reducing the ability of interparticle bridging. With fewer particle-to-particle bridges, the flocs that formed were smaller, and therefore more difficult to separate, due to their lower terminal settling velocities.

---

1 Recovery = (Coal weight in flocs / coal weight in feed) x 100%
2 Upgrading = (Feed ash % - floc ash % / feed ash %) x 100%

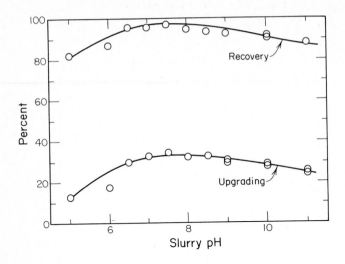

Fig. 2.   The Effect of Slurry pH on Coal Recovery and Upgrading.

At high pH values (pH greater than 9), the lower recovery was a result of excess negative charges on the coal surface. Regardless of how fresh the starting coal, some oxidation usually takes place during grinding and handling. The presence of the negative surface charge would reduce the adsorption of the negatively charged flocculant on the coal particles. Obviously, if no flocculant absorbed on the particle surface, it could not be incorporated into a floc, thus reducing the recovery. The lower upgrading seemed to contradict what is currently known about the process. It would seem that if the coal particles were negatively charged along with the shale minerals, flocculant absorption would be even more selective, promoting a higher floc grade, not a lower one. This phenomenon requires further investigation.

Effect of Solids Content

The effect of solids content on selective flocculation is shown in Fig. 3. This figure shows that as the solids content of the slurry was increased, the recovery and upgrading decreased. The high solids content during flocculation produced a higher rate of floc formation, a condition which is known to produce finer flocs (ref. 9). The smaller flocs have a lower terminal velocity, so not all of the flocs settle in the allotted time. The reduction in upgrading was caused by entrainment of ash-forming minerals during separation. Although the flocs were small and had less material entrapped within them, there was a large amount of mineral matter in the water surrounding the flocs which was not separated when the dispersed phase was decanted from the floc phase. In the range of values studied, the "optimum" appeared to be one percent solids by weight. However, operation at this low of a solids content was not practical. Therefore, 5 percent by weight was recommended since the difference in process performance was almost negligible.

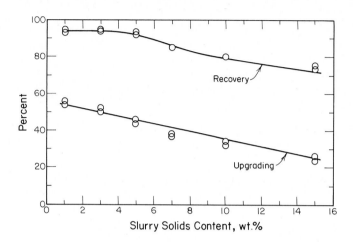

Fig. 3. The Effect of Weight Percent of Solids in Slurry on Process Performance.

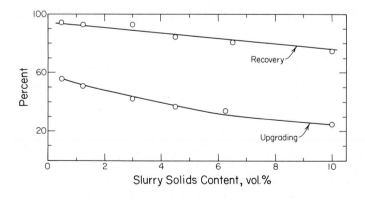

Fig. 4. Process Performance as a Function of Volume Percent Solids.

It is interesting to note that when the data is transformed to volume percent, using the method of Mayer (ref. 10) and plotted in Fig. 4, the relationship is almost linear. This is due to the fact that the process depends on the amount of surface area available for flocculant absorption,

324

and not strictly on the mass of coal present. The volume fraction of coal in
the slurry is more closely related to the total surface area and the number of
particles in the slurry than the mass fraction of coal.

Effect of Polymer Stock Solution Concentration

The effect of the polymer stock solution concentration, as shown in Fig. 5,
was to increase the upgrading and decrease the recovery as the concentration
was increased. These effects were attributable, again, to the rate of floc
formation. At higher stock solution concentrations, the flocs formed faster,
and therefore were smaller as a result of the mixing conditions employed in
these test. These results are similar to those obtained by Kogan, et al (ref.
9) in their study using synthetic coal feed.

This is one of the parameters for which there are two optima. If the
desired goal was a high recovery, then the lower stock solution concentration
would be best. However, in this case, the desire was to maximize the
upgrading, so a high stock solution concentration was selected. Based on the
data presented, a 0.5 weight percent polymer solution appeared to be the
"optimum", since further increase in the concentration produced no change in
process performance.

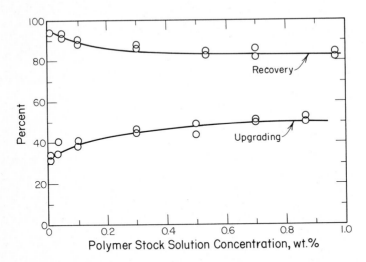

Fig. 5. The Effect of Polymer Stock Solution Concentration on Process
Performance.

Effect of Shear Rate

There is some disagreement as to whether the shear rate (velocity gradient)
is a valid parameter for flocculation processes. Other researchers disagree
with the use of the velocity gradient G as a flocculation parameter for

various reasons. Cleasby (ref. 11) argues that the use of G is only valid if the particles being flocculated are smaller than the Kolmogoroff microscale. The Kolmogoroff microscale is an internal scale of local turbulence (ref. 16) which is defined as:

$$\eta = (\nu^3/\varepsilon)^{1/4}$$

where: η = Kolmogoroff microscale
ν = Kineratic viscosity
ε = Energy Dissipation rate

He advocated the use of the energy dissipated per unit volume raised to the two-thirds power for systems in which the particles are larger than the microscale. Others argue that in a backflow mixer, the calculated velocity gradient is based upon the impeller tip speed, and this shear rate is not necessarily the average shear rate in the mixer (ref. 12). There are two reasons why the velocity gradient was used, in spite of these objections. First, the actual physical phenomenon that is responsible for floc breakage is shear stress, which is directly related to the shear rate, and G is an acceptable estimation of the shear rate. Secondly, the authors agree with Argman and Kaufman (ref. 13) that G is a parameter that is easily obtainable and familiar to most engineers.

The velocity gradient expression is based on the basic shear rate equation (ref. 14) and the following equation is obtained:

$$G = (N_p D_i^5 \rho/g_c \mu V)^{1/2} N^{3/2}$$

Where N_p = Power number

D_i = Impeller diameter

N = Revolutions per second

g_c = A dimensional constant

ρ = Fluid density

and μ = Viscosity

The complete derivation of this expression is presented elsewhere (ref. 9).

The effect of shear rate on the process performance is shown in Figure 6. It can be seen that by increasing the shear rate, a slight increase in process performance can be obtained. However, the effect is so small that the additional costs connected with the incresed shear rate would not seem to be worthwhile. Therefore, the "optimum" shear rate would seem to be the lowest investigated, 170 sec^{-1} under the current test conditions.

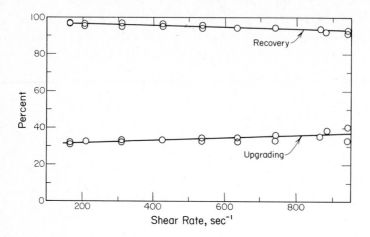

Fig. 6   The Effect of Shear Rate on Coal Recovery and Upgrading.

Effect of Rate of Polymer Addition

The effects of the rate of polymer addition, shown in Fig. 7, were to increase the upgrading and decrease the recovery as the rate of addition decreased (increased time of addition).   These effects were the result of the longer addition time promoting floc breakup and re-dispersion, followed by re-flocculation.   This can lead to what some authors (ref. 15) have called secondary adsorption.   In secondary adsorption, the polymer "wraps itself" around a floc fragment, producing stable floc fragments that are smaller

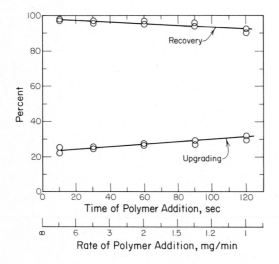

Fig. 7.   The Effect of Rate of Polymer Addition on Process Performance.

than the original floc. As mentioned earlier, small flocs have lower terminal velocities, which would reduce their recovery, but allow for higher upgrading by reducing the amount of mineral matter entrapped within the floc structure.

The effects of polymer addition time on process performance were rather small. The standard condition of one minute addition time is apparently as good as any other value. Additionally, this value has some experimental convenience, so it was used as the "optimum" value for this parameter.

Effect of Polymer Dosage

Fig. 8 shows the effects of the polymer dosage on selective flocculation. The effects were to increase the recovery and to decrease the upgrading as the dosage was increased. The high recovery was a result of the larger amount of polymer-polymer bridges that were formed. This produced larger flocs with higher settling velocities, thus promoting a higher recovery. However, since the flocs were larger, there was a larger amount of gangue minerals trapped within them, thus reducing the floc grade. Since the polymer dosage had such a strong effect on the floc grade, especially at lower dosages, the "optimum" would seem to be a low dosage, 0.02 kg/ton (0.02 mg/g).

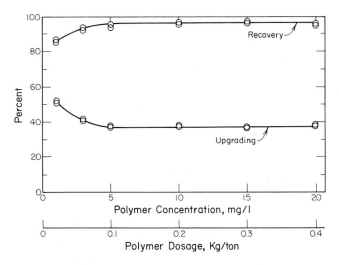

Fig. 8. The Effect of Polymer Dosage on Coal Recovery and Upgrading.

Effect of Polymer Dispersion Time

The effects of polymer dispersion time, as shown in Fig. 9, were to reduce the recovery and increase the upgrading with increased dispersion time. These effects are similar to those seen for the rate of addition, and are the results of the repeated breakup and reformation of the flocs which lead to the formation of fine flocs. As mentioned earlier, the fine flocs produce higher upgrading, but lower recovery due to their lower terminal velocity. In this figure, the effects at two levels of shear are shown (425 and 740 sec^{-1}). The effect of the higher shear rate was to enhance the upgrading, without significantly changing the recovery. For these reasons, a "high" shear rate for polymer dispersion, along with relatively long dispersion times, should be used. The "optimum" dispersion time seems to be 90 seconds, since the difference between 90 and 120 seconds was relatively small.

Effect of Conditioning Time

The effect of floc conditioning time is shown in Fig. 10. It can be seen that some conditioning of the flocs is necessary to promote both high upgrading and recovery. The increase in upgrading is believed to be due to the gentle shearing which causes the floc structure to collapse to a more compact structure, while squeezing out gangue minerals. The increase in recovery was the result of the number of collisions between independent particles (or small flocs) and larger flocs. This allows the smaller flocs and independent particles to be incorporated into the larger flocs.

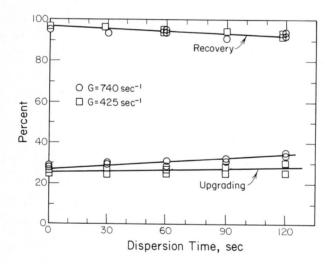

Fig. 9. Coal Recovery and Upgrading as a Function of Polymer Dispersion Time.

The "optimum" conditioning time (at this shear rate of 170 sec^{-1}) seems to be approximately 2.5 to 3 minutes. After this time, there was no additional improvement in process performance.

Results Using the Optimized Parameters

Once the "optimum" values for the various parameters were determined, it was necessary to ascertain whether or not the process did, indeed, perform optimally using these conditions. Listed in Table 3 are the "optimum" values for the parameters investigated.

To test these conditions, two sets of selective flocculation experiments were performed: one using a single-step selective flocculation, and the other a two-step selective flocculation process. In the two-step process, the flocs were diluted to 5% solids, and redispersed for 10 minutes at a shear rate of 1100 sec $^{-1}$. The high shear rate was used to ensure that all the flocs were ruptured, and to promote breakage of the flocculant used in the first stage.

TABLE 3.
Optimum Values for the Mixing Parameters.

PARAMETER	VALUE
Polymer dosage	0.02 mg/g
Time of polymer addition	90 sec.
Slurry solids content	5 wt%
Slurry pH	7.5
Polymer stock solution concentration	0.5 wt%
Polymer dispersion time	60 sec.
Shear rate	740 sec^{-1}
Floc conditioning time	3 min.

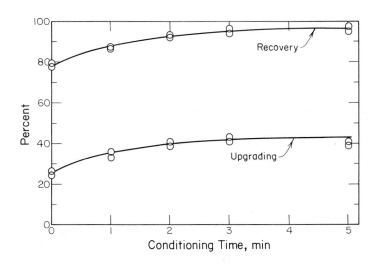

Fig. 10. The Effect of Floc Conditioning Time on Coal Recovery and Upgrading.

The results of these experiments are shown in Tables 4 and 5. In order to test the applicability of the process to industrial use, the coal used in these tests was first precleaned using a heavy liquid (tetrachloroethylene) separation at a specific gravity of 1.62. These results show that with one step of selective flocculation, about 67% of the ash-forming minerals can be

TABLE 4.
Process Performance Using Optimized Parameters In a Single-Step Flocculation.

Product	Weight %	ANALYSIS Ash %	Coal %	DISTRIBUTIONS Ash %	Coal %
Flocs	87.0	5.59	94.41	32.8	96.4
Dispersed	13.0	76.42	23.58	67.2	3.6
Feed	100.0	14.81	85.19	100.0	100.0
Flocs	87.7	5.45	94.55	33.5	96.7
Dispersed	12.3	77.24	22.76	66.5	3.3
Feed	100.0	14.27	85.73	100.0	100.0
Flocs	85.4	5.72	94.28	32.8	94.6
Dispersed	14.6	68.44	31.56	67.2	5.4
Feed	100.0	14.88	85.12	100.0	100.0

TABLE 5.
Process Performance Using Optimized Parameters in a Two-Step Selective Flocculation.

Product	Weight %	ANALYSIS Ash %	Coal %	DISTRIBUTIONS Ash %	Coal %
Flocs	80.9	2.73	97.27	14.9	92.4
Dispersed 1	11.9	71.78	28.22	57.6	3.9
Dispersed 2	7.2	56.53	43.47	27.5	3.7
Feed	100.0	56.53	43.47	27.5	3.7
Flocs	78.9	2.85	97.15	15.1	90.0
Dispersed 1	10.5	73.24	26.76	51.8	3.3
Dispersed 2	10.6	46.34	53.66	33.1	6.7
Feed	100.0	14.85	85.15	100.0	100.0
Flocs	80.5	2.77	97.23	15.4	91.5
Dispersed 1	12.1	72.12	27.88	60.4	3.9
Dispersed 2	7.4	47.29	52.71	24.2	4.6
Feed	100.0	14.45	85.55	100.0	100.0

removed, while retaining 95% of the feed coal. The mineral content of the coal was reduced to 5.5%, very close to the super-clean level. When a second step was employed, the mineral content was reduced to 2.7%, while retaining 90% coal recovery.

CONCLUSIONS AND RECOMMENDATIONS

It is obvious that the selective flocculation process is capable of producing super-clean coal. The conditions presented in Table 3 yielded excellent results using a single step selective flocculation. A pre-cleaned coal with an ash content of 14.7% was reduced to 5.5% with 95% of the coal recovered. When a second step of selective flocculation was used, the ash content of the cleaned coal was reduced to 2.7%, with a coal recovery of 90.2%.

Further work is required to completely optimize the process. The conditions presented here represent a technical optimum, which does not guarantee that the conditions are also economical. At this point, an economic analysis of the process is needed to determine whether or not the process is capable of producing super-clean coal in a cost effective manner. Gross economic assessment, however, indicates that this process would be highly viable for commercial production.

ACKNOWLEDGEMENTS

The Fellowship to Keith Driscoll provided by the Ohio Mining and Mineral Resource Research Institute is gratefully acknowledged. The authors would like to thank Babcock and Wilcox Company, Alliance Research Center, for their help in obtaining the coal samples and their interest in, and support of, this project.

REFERENCES

1 Y.A. Attia, Development of a Selective Flocculation Process for a Complex Copper Ore, Int. J. Min. Pros., Vol. 4, 1977, pp. 209-225.
2 C. Clauss, E. Appleton, and J. Vink, Selective Flocculation of Cassiterite in Mixtures with Quartz Using a Modified Polyacrylamide Flocculant, Int. J. Min. Proc., Vol. 3, No. 1, 1976, pp. 27-34.
3 V.I. Motzgovi, Application of the Sodium Salt of Sulfonated Polystyrene for Selective Coagulation of Coal Suspensions, IV Vyssh. Zavd, Gorn, ZH 4, 1969.
4 Y. Attia, S. Yu, and S. Vecci, Selective Flocculation of Upper Freeport Coal with a Totally Hydrophobic Polymeric Flocculant, in: Y.A. Attia (Ed.), Flocculation in Biotechnology and Separation Systems, Elsevier, 1987, pp. 547-64.
5 A.F. Banks, Selective Flocculation-Flotation of Slimes from a Sylvinite Ore, in: Benficiation of Mineral Fines, NSW Workshop, 1979.
6 R. Sisselman, Cleveland Cliffs Takes the Wraps Off Revolutionary New Tilden Iron Ore Process, Eng. Min. J., V 176, No. 10, pp. 79-84.
7 S. Yu, and Y. Attia, Review of Selective Flocculation in Mineral Separations, in: Y.A. Attia (Ed.) Flocculation in Biotechnology and Separation Systems, Elsevier, 1987, pp. 601-37.
8 Y.A. Attia, H.N. Conkle and S.V. Krishnan, Selective Flocculation Coal Cleaning for Coal Slurry Preparation, 6th Int. Coal Slurry Combustion and Tech. Symp., Orlando, Florida, 1977.
9 V. Kogan, K. Driscoll, and Y. Attia, Polymer Mixing in Selective Flocculation, in: Y.A. Attia (Ed.), Flocculation in Biotechnology and Separation Systems, Elsevier, 1987, pp. 321-34.

10 C.W., Dell, The Composition of Suspensions, Colliery Engineering, 1961, pp. 401-5.
11 J.L. Cleasby, Is Velocity Gradient a Valid Flocculation Parameterπ, Journal of Environmental Science, Vol. 110, No. 5, 1984, pp. 875-97.
12 R.S. Brodkey, Private Communication, Spring, 1986.
13 Y. Argman, and W. Kaufman, Turbulence in Orthokinetic Flocculation, SERL Report No. 68-55, Sanitary Engineering Research Laboratory, University of California, Berkeley, 1968.
14 R.H. Perry and C.H. Chilton, Chemical Engineers' Handbook, 5th edn., McGraw-Hill, New York, 1973, pp. 19-26.
15 C.R. O'Melia, A Review of the Coagulation Process, Pub. Works, Vol. 100, No. 5, 1969, pp. 87-98.

Interfacial Phenomena in Biotechnology and Materials Processing,
edited by Y.A. Attia, B.M. Moudgil and S. Chander
Elsevier Science Publishers B.V., Amsterdam, 1988 — Printed in The Netherlands

INDUCTION TIME MEASUREMENTS FOR A COAL FLOTATION SYSTEM

J.L. YORDAN and R.H. YOON

Department of Mining and Minerals Engineering, Virginia Polytechnic Institute
and State University, Blacksburg, Virginia 24061

SUMMARY

While contact angle offers a thermodynamic criterion for bubble-particle
adhesion, induction time can provide kinetic information regarding the
drainage of the disjoining film between a bubble and a particle. For a
hydrophobic particle to adhere to the surface of a bubble, the contact time
must be longer than the induction time. An improved version of the induction
timer is described in this paper and its utility is discussed. It is shown
that measurement of induction time is useful in studying the effect of coal
oxidation on flotation and in determining the optimum flotation conditions in
terms of pH and frother and collector additions. In addition, the induction
time measurements conducted as a function of temperature give the activation
energy for bubble-particle adhesion, which may be useful in providing
information on the role of flotation reagents.

INTRODUCTION

Froth flotation is based on the elementary process of bubble-particle
adhesion. Contact angle measurements show whether the adhesion is thermodynamically possible or not, but fail to describe the dynamic nature of the
process. When a particle collides with a bubble during flotation, it will
have a finite contact time on the surface of the bubble. If the contact time
is not long enough to drain the disjoining water film between the particle and
the bubble, bubble-particle adhesion is not possible even if it is thermodynamically possible. The kinetics of film drainage is determined by factors
such as viscosity, temperature, particle size, bubble size, etc.

Induction time is defined as the minimum time required to thin the
disjoining film between a particle and a bubble. The technique of measuring
induction time was first conceived by Sven-Nillson (ref. 1), and implemented
by Eigeles and Volova (ref 2.) in mineral flotation systems. Recently, Yordan
and Yoon (ref. 3) have used an improved apparatus to measure induction times
for the quartz-amine flotation system. They have shown excellent correlations
between measured induction times and flotation results obtained under various
conditions.

In the present work, induction time measurements have been carried out on
a coal flotation system as a function of frother and collector additions, pH
and degree of coal oxidation.

MATERIALS AND PROCEDURE

Coal Samples

A run-of-mine Elkhorn seam coal containing 15.5% ash was used in the
present work without further cleaning. Before each measurement, the –1/4-inch
fraction was hand-ground in an agate mortar, pulverized in a coffee grinder
and screened to obtain –212+150– and –150+100-micron fractions.

Reagents

Polypropylene glycol (MW 425) supplied by Aldrich Chemical Company was
used without further purification. Kerosene was used as the collector.
Research grade potassium chloride (KCl), supplied by Fisher Scientific
Company, was used in the salt flotation tests, and hydrochloric acid (HCl) and
sodium hydroxide (NaOH) were used for pH control. Double-distilled water,
prepared in an all-glass still, was used in all of the experiments. Ultra-
pure nitrogen (99.999%) from AIRCO Industrial Gas Company was used to produce
the bubbles for the microflotation tests.

Equipment

An induction time apparatus has been used in the present work which has a
sensitivity limit of 0.1–0.15 milliseconds. The basic unit is similar to the
one used by Eigeles and Volova (ref. 2) and Trahar (ref. 4), but it operates
with a microcomputer and has a greater sensitivity. A schematic representation of the induction time apparatus is shown in Figure 1. The microflotation

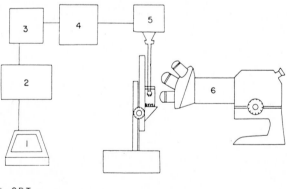

1 - CRT
2 - COMPUTER
3 - INTERFACE
4 - ELECTRONIC PULSE AMPLIFIER
5 - ELECTROMECHANICAL ACTUATOR
6 - MICROSCOPE

Fig. 1. Schematic diagram of the induction time apparatus.

tests were made using a Partridge and Smith (ref. 5) type flotation cell.
Bubbles were produced by sparging ultra-pure nitrogen gas through a medium-
porosity glass frit at the bottom of the cell. The gas flow was monitored
using a flowmeter. Gentle agitation was provided by means of a teflon-coated
magnetic stirring bar.

Procedure

For the induction time measurements, two grams of coal particles were
conditioned in a beaker containing 100 ml of collector or frother solution.
After pH adjustment, the coal suspension was agitated for 15 minutes before it
was transferred to a rectangular optical cell to form a bed of particles.
The cell, containing approximately 20 ml of the reagent solution, was then
placed on the moving stage of a microscope for measurement.

Inside the cell, an air bubble approximately 2 mm in diameter was formed
at the tip of a glass capillary tube using a microsyringe, and left to stand
for one minute to reach equilibrium before making contact with the particle
bed. At a preset contact time, the bubble was lowered to the particle bed ten
times and examined each time through the microscope to see if any particles
were picked up by the bubble. If the contact time was too short, no particles
attached themselves to the bubble. The experiments were therefore repeated,
changing the contact time incrementally. In this manner, the minimum contact
time, for which at least one particle was actually picked up for five of ten
contacts, was determined. This critical time was taken in the present work to
be the induction time. Most of the tests were carried out at ambient
temperature, except for those studying the effect of temperature.

In each microflotation test, approximately 1 gram of sample was used with
approximately 75 ml of collector solution at a gas flow rate of 45 ml/min STP.
The recovery after one minute of flotation time was taken as a convenient
measure of floatability.

RESULTS AND DISCUSSION

Effect of Oxidation Time

Many investigators (ref. 6, 7) have studied the effect of oxidation on
the flotation of coal. In most of these investigations, coal samples were
oxidized in a furnace at elevated temperatures to simulate the oxidation that
takes place in the field. It should be pointed out, however, that the coals
subjected to coal cleaning processes are seldom subjected to dry oxidation at
high temperatures. The oxidation mechanisms and the reaction products can be
significantly different from those of the low-temperature oxidation that takes

336

place in moist or wet environments. The kinetics of the oxidation, in
particular, can be quite different.

In an effort to obtain information more relevant to coal preparation
practice, a run-of-mine coal sample was subjected to wet oxidation for a
prolonged period of time while periodically taking samples. A 50-gram coal
sample (-150+100 microns) was initially placed in a 250-ml beaker containing
200 ml of distilled water at pH 6. The beaker was exposed to the ambient, so
that atmospheric oxygen could freely diffuse into the suspension. A small
amount of sample was taken periodically and subjected to induction time
measurements and microflotation experiments.

The induction time measurements were carried out without any reagents,
while the flotation experiments were conducted in the presence of 10^{-3} M KCl
solutions. The results are given in Figure 2. As shown, the induction time
increased steadily with time, and then reached a plateau of 5.6 milliseconds
after approximately 16 days. These results indicate that as the oxidation
progresses, the displacement of the wetting film by an air bubble from the
surface of the coal becomes increasingly difficult. The reason is, of course,
that oxidation produces hydrophilic species on the coal surface. The most
likely oxidation products are carboxylic and phenolic groups (ref. 7).

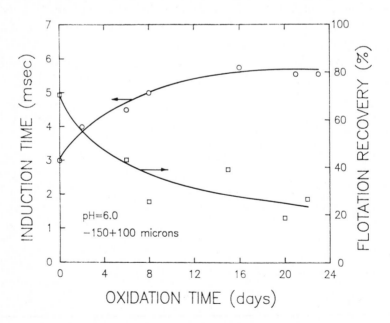

Fig. 2. Effect of oxidation time on the induction time and flotation recovery
of an Elkhorn seam coal (-150+100 microns) at pH 6.0.

The results of the salt flotation experiments, also shown in Figure 2, are in close agreement with those of the induction time measurements, i.e., the flotation recovery decreases steadily as the induction time increases. The flotation recovery was maximum (70%) when the coal was fresh, and then decreased steadily as the oxidation progressed. After 20 days of ambient oxidation, the floatability was reduced substantially to approximately 20%. As shown by Laskowski (ref. 8), the salt flotation technique is sensitive to changes in the degree of coal oxidation. Sun (ref. 9) showed that wet-oxidation of coal reduced the coal recovery in batch flotation experiments. More recently, Gutierrez and Aplan (ref. 10) also showed that wet-oxidation of a bituminous coal caused significant reductions in contact angle.

Effect of Kerosene Addition

Figure 3 shows the results of induction time measurements conducted on the -212+150-micron fraction of the Elkhorn seam coal at pH 6.0 as a function of kerosene addition. The induction time is shown to decrease sharply with increasing kerosene addition. It is likely that the droplets of the hydrocarbon oils adsorbing on the coal surface bridge the hydrophobic sites, masking the hydrophilic sites in-between. The net result is that the hydrocarbon adsorption makes the coal surface more hydrophobic and causes the disjoining film to be more labile, making it easier for the approaching air bubble to displace the film.

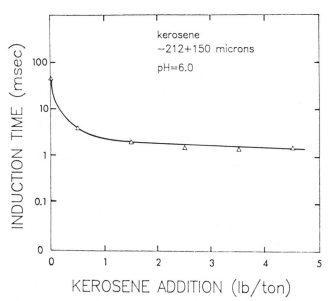

Fig. 3. Effect of kerosene addition on the induction time of an Elkhorn seam coal (-212+150 microns) at pH 6.0.

338

Note that a kerosene addition beyond about 1.5 lb/ton does not further reduce the induction time significantly. It is also interesting to note that the average consumption of oily collectors in U.S. coal preparation plants is 1.7 lb/ton (ref. 11), which is remarkably close to the results obtained from the induction time measurements.

Figure 4 shows the results of induction time measurements and microflotation experiments conducted as a function of pH. The experiments were carried out using 1.5 lb/ton of kerosene and no frother. It is shown that the flotation maximum occurs at the pH where the induction time becomes minimum, i.e., pH 7-8. It was observed by Derjaguin and Shukakidse (ref. 12) and Fuerstenau (ref. 13) that the flotation maxima of various minerals occur at pH values where the minerals have minimum zeta-potentials. This phenomenon was explained by the fact that at such pH values, the electrostatic repulsion between the mineral and the bubble should become minimum. However, the zeta-potential measurements conducted on the Elkhorn seam coal (containing 15.5% ash) used in the present work showed that its iso-electric point (i.e.p.) occurs at pH 3.5-4.0, which is far removed from the pH of maximum flotation and the minimum induction time observed in the present work. Perhaps surface forces other than the zeta-potential play a more important role in this case.

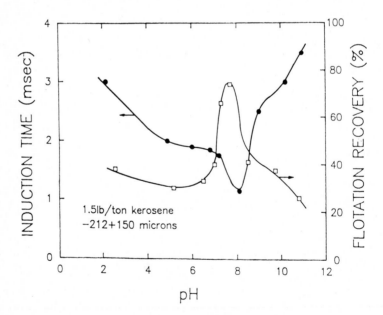

Fig. 4. Effect of pH on the induction time and flotation recovery of an Elkhorn seam coal (-212+150 microns) for a kerosene addition of 1.5 lb/ton.

Effect of PPG Addition

Figure 5 shows the results of induction time measurements conducted in the presence of PPG at pH 6.0. As shown, the induction time decreases steadily with increasing frother concentration, suggesting that the presence of PPG in the solution favors the bubble-particle adhesion process. Adsorption studies carried out by many investigators (ref. 14, 15) have shown that nonionic surfactants adsorb on coal significantly via hydrophobic bonding. The results given in Figure 5 indicate that PPG does adsorb on coal and that its adsorption makes the surface more hydrophobic, thereby causing the disjoining film to become more labile.

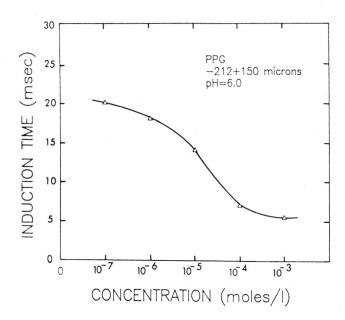

Fig. 5. Effect of PPG concentration on the induction time of an Elkhorn seam coal (-212+150 microns) at pH 6.0.

Induction time measurements have also been carried out as a function of pH in the presence of 10^{-3} moles of PPG solution. No collector was used in these tests. The results shown in Figure 6 are similar to those obtained in the presence of kerosene without PPG (Figure 4). The flotation reaches a maximum at pH 7-8, at which the induction time reaches a minimum. It is interesting to note that the optimum flotation pH values in the presence of kerosene and PPG are the same.

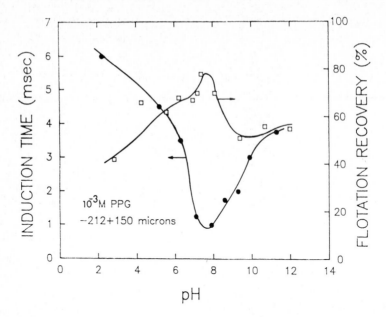

Fig. 6. Effect of pH on the induction time and flotation recovery of an Elkhorn seam coal (-212+150 microns) at 10^{-3} M PPG.

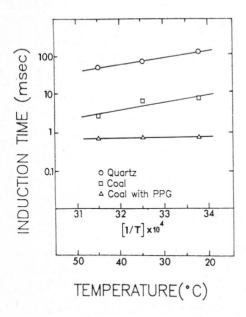

Fig. 7. Effect of temperature on the induction time of an Elkhorn seam coal (-212+150 microns) and quartz (-212+150 microns) at pH 7.6 in the presence and absence of PPG.

Effect of Temperature

Figure 7 shows the results of induction time measurements conducted on Elkorn seam coal as a function of temperature at pH 7.6. The measurements were conducted in the absence and presence of 10^{-3} M PPG solution. One can see that the induction time decreases significantly in the presence of PPG. In the absence of PPG, the induction time decreases with increasing temperature. The effect of temperature becomes less significant in the presence of PPG. Also shown in Figure 7 are the results obtained with quartz which may be representative of a typical mineral matter present in coal. It can be seen that the induction times for quartz are an order of magnitude larger than those of coal without PPG, and two orders of magnitude larger than those of coal in the presence of PPG.

Using the data given in Figure 7, one can calculate the activation energy for bubble-particle adhesion using the Arrhenius-type equation:

$$t = t_o \exp(E/kT), \tag{1}$$

in which t is the induction time, t_o is a constant, E is the activation energy, k is the Boltzman constant and T is the absolute temperature. Rearranging Eq. (1), one can obtain a convenient expression:

$$\log t = E/2.3kT + \log t_o, \tag{2}$$

in which E/2.3k is the slope of the log t vs (1/T) plot. Table 1 gives the values of E calculated from the slopes of the plots shown in Figure 7. The activation energy for the coal-air bubble adhesion is significantly less than for quartz. In the absence of PPG, the activation energy is 3.87 kcal/mole. In the presence of 10^{-3} M PPG, however, the activation energy is reduced to 1.27 kcal/mole. The difference between the two activation energy values, i.e., 2.6 kcal/mole, may have been provided by the free energy of PPG adsorption. Work is currently in progress at Virginia Tech to determine the free energy of frother adsorption.

TABLE 1

Activation Energies for Bubble-Particle Adhesion

SYSTEM	ACTIVATION ENERGY (kcal/mole)
Pure Quartz	5.07
Elkorn Seam Coal	3.87
1×10^{-3} M PPG-Coal	1.27

342

CONCLUSIONS

1. The induction time apparatus used in the present work has been found to be useful in studying the chemical and hydrodynamic aspects of a coal flotation system.

2. It has been shown using induction time measurements that the natural floatability of coal is seriously deteriorated by low-temperature wet oxidation. The flotation recovery is reduced from 70% for fresh coal to 20% after 20 days of oxidation, while the induction time is reduced concomitantly.

3. Induction time measurements have been shown to be useful in determining optimum flotation conditions in terms of pH and collector addition. For the coal used in this work, the optimum flotation pH occurs at pH 7-8. The optimum kerosene addition has been found to be 1.5 lb/ton.

4. Activation energies of the bubble-particle adhesion process have been determined. The values range between 3.87 and 1.27 kcal/mole at PPG concentrations ranging between 0 and 1×10^{-3} M. This reduction in the activation energy has been related to the free energy of collector adsorption.

REFERENCES

1 I. Sven-Nillson, Effect of Contact Time between Mineral and Air Bubbles on Flotation, Kolloid Z., 69 (1934) 230.
2 M.A. Eigeles and M.L. Volova, Kinetic Investigation of Effect of Contact Time, Temperature and Surface Condition on the Adhesion of Bubbles to Mineral Surfaces, in: Proceedings, 5th International Mineral Processing Congress, IMM, London, 1960, pp. 271-284.
3 J.L. Yordan and R.H. Yoon, Induction Time Measurements for the Quartzamine Flotation System, SME-AIME Annual Meeting, New Orleans, Louisiana, Preprint No. 86-105 (1986).
4 W. Trahar, Private communication (1983).
5 A.C. Partridge and G.W. Smith, Small-sample Flotation Testing, Trans. IMM, 80 (1971) C199.
6 W.W. Wen and S.C. Sun, An Electrokinetic Study on the Oil Flotation of Oxidized Coal, Sep. Science and Technology, 16(10) (1981) 1491-1521.
7 D.W. Fuerstenau, G.C.C. Yang and J.S. Laskowski, Oxidation Phenomena in Flotation, Part 1, Coal Preparation, 4 (1987) 183-191.
8 J.S. Laskowski, Particle-bubble Attachment in Flotation, Miner. Sci. Engineering, 6(4) (1974) 223.
9 S.C. Sun, Hypothesis for Different Floatabilities of Coals, Carbons and Hydrocarbon Minerals, Trans. AIME, 8 (1954) 67-75.
10 J.A. Gutierrez and F.F. Aplan, The Effect of Oxygen on the Hydrophobicity and Floatability of Coal, Colloids and Surfaces, 12 (1984) 27.
11 F.F. Aplan, Coal Flotation, in: M.C. Fuerstenau (Ed.), Flotation, A.M. Gaudin Memorial Volume, AIME, New York, (1976).

12 B.V. Derjaguin and N.D. Shukakidse, Dependence of the Floatability of Antimonite on the Value of the Zeta-potential, Trans. IMM, 70 (1961) 569-600.
13 D.W. Fuerstenau, Correlation of Contact Angles, Adsorption Density, Zeta Potentials and Flotation Rate, Trans. AIME, 208 (1957) 1365-1367.
14 D.W. Fuerstenau and Pradip, Adsorption of Frothers at Coal/Water Interfaces, Colloids and Surfaces, 1 (1982) 229-243.
15 M.S. Celik and R.H. Yoon, Adsorption of Nonionic Surfactants on Coal, in: Proc. Surfactants in Mineral and Materials Systems, American Chemical Society National Meeting, New York, New York, April, 1986; in press.

Interfacial Phenomena in Biotechnology and Materials Processing,
edited by Y.A. Attia, B.M. Moudgil and S. Chander
Elsevier Science Publishers B.V., Amsterdam, 1988 — Printed in The Netherlands

POSSIBILITY OF USING STARCHES IN SELECTIVE FLOCCULATION OF A RUTILE ORE

A. MARABINI, M. BARBARO and A. FALBO

C N R - Istituto Trattamento Minerali - Via Bolognola, 7 - 00138 Roma - ITALY

SUMMARY

In the present work, dispersin and flocculating efficiency of principal starch components has been examined on pure rutile, ilmenite, quartz, pyrope and almandite, in order to evaluate the possibility of separating rutile and ilmenite from each other and from gangue minerals.

Sedimentation tests were performed on individual minerals at different pH and reagent concentration. Electrophoretic mobility of minerals was determined as a function of pH, to ascertain if there is any correlation between surface charge and flocculating or dispersing power.

An hypothesis on the mechanism of flocculating and dispersing action has been suggested. The chemical reaction involving metallic ions and starch molecules seems to be a necessary condition for flocculation and dispersion.

Flocculation could be due to the formation of surface chelates, followed by hydrogen bridging between adsorbed molecules. Dispersion is attributed to the formation of hydrophilic, surface complexes.

On the basis of experimental results, pH and concentration conditions have been established which should permit separation of the minerals by selective flocculation and/or dispersion.

INTRODUCTION

The work reported here was performed within the context of a larger research project designed to establish methods of extracting rutile from a major occurrence of eclogites in northern Italy. The ore itself contains pyroxenes, amphiboles, garnets and titaniferous minerals including rutile but also ilmenite and titanite.

The recoveries and grades achieved so far with such techniques as flotation using phosphate reagents and magnetic separation are not such as to guarantee the economic viability of a plant, at present market prices for concentrates. The reasons for the low efficiency of the physical process investigated are mainly the poor selectivity both of flotation with phosphate collectors and of magnetic separation in respect to the major gangue minerals, and the ineffectiveness of the two processes in rutile-ilmenite separation.

Another problem is the finely divided state of the rutile which is present as crystals whose average size is between 30 and 40 microns. Very fine grinding is thus required to ensure an adequate degree of liberation of the rutile from the gangue minerals and ilmenite. Under such conditions part of the ore is overground, being reduced to only a few microns, so special separation processes are needed.

It was therefore decided to study the possibility of employing selective flocculation for separating the various minerals. To this end, tests were performed with natural polymers of the polysaccharide type, to check on their dispersant and/or flocculant efficiency in respect of individual minerals, with a view to separating these by differential gravity sedimentation or by flotation of the valuable minerals as multiparticle aggregates.

Previous studies (ref. 1) had demonstrated that natural polymer reagents, such as tannins, could be used successfully for selective flocculation of rutile. Starches, like tannins, are high molecular weight polymers (10^5-10^6) characterized by the presence of hydrated polar groups -OH and CH_2OH. They are polysaccharides formed from molecules of D(+) glucose linked by glucosidic bonds, and are the result of chlorophyllic photosynthesis (ref. 2).

The main constituents of starch, amylose (AL) and amylopectin (AP), differ from one another inasmuch as AL has a linear structure due to condensation of glucose molecules held together by 1 4 acetal bonds (Fig. 1), while AP, in addition to linear portions, has a branched structure with branching points due to 1 6 bonds (Fig. 2).

Starches are, of course, employed as modifiers in flotation: in particular, they act as depressants for phosphates, carbonates, iron oxides, monazite and other minerals (ref. 3). Various interpretations have been put forward to explain the way they act. Some workers (ref. 4) speak of starch-collector co-adsorption on the mineral surface, and some (ref. 5) consider there is competition between starch and collector for the surface, while others (ref. 3) are of the opinion that there is a synergic collector-starch effect. According to this latter hypothesis the starch does not compete with the

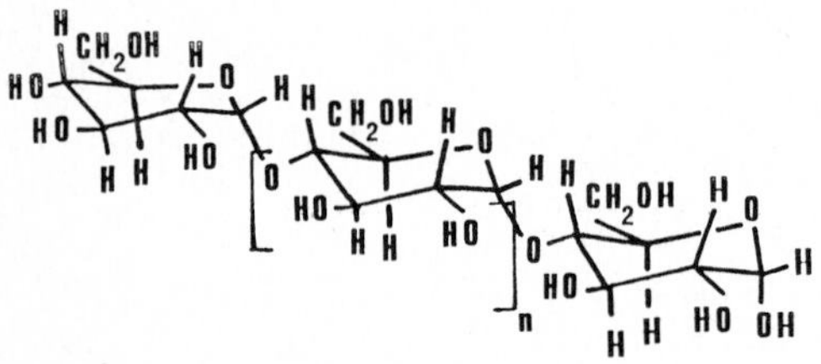

Fig. 1. Structure of amylose.

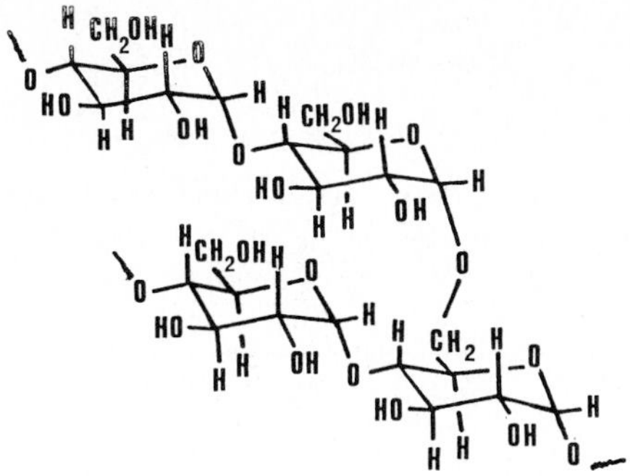

Fig. 2. Structure of amylopectin.

collector for adsorption on the particle but forms clathrates with it, including the collector adsorbed in its helices: hydrophobic inside and hydrophilic outside. In some instances the starches can be utilized for their flocculant action, as is the case with iron oxide minerals.

Iwasaki (ref. 6) refers to the action of starches as selective flocculants which can be employed in the recovery of magnetite or taconite fines.

The Tilden plant in north Michigan floats quartz with amine collectors and depresses iron oxides by flocculation with starch (ref. 7). This is one of the more significant examples of the industrial application of selective flocculation (ref. 8).

Studies have also been made of other separation processes which employ starch flocculation together with flotation (ref. 9) in some cases and with magnetic separation in others (ref. 10). Numerous workers (refs. 11-14) have studied the way starches interact with mineral surfaces. However, no explanation has as yet been found for hoe starches can act.

The work reported here concerns the flocculating and/or dispersing efficiency of AL and AP on pure rutile, ilmenite, quartz, pyrope and almandite, which represent the main minerals present in an Italian ore; moreover, a mechanistic hypothesis has been given.

EXPERIMENTAL

Minerals

The following minerals, free from any marked impurities were used: rutile from Florida, ilmenite and almandite from Norway, pyrope from Arizona and synthetic quartz of C. Erba Co. These were wet ground in a porcelain mill and the plus 15 micron fraction was removed by sedimentation. The minus 15 micron fraction was kept as pulp.

Reagents

The amylose and amylopectin were pure chemicals made by the Calbiochem-Behring Company.

Sedimentation tests

Two grams of mineral and 70 ml of water were treated in a 100 ml cylinder with appropriate concentrations of starch and the pH was adjusted by means of NaOH and HCl. After fifteen minutes shaking, the solution was allowed to settle for a given time, established via preliminary tests in such a way that in the absence of starch the quantity of mineral in the upper 70 ml of liquid was about 1 g, namely 50% of the mineral used in the test. The upper 70 ml of solution was drawn off by a peristaltic pump and dried to constant weight. The remaining 30 ml was also dried to constant weight.

Flocculating power was calculated by the formula:

$$F = \frac{P_o - P}{P_o} \times 100 \tag{1}$$

and dispersing power by the formula:

$$D = \frac{P - P_o}{P_w - P_o} \times 100 \tag{2}$$

where P_o is the weight of mineral in the upper 70 ml of liquid in the absence of polymer, P is the weight of mineral in the upper 70 ml of liquid in the presence of polymer, and P_w represents 70% of the total weight of mineral introduced into the dispersion.

Technique and apparatus

The zeta potential was measured using a Rank Brothers apparatus, Rank Bros Mark II, with a flat cell (a rectangular cross section, 10x1 mm of fused silica) and two platinum electrodes.

In the experiments, high-purity water from a millipore water system Milli-Q, was used.

Conditioning time of both suspensions was 15' with a stirrer of approximately 500 rpm.

Each electrophoretic mobility measurement was derived from an average of about twenty readings at 25^oC.

The reagents used to vary the pH were NaOH and HCl, while the ionic strength was 10^{-3} M NaCl.

The pH of suspensions was measured by means of direct reading pH meter (Orion model 701-A).

Electrophoretic Mobility is readily computed by the known formula:

$$EM = \frac{\mu}{t} \quad \frac{v}{L} \tag{3}$$

where μ is the distance over which the particle is tracked, in microns; L is
the distance between walls of the electrophoresis cell, in centimeters; v is
the voltage applied, in volts; and t represents the average time required to
track one particle over a quarter-scale division, in seconds.

RESULTS AND DISCUSSION

Figures 3 to 7 report the flocculating and dispersing power exerted by
amylose (AL) and amylopectin (AP) on the minerals examined (rutile, ilmenite,

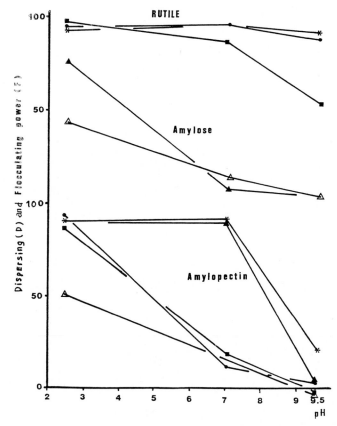

Fig. 3. Flocculating and dispersing power of amylose and amylopectin on pure
rutile versus pH at different concentrations (g/l).

$\triangle = 5 \times 10^{-4}$ g/l; $\blacktriangle = 1 \times 10^{-3}$ g/l; $\blacksquare = 2,5 \times 10^{-3}$ g/l; $\bullet = 5 \times 10^{-3}$ g/l;
$* = 1 \times 10^{-2}$ g/l.

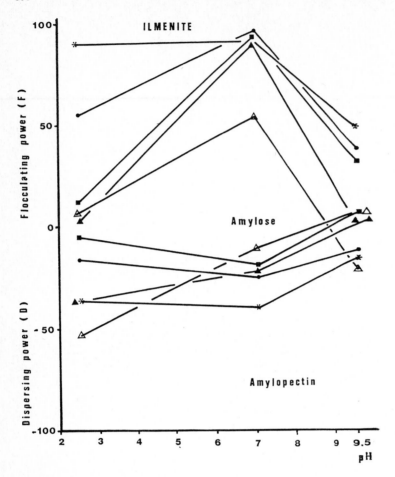

Fig. 4. Flocculating and dispersing power of amylose and amylopectin on pure ilmenite versus pH at different concentrations (g/l).

$\triangle = 5 \times 10^{-4}$ g/l; $\blacktriangle = 1 \times 10^{-3}$ g/l; $\blacksquare = 2,5 \times 10^{-3}$ g/l; $\bullet = 5 \times 10^{-3}$ g/l; $\ast = 1 \times 10^{-2}$ g/l.

quartz, almandite and pyrope) at three different pH values (2.5, 7 and 9.5) and five concentration levels (5×10^{-4} g/l to 1×10^{-2} g/l). Figures 8 to 12 illustrate experimentally determined zeta potentials of these minerals, the aim being to ascertain whether there is any correlation between surface charge and flocculating power. These values are in good agreement with the literature (ref. 15).

Table 1 provides an indication of the flocculating (F) and dispersing power (D), of amylose and amylopectin at all pH and concentration values examined. The plus and minus signs (+ and -) are used to indicate floccu-

TABLE 1

POSSIBILITY OF SEPARATION BY SELECTIVE FLOCCULATION

AMYLOSE

CONCENTRATION	pH = 2.5					pH = 7					pH = 9.5				
	R	I	Q	A	P	R	I	Q	A	P	R	I	Q	A	P
5.10^{-4}	+	+	-	--	--	+	++	+	+	--	+	-	0	+	--
1.10^{-3}	++	+	-	--	--	+	+++	0	0	--	+	0	0	++	--
$2.5.10^{-3}$	+++	+	-	0	--	+++	+++	-	+	--	++	+	0	+++	-
5.10^{-3}	+++	++	-	+	--	+++	+++	0	+	--	+++	+	0	+++	-
1.10^{-2}	+++	+++	-	+++	--	+++	+++	-	+	--	+++	++	0	++	-

AMYLOPECTIN

CONCENTRATION	pH = 2.5					pH = 7					pH = 9.5				
	R	I	Q	A	P	R	I	Q	A	P	R	I	Q	A	P
5.10^{-4}	++	--	-	--	--	+	-	0	-	--	-	+	+	-	--
1.10^{-3}	+++	-	-	--	--	+++	-	0	-	--	-	+	+	-	--
$2.5.10^{-3}$	+++	-	-	--	--	+	-	0	-	--	0	+	0	-	-
5.10^{-3}	+++	-	-	-	--	+	-	0	-	--	+	-	0	-	-
1.10^{-2}	+++	-	-	+	--	+++	-	0	-	--	+	-	0	-	-

R = Rutile; I = Ilmenite; Q = Quart; A = Almandite P = Pyrope

+ = Flocculating power

- = Dispersing power

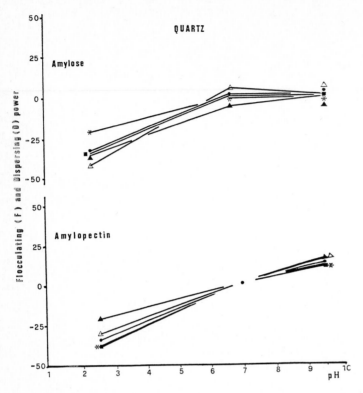

Fig. 5. Flocculating and dispersing power of amylose and amylopectin on pure quartz versus pH at different concentrations (g/l).

Δ = 5×10^{-4} g/l; $\blacktriangle$ = 1×10^{-3} g/l; $\blacksquare$ = $2,5 \times 10^{-3}$ g/l; $\bullet$ = 5×10^{-3} g/l;

$\ast$ = 1×10^{-2} g/l.

lating and dispersing power, respectively. A single + means that flocculating power is less than 50% and ++ signifies it is between 50 and 75%,
while +++ places it in the 75 to 100% bracket. The minus sign is used in
the same manner to signify dispersing power.

It would appear from the Table that it is theoretically possible to
separate rutile and ilmenite from gangue minerals, by means of amylose and
amylopectin. In detail, amylose and amylopectin act as flocculants for
rutile and dispersants for pyrope and to some extent for almandite, but
they are ineffective where quartz is concerned. It should be pointed out
that amylose has a more pronounced effect than amylopectin.

The flocculating and dispersing powers of both reagents are now examined
on a mineral-by-mineral basis.

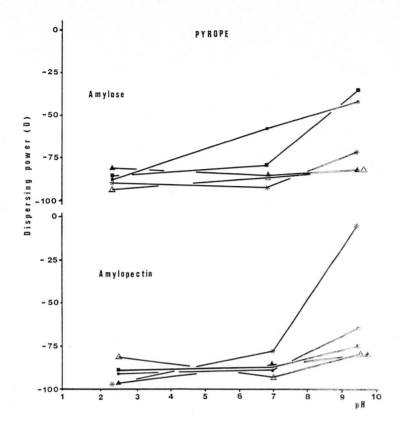

Fig. 6. Flocculating and dispersing power of amylose and amylopectin on pure pyrope versus pH at different concentrations (g/l).

$\blacktriangle$ = 5×10^{-4} g/l; $\blacktriangle$ = 1×10^{-3} g/l; $\blacksquare$ = $2,5 \times 10^{-3}$ g/l; $\bullet$ = 5×10^{-3} g/l; $*$ = 1×10^{-2} g/l.

Rutile

Rutile is flocculated by amylose and amylopectin throughout the pH range, though the effect is greatest at acid pH. Amylose is more efficient than amylopectin. Flocculating power starts to fall off at pH in excess of 7, the decline being more marked in the case of amylopectin. With both reagents, flocculating power increases with concentration (Fig. 3).

Surface charge would appear to exert a minimum influence. In fact, the efficiency of the two reagents at concentrations greater than 1×10^{-3} g/l remains unchanged at pH values above and below the zero point of charge, despite the marked accumulation of electrons on both amylose and amylopectin at all pH values. The accumulation of negative charges at acid pH is attributable to the electron-rich nature of the external hydroxyls of the glucose

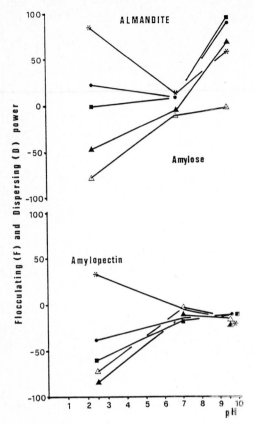

Fig. 7. Flocculating and dispersing power of amylose and amylopectin on pure almandite versus pH at different concentrations (g/l).

$\triangle$ = 5×10^{-4} g/l; $\blacktriangle$ = 1×10^{-3} g/l; $\blacksquare$ = $2,5 \times 10^{-3}$ g/l; $\bullet$ = 5×10^{-3} g/l; $*$ = 1×10^{-2} g/l.

units, while at alkaline pH the accumulation is attributable to their strong dissociation (refs. 13, 16, 17).

It would appear very likely, therefore, that the flocculating action of amylose and amylopectin is due to chemical interaction.

It is a well-known fact that titanium, a transition element, has a strong tendency to form coordination compounds, mainly with oxygen-containing ligands (ref. 18). It is also known that titanium dioxide can react with starches (ref. 19) and form films that are not very soluble in water. The formation of chelating compounds with the hydroxyls of the glucosidic units could explain this phenomenon. It is also known from flotation and flocculation studies (refs. 1, 20, 21) that titanium ions on the surface of rutile have a tendency to form stable coordination compounds with oxygen-containing ligands.

The greater flocculating power at acid pH is probably correlated with the enhanced coordinating ability of titanium in this pH range (ref. 1). Moreover, electrostatic forces of attraction may exert a synergistic effect at acid pH (Fig. 8).

The decrease in flocculating action with increasing pH may be attributed not only to less intense chemical interaction between titanium ions and the ligands, but also to greater solvation of the surface chelate, and/or to electrostatic repulsion.

The fact that amylopectin is less effective than amylose becomes more evident when conditions are not so favourable (lower concentrations and alkaline pH). It is probably explained by the smaller concentration of hydroxyls available for complexing and the decreased bridging tendency of branched molecules.

The rise in flocculating power with increased concentration is readily explained on the basis of chemical equilibria.

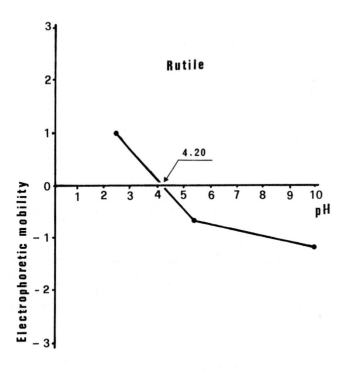

Fig. 8. Electrophoretic mobility of rutile as a function of pH.

Ilmenite

Amylose flocculates ilmenite, whilst amylopectin has a dispersing effect. Flocculating power of amylose peaks at pH 7, at which value the dispersing power of amylopectin is greatest. However, at high concentrations, amylose is strongly flocculating at all pH values below neutral (Fig. 4).

This indicates that with both reagents, complexation is favoured at neutral pH (ref. 13), though a synergistic effect of electrostatic attractive forces can be observed at pH values below neutral (Fig. 9) and at high concentrations.

The difference in the behaviour of amylose and amylopectin is probably attributable to their different molecular structure. Namely, the linear structure of amylose, which allows hydrogen bridging between adsorbed molecules and consequent flocculation (ref. 23), while the branched structure of amylopectin does not favour such packing; indeed, it tends to expose strongly hydrophilic hydroxyl groups of side chains to the water, the result being complex solvation (refs. 12, 22).

Here, too, the difference in the absolute value of that action (F-100%; D-40%) is attributable to the smaller availability of bonding groups in the amylopectin.

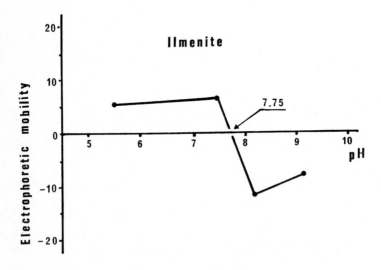

Fig. 9. Electrophoretic mobility of ilmenite as a function of pH.

Quartz

As is evident from Fig. 5, amylose and amylopectin have virtually no effect on quartz (ref. 12) probably because of the lack of chemically-active metal sites that can react with the starch hydroxyls.

The slight dispersing effect exerted by both reagents at acid pH may well be due to a small amount of physical adsorption. This is possible because the surface electric charge of quartz is virtually nil at this pH (Fig. 10). Surface hydrophilicity is increased as a result of this adsorption.

Pyrope

Amylose exerts strongly dispersing action on pyrope, $Mg_3Al_2(SiO_4)_3$, throughout the pH range (Fig. 6), while amylopectin acts similarly except at alkaline pH and maximum concentration (1×10^{-2} g/l). This dispersing action may be attributable to the particular nature of the complexes presumably formed between the magnesium ions and starches (ref. 5).

Indeed, it is known that these complexes are strongly solvated owing to the more ionic character of the bond between the hydroxyls and the metal. This idea fits in well with what is known (ref. 19) about solution complexes between carbohydrates and the divalent ions of the alkaline earth metals (e.g., lactose, $CaCl_2 \cdot nH_2O$).

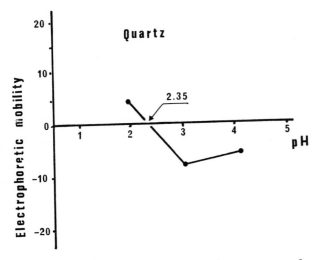

Fig. 10. Electrophoretic mobility of quartz as a function of pH.

The decline in the dispersing effect of amylopectin at alkaline pH and high concentration is attributable to a lesser degree of adsorption, perhaps due to electrostatic repulsion (Fig. 11), which is more evident in the case of this reagent because of its lower complexing ability.

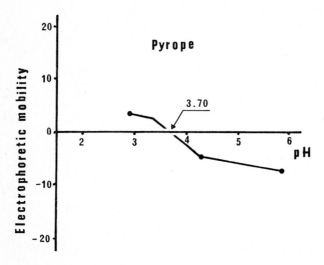

Fig. 11. Electrophoretic mobility of pyrope as a function of pH.

Almandite

Where almandite, $Fe_3Al_2(SiO_4)_3$, is concerned, the flocculating and dispersing power of amylose and amylopectin depends markedly on pH and concentration (Fig. 7).

However, it can be assumed that there is a chemical interaction between amylose and amylopectin on the one hand and the ferrous ions of the mineral on the other, owing to the marked flocculating and dispersing effect.

Though it is difficult to interpret the experimental results obtained with amylose, the behaviour of amylopectin, which disperses the mineral at acid pH and is ineffective at values higher than 7, can be explained in the same was as for the other minerals, namely, hydrophilization of surfaces in the 2.5 to 7 pH range and electrostatic repulsion at higher pH values.

Possibility of separation by selective flocculation

It would appear from the results obtained that rutile can be separated from ilmenite and both of these from gangue minerals by means of amylose and amylopectin.

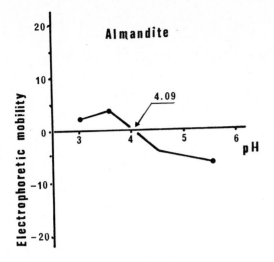

Fig. 12. Electrophoretic mobility of almandite as a function of pH.

The possibility of selectively flocculating and/or dispersing one mineral from another is evident from Table 1 where the efficiency of the two reagents is expressed by the number of plus or minus signs accorded to each. It would seem that selective flocculation is feasible by adopting certain pH and amylose and amylopectin concentration values.

With amylose, the most favourable condition for separation of rutile by selective flocculation from all the other minerals occurs at pH 2.5 and a concentration around 2.5×10^{-3} g/l. Where ilmenite is concerned, instead, the most favourable condition is pH 7 and a concentration of about 1×10^{-3} g/l. Rutile can also be selectively flocculated from all the other minerals by amylopectin at pH 2.5 and concentrations between 1×10^{-3} and 1×10^{-2} g/l.

CONCLUSIONS

A study has been made of the flocculating and/or dispersing action of amylose and amylopectin on rutile, ilmenite, quartz, almandite and pyrope, namely the minerals present in an Italian ore with a highly intergrown structure. The pH and concentration values have been ascertained at which it should be possible to separate rutile and ilmenite from one another and from the accessory minerals by means of amylose and amylopectin.

The experimental data have been interpreted by considering chemical surface adsorption reactions, the structural effects of amylose and amylopectin and the surface charge of the minerals.

It is assumed that flocculating and dispersing power are both attributable to strong chemical interaction, with the formation of surface complexes between the metal ions and the starch hydroxyls.

According to this interpretation, flocculation results from the formation of coordination chelates on the surface followed by the formation of hydrogen bridging bonds between the glucosidic chains adsorbed on the surfaces of several particles. Dispersion is also correlated with the formation of complexes which are highly ionic and thus strongly hydrophilic.

The influence of the surface charges is also evident, when there is no strong chemical interaction.

ACKNOWLEDGEMENTS

The authors wish to thank M. Esposito, S. Quaresima and B. Passariello for performing the experimental flocculation and zeta potential tests.

REFERENCES

1 G. Rinelli and A.M. Marabini, A new reagent system for the selective flocculation of rutile, XIII International Mineral Processing Congress, Panstwowe Wydawnictwo Naukowe, Warszawa, Vol. I, 1979, 304.
2 T. Morrison and R.N. Boyd, Chimica organica, Ambrosiana, Milano, 1976.
3 H.S. Hanna and P. Somasundaran, Flotation of salt-type minerals, in Flotation, A.M. Gaudin Memorial, Vol. I, AIME Publ., New York, 1976.
4 P.M. Afenya, Adsorption of xanthate and starch on synthetic graphite, International Journal of Mineral Processing, 9 (1982) 303-319.
5 K. Khosla and A.K. Biswas, Mineral collector starch constituent interactions, Colloids and Surfaces, 9 (1984) 219-235.
6 I. Iwasaki et al., The use of starches and starch derivatives as depressant and flocculants in iron ore benefication, Trans. AIME, 244 (1969), 98.
7 A.F. Colombo and D.W. Frommer, Cationic flotation of Mesabi range oxidized taconite, in Flotation, A.M. Gaudin Memorial, Vol. II, AIME Publ., New York, 1976, 1285.
8 H.D. Jacobs and A.F. Colombo, Cationic flotation of a hematic oxidized taconite, Bureau of Mines Report of Investigations RI (1981) 8505.
9 S.C. Termes, R-L. Wiefong and P.E. Richardson, Insoluble cross linked starches xanthate as selective flocculant for sulfide minerals, 1982 AIME Annual Meeting, Dallas, Texas, February 14-18, 1982.
10 M. Zuelta, L.V. Gutierrez and J.A. Matar, Use of starch in selective flocculation of low grade hematite ore and high ash-content coal, IMM-CNR Conference, Rome, September 1984.
11 K. Rao Hanumantha and K.S. Narasimhan, Selective flocculation applied to Barsuan iron ore tailing, International Journal of Mineral Processing, 14 (1985) 67-75.
12 R. Hovot, Beneficiation of iron by flotation - review of industrial and potential applications, International Journal of Mineral Processing, 10 (1983) 183-204.
13 N.K. Khosla et al., Calorimetric and other interaction studies on mineral-starch adsorption systems, Colloids and Surface, 8 (1984) 321-336.
14 B. Gururay et al., Dispersion-flocculation studies on hematite clay systems, International Journal of Mineral Processing, 11 (1983) 285-302.

15 P. Ney, Zeta-Potentiale und Flotierbarkeit van Mineralen, Springer-Verlag, Wien, New York, 1973.

16 R.L. Whistler and E.F. Paschall, Starch, chemistry and technology, Academic Press, Vol. 1, 455; Vol. 2, 149 (1965).

17 P. Somasundaran, Adsorption of starch and oleate and interaction between them on calcite in aqueous solutions, J. Coll. Interf. Sci., Vol. 2, 31(4) (1969) 557.

18 D.D. Perrin, Organic complexing reagents, structure, behavior, and application to inorganic analysis, Intersci. Publ., New York, 1964, 314.

19 R.L. Whistler and E.F. Paschall, Starch, chemistry and technology, Academic Press, Vol. 1, 309-329 (1965).

20 A.M. Marabini, M. Barbaro and M. Ciriachi, Calculation method for selection of complexing collectors, Trans. Inst. Min. Metall., Sec. C, 92, 1983.

21 Y.A.I. Attia and J.A. Kitchner, Development of complexing polymers for the selective flocculation of copper minerals, Proc. of the 11th Int. Min. Proc. Congr., Cagliari, April 20-26, 1975, 1233.

22 Pradip and D.W. Fuerstenau, The effect of polymer adsorption on the wettability of coal, Flocculation in Biotechnology and Separation Systems, Y.A. Attia (Ed.), Elsevier Sci. Publ. B.V., Amsterdam, 1987.

23 Y.A. Attia et al., Investigation of polymer adsorption with electron microscopic techniques, Flocculation in Biotechnology and Separation Systems, Y.A. Attia (Ed.), Elsevier Sci. Publ. B.V., Amsterdam, 1987.

Interfacial Phenomena in Biotechnology and Materials Processing,
edited by Y.A. Attia, B.M. Moudgil and S. Chander
Elsevier Science Publishers B.V., Amsterdam, 1988 — Printed in The Netherlands

MICROBUBBLE FLOTATION OF FINE PARTICLES

R.H. YOON, G.T. ADEL, G.H. LUTTRELL, M.J. MANKOSA and A.T. WEBER

Department of Mining and Minerals Engineering, Virginia Polytechnic Institute
and State University, Blacksburg, Virginia 24061

ABSTRACT
Hydrodynamic studies carried out in the past have suggested that the
capture of fine particles by air bubbles can be improved by decreasing the
bubble size. This concept has recently been applied in column flotation as a
means of improving flotation kinetics. In the present work, a series of
computer simulations have been conducted to better define the effects of
various parameters on flotation column performance. The results of these
analyses suggest that the best way to obtain high flotation recoveries is to
use tall columns and small air bubbles. In addition, the simulations indicate
that countercurrent wash water addition can greatly improve product quality.
On the basis of these simulation results, microbubble column flotation tests
have been conducted. Excellent results have been obtained with a variety of
materials, including coal and kaolin.

INTRODUCTION

Froth flotation is a process commonly employed for the selective separation
of fine particles from unwanted gangue. Because of its simplicity and
relatively low operating cost, flotation is now widely used throughout the
mineral processing industry. Although the process itself is conceptually very
simple, the fundamental principles that govern its behavior are complex and
not well understood at the present time. Rapid advances in flotation theory
have been hampered by the large number of interdependent variables which
determine flotation response (ref. 1).

The phenomena governing flotation can be roughly classified as either
chemical or physical in nature. Chemical phenomena are controlled by reagent
additions and the surface characteristics of the materials to be separated.
Physical phenomena are largely controlled by the hydrodynamic interactions
between bubbles and particles, which are determined by such factors as bubble
size, particle size, turbulence imparted by the flotation machine, etc. In
past decades, most of the major advances in flotation technology have been in
the area of flotation chemistry. On the other hand, physical phenomena,
although equally important, have received much less attention. For example,
only a limited amount of success has been achieved in the development of
phenomenological process models that can be used to predict flotation

response. One advantage of such a model is that it can be used to examine process variables independently so that the effects of changes in these variables can be adequately assessed.

The purpose of this paper is to describe the microbubble column flotation process developed at Viginia Tech. The fundamental theory behind the process and the process model will be described along with the results obtained with a bench-scale, continuous column that was 5 cm in diameter.

THEORETICAL BASIS

Of the various subprocesses which contribute to the overall rate of flotation, the elementary step of particle capture by a rising bubble may be considered the most important. It is this step which eventually determines how operating parameters affect flotation performance.

From a fundamental viewpoint, the process of particle capture may be described by:

$$P = P_c P_a (1-P_d) \tag{1}$$

where P is the overall probability of particle capture, P_c is the probability of collision between a bubble and particle, P_a is the probability of adhesion after a particle has collided with a bubble, and P_d is the probability of particle detachment. For particles smaller than approximately 100 microns, detachment is negligible ($P_d = 0$) and need not be considered (ref. 2).

Many investigators have shown that a decrease in P_c with decreasing particle size is largely responsible for the difficulty in floating fine particles. In the present work, this relationship has been quantified by considering the case of an isolated bubble rising through a suspension of particles. As the bubble rises, a flow pattern develops around the bubble that can be represented by an infinite series of streamlines. The critical streamline represents the trajectory of a finite-sized particle that just grazes the surface of the bubble. Thus, only those particles that are inside the critical streamline will collide with the bubble, while those that are outside will miss the bubble. The probability of collision is determined from the ratio of the area inside the critical streamline at an infinite distance away from the bubble to the cross-sectional area of the bubble. The details of this analysis will be discussed elsewhere (ref. 3). Based on this concept, a mathematical model has been developed for P_c as a function of bubble size (D_b) and particle size (D_p), as follows:

$$P_c = \left[\frac{D_p}{D_b}\right]^2 \left[\frac{3}{2} + \frac{4Re^{.72}}{15}\right] \qquad (2)$$

where Re is the Reynolds number of the bubble. This expression is valid for rigid spheres having a diameter ratio $D_p/D_b < 0.1$ and bubbles with Reynolds numbers between 0 and 100. Eqn. (2) clearly shows that at very small Reynolds numbers, P_c is directly proportional to the square of the particle size and inversely proportional to the square of the bubble diameter. Thus, for the flotation of fine particles, there must be a corresponding decrease in bubble size to maintain an adequate probability of collision.

Once the collision occurs, it may or may not result in the formation of a stable bubble-particle aggregate. This, of course, depends on the hydrophobicity of the particle. For a perfectly hydrophobic particle, the value of P_a is unity and, therefore, P is determined directly by eqn. (2). For particles of lesser hydrophobicity, P_a can be determined from induction time measurements, as has been described elsewhere (ref. 4).

Once P is known, the first-order rate constant for bubble-particle attachment (k) can be evaluated from the following:

$$k = \frac{3PV_g}{2D_b} \qquad (3)$$

in which V_g is the superficial gas velocity (ref. 5). If all other parameters are held constant, this equation indicates that k increases as P and V_g increase, and decreases as D_b increases. It should also be noted that since P is approximately proportional to the inverse square of bubble diameter when Re is small, eqn. (3) suggests that k is proportional to the inverse cube of bubble diameter. Thus, by decreasing bubble size by half, an eight-fold increase in particle attachment is predicted. This analysis agrees well with experimental findings (ref. 6,7) which suggest that the recovery of fine coal particles increases rapidly with decreasing bubble size.

FLOTATION COLUMN MODELING

Although eqn. (3) is useful for describing the rate at which particles become attached to bubbles, it does not completely describe the flotation process. The overall effectiveness of flotation is determined by a combination of both rate and transport terms. For example, poor flotation may result when the rate of bubble-particle attachment is high, but the transport of the resultant bubble-particle aggregate is slow. Therefore, in order to adequately model the flotation process, both rate and transport terms must be

considered. Transport is particularly important for gangue particles, since
the bubble-particle attachment rate does not play the major role in the
recovery of this type of particle.

In the present work, the flotation column has been modeled by describing
the flow pattern along the length of the column by a series of sections, each
representing different flow conditions. Each section has been subdivided into
one or more well-mixed zones, the number of which depends on the height of the
column. For each zone, a mass (or volume) balance has been applied to each
particulate phase present in the column. For the case of flotation, the
different phases which must be considered include air, unattached solids and
solids attached to air bubbles. Particulate solids can be further classified
as either valuable particles or gangue particles, with each having different
rate constants for bubble-particle attachment and different settling
velocities depending on the particle size. Perfect liberation of valuable
particles from the gangue particles has been assumed, which makes the model
suitable only for cases of very finely ground particles. Factors which have
not been considered include particle agglomeration, bubble coalescence, bubble
loading and bubble-particle detachment. Simple modifications are presently
being incorporated into the present form of the model to account for these
effects.

Model equations have been derived by applying a mass (or volume) balance
around each zone. For a given zone, four transport and two rate terms are
possible. Transport terms include the flows of material into the zones
directly above and below from the zone under consideration, and the flows of
material from the zones directly above and below into the zone under
consideration. Transport terms are used to describe the movement of particles
(or bubbles) due to volumetric flows, settling (or rising) velocities and
axial mixing. Rate terms, which can be quantified using the rate constant
given in eqn. (3), describe the disappearance from or appearance into a
particulate phase due to bubble-particle attachment. The rate of change, or
accumulation, of mass or volume in any zone is given by the difference between
input and output terms due to transport and rate. Steady-state conditions are
achieved when the accumulation becomes equal to zero.

Using this type of analysis, one can develop expressions that describe the
movement of the various particulate phases through the column in both time and
space. A complete derivation of these expressions has been reported in detail
elsewhere (ref. 8). The total mass flow of valuable particles is determined
by the summation of the mass of attached and unattached valuable particles
reporting to either the product or reject streams of the flotation column.
Combining this solution with a similar analysis for the gangue particles

allows both the recovery and grade of the product and of the reject streams to be calculated.

Simulation Results

Using the steady-state solution to the column model, a series of simulations were conducted to determine the effects of various operating and design parameters on the product recovery and ash obtained for the flotation of fine coal. The conditions under which these simulations were conducted are given as follows:

Superficial Feed Velocity = 15.4 cm/min
Superficial Gas Velocity = 19.7 cm/min
Feed Percent Solids   =   5.3%
Feed Percent Ash   = 36.4%
Mean Particle Size   =   5.5 microns.

The unknown parameters of froth film thickness (F) and particle collection probability for coal (P), as determined from experimental data, were 4.9 microns and 0.00028, respectively (ref. 9). The particle collection probabilty for mineral matter was assumed to be zero.

As a result of several simulations, it was determined that column length-to-diameter ratio, bubble diameter and wash water addition rate had the largest influence on product grade and recovery. The effects of these three parameters are illustrated in Figures 1-4. Figure 1 shows the relationship between recovery and bubble diameter (D_b) for various column length-to-diameter ratios (L/D) in the absence of wash water. As can be seen, recovery increases significantly as bubble size is reduced. This is a direct result of the increase in the number of bubbles and the probability of collection (P) with decreasing bubble size. The latter effect has been discussed in detail elsewhere (ref. 5). An increase in recovery is also observed as L/D increases, although this increase is not as significant as that produced by decreasing the bubble size. This effect is primarily a result of the increased residence time which provides more opportunity for bubble/particle collision.

A corresponding relationship to that discussed above is illustrated in Figure 2 for product ash. As shown, the ash content tends to decrease with decreasing bubble size down to a diameter of approximately 250 microns, after which the ash content tends to increase. The initial decrease in ash content with decreasing bubble diameter is largely due to the increase in P discussed previously. Since the mineral matter in this simulation has been considered

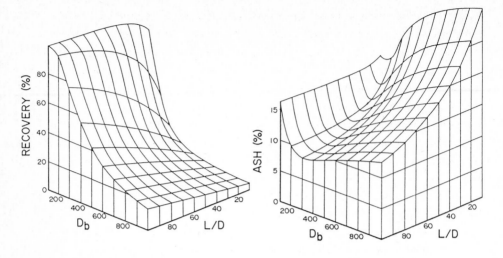

Fig. 1. Recovery as a function of bubble diameter for different L/D ratios in the absence of counter-current wash water.

Fig. 2. Product ash as a function of bubble diameter for different L/D ratios in the absence of counter-current wash water.

to have no floatabilty, an increase in P should increase the selectivity. Below a bubble size of approximately 250 microns, however, the increased water recovery due to the large number of bubbles increases the nonselective entrainment of mineral matter and the ash content of the product increases.

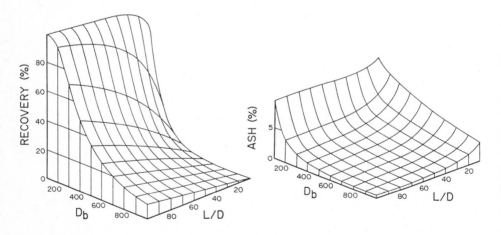

Fig. 3. Recovery as a function of bubble diameter for different L/D ratios in the presence of counter-current wash water.

Fig. 4. Product ash as a function of bubble diameter for different L/D ratios in the presence of counter-current wash water.

Thus, the simulations predict that operating at a bubble size between 200 and 300 microns should give the best conditions for obtaining a low-ash product at a relatively high recovery. The results shown in Figure 2 also suggest that increasing the column L/D ratio has a significant effect on the product ash. In fact, under the conditions employed for this simulation, increasing the L/D ratio has a more significant effect on product quality than on recovery.

When countercurrent wash water is added just below the froth/pulp interface, little change is seen in the shape of the recovery curve as shown in Figure 3, although the recovery values are somewhat lower. In this case, wash water was added at a superficial velocity of 2 cm/min which lowered the recovery values approximately 5%. The effect of wash water on product ash, however, is considerably more significant. As shown in Figure 4, the use of countercurrent wash water reduces the product ash content for larger values of D_b and L/D to less than 1%. If a typical superficial wash water velocity of approximately 20 cm/min is used, nearly all ash entrainment can be suppressed.

The dynamic solution to the column model was used to determine the time required to achieve steady-state from a start-up condition. This can also be used as an indication of the time needed for the column to stabilize when it encounters a disturbance. As shown in Figure 5, the recovery appears to reach steady-state quite rapidly (i.e., < 5 min.). However, the product ash content

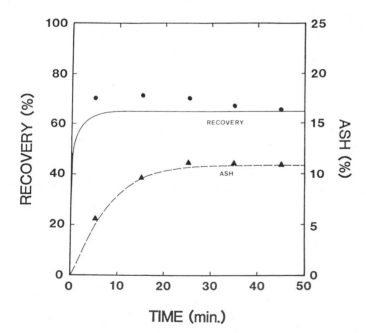

Fig. 5. Simulated (lines) and experimental (points) recovery and product ash as functions of time.

requires nearly 20 minutes to achieve steady-state. The simulated results were found to be in good agreement with experimental results produced in a 2.5-cm laboratory column. The implication of this finding is that if the column is operated for a short time, biased results can be produced which will show a much cleaner product than can be achieved under steady-state conditions. Thus, it is important to monitor the product grade rather than the recovery when determining steady-state for flotation column testing.

EXAMPLES

Production of Superclean Coal

Coal samples from the Elkhorn No. 3 seam, the Upper Cedar Grove seam and the Pittsburgh No. 8 seam were subjected to microbubble column flotation after being finely pulverized to improve the liberation of mineral matter. These tests were performed in an attempt to demonstrate the ability of the microbubble column process to produce superclean coals containing less than 2% ash. All samples were prepared by crushing the run-of-mine coal to -6 mm using a laboratory jaw crusher. Samples were then split into representative lots of 300 grams each, placed in air-tight containers, and stored at -20°C in a freezer. Prior to flotation, samples were passed through a laboratory hammer mill and dry-ground to -100 mesh. This procedure was followed by wet-grinding at 40% solids in a 13.3-cm diameter stirred ball mill for 30 minutes with 3.2-mm diameter stainless steel grinding media. The mean product size of the mill product was found to be approximately 5 microns, as determined using an Elzone 80-XY particle size analyzer. After grinding, samples were diluted to 5% solids by weight in a conditioning sump. A kerosene addition corresponding to 0.7 kg/ton of feed coal was utilized in all of the experiments, and Dowfroth M-150 was added directly into the bubble generation circuit at a rate of 2.7 kg/ton. All flotation tests were conducted in a 5-cm diameter flotation column with an L/D ratio of 37. The flow rates of the feed slurry and countercurrent wash water were held constant at 0.05 l/min and 0.80 l/min, respectively. The results of these tests are listed in Table 1.

As shown, the microbubble flotation process was able to consistently produce superclean coal (< 2% ash) from a variety of coal seams. At the same time, recovery was maintained near or above 70%. It is also interesting to note that although nothing was done specifically to prevent the flotation of pyrite, the microbubble process appears to inherently reject pyritic sulfur, as indicated by the total sulfur values in Table 1. In fact, for the Pittsburgh No. 8 seam, nearly 30% of the total sulfur was removed.

TABLE 1

Production of Superclean Coal

Coal Seam	Feed Ash %	Feed Sulfur %	Product Ash %	Product Sulfur %	Yield %	Recovery %
Elkhorn No. 3	9.12	0.81	1.73	0.75	78.8	84.7
Upper Cedar Grove	7.80	0.81	1.87	0.69	64.4	68.5
Pittsburgh No. 8	5.10	1.46	1.87	1.05	72.9	75.3

Additional tests were conducted to determine the feasibility of the microbubble flotation process for producing an ultraclean coal (< 0.8% ash). These tests were carried out using an Elkhorn No. 3 seam coal which had been cleaned to approximately 1% ash in a heavy media cyclone at 1.3 specific gravity. A -20 mesh "as-received" sample was pulverized in a laboratory ball mill to -100 mesh, followed by grinding to a mean size of approximately 5 microns in a stirred ball mill. Microbubble experiments were conducted at two different froth heights, and the results are given in Table 2. At a froth height of 30 cm, an ultraclean coal containing 0.45% ash was obtained with a combustible recovery of over 98%. By increasing the froth height to 45 cm, the ash content was further reduced to 0.41%, although the recovery dropped to 74%. It appears that much of the pyritic sulfur was removed from the original feed sample since little change in the total sulfur was observed.

TABLE 2

Production of Ultraclean Coal

Froth Height	Feed Ash %	Feed Sulfur %	Product Ash %	Product Sulfur %	Yield %	Recovery %
30 cm	1.03	--	0.45	--	98.1	98.6
45 cm	0.93	0.58	0.41	0.57	73.6	74.0

Refuse Coal Flotation

The processing of refuse material is also considered to be a potential application for microbubble flotation. A -100 mesh refuse sample from the Coalburg coal seam, West Virginia, was obtained for microbubble testing. This particular sample was chosen since the preparation plant which was processing this coal was unable to produce an acceptable product from the -100 mesh material using conventional flotation techniques. This material accounts for approximately 6-7% of the raw coal entering the plant, and is currently being discarded as plant reject.

The sample, containing approximately 40% ash, was received in slurry form at 30% solids. Upon conditioning with 0.2 kg/ton of kerosene, the sample was fed without dilution directly into the column at a rate of 0.69 liters/min. A countercurrent wash water rate of 0.8 liters/min and an air flow rate of 1.3 liters/min were employed. Dowfroth M-150 was added into the bubble generation circuit at a level corresponding to 0.2 kg/ton of raw coal. Flotation tests were carried out using a 5-cm diameter microbubble column with an L/D ratio of 37.

The results of the microbubble flotation test are given in Table 3, along with results from a conventional flotation test conducted using a Denver Model D-12 flotation machine. All experimental conditions were held constant in both tests, although the air flow rate of the conventional test had to be increased to 6 liters/min in order to obtain a froth layer of adequate stability. Both experiments were conducted at the same mean residence time. As shown, the microbubble process produced a clean coal containing less than 7% ash, with a combustible coal recovery of over 90%. The ash content of the product produced by the conventional flotation machine was very poor (32.7% ash), despite having a coal recovery below that of the microbubble column (79.3% versus 92.4%). Since the feed sulfur content was low, neither method seemed to have a large impact on sulfur reduction.

TABLE 3

Flotation of the Coalburg Seam Coal (-100 Mesh)

Test	Feed Ash %	Feed Sulfur %	Product Ash %	Product Sulfur %	Yield %	Recovery %
Microbubble	39.20	0.72	6.77	0.65	60.3	92.4
Conventional	37.98	0.72	32.71	0.74	73.1	79.3

Removal of Impurities from Kaolin

In order to study the effectiveness of microbubble flotation for removing anatase from kaolin, a series of tests were conducted on a Middle Georgia clay sample. Test samples consisting of 2000-gram lots were blunged for 5 minutes in a Waring blender at 60% solids using 2.7 kg/ton of sodium silicate and 1.4 kg/ton of ammonium hydroxide. The samples were then diluted to 45% solids and conditioned for 5 minutes with 1.5 lb/ton of potassium octylhydroxamate which was used as a collector for anatase (ref. 10). As shown in Table 4, three tests were conducted at various feed flow rates. A gas flow rate of 1200 ml/min and a wash water flow rate of 250 ml/min were used in all three tests. This resulted in a mean residence time of under 5 minutes for all tests. The results clearly show that very high clay recoveries (i.e., > 96%) were obtained in all cases, while the TiO_2 content of the kaolin product was extremely low (i.e., < 0.25%). In addition, it can be seen that the TiO_2 content decreased from 0.25% to 0.19% with an increase in feed rate from 22 ml/min to 60 ml/min. These results are substantially better than what is obtainable with conventional flotation.

TABLE 4

Removal of Impurities from Kaolin

Test No.	Feed Rate (1/h)	Feed % TiO_2	Product % TiO_2	Clay Recovery
1	1.32	1.46	0.25	96.2
2	2.40	1.46	0.20	96.3
3	3.60	1.46	0.19	96.1

CONCLUSIONS

1. The fundamental hydrodynamic theory behind the microbubble flotation process has shown that small bubbles provide improved recovery and selectivity in the flotation of fine particles.

2. Using the theoretically determined rate constant for bubble-particle attachment, a population balance model for the flotation of fine coal in a column has been developed. The model is well suited for independently assessing the effects of various operating parameters on column

performance. Furthermore, the model should prove to be useful for process optimization and design since it was developed from first principle considerations.

3. The column simulation results suggest that the best means for obtaining high flotation recoveries is through the use of tall columns and small bubbles. Once the maximum possible recovery has been obtained, improvements in product grade are best achieved by preventing the entrainment of fine gangue particles into the froth. Of the various operating parameters studied, the addition of countercurrent wash water appears to be the most effective means of preventing entrainment.

4. A bench-scale microbubble flotation column was found to perform well for a variety of applications including the production of superclean coal (< 2% ash), the production of ultraclean coal (< 0.8% ash) and the removal of anatase from kaolin clay.

ACKNOWLEDGMENTS

The authors would like to acknowledge the United States Department of Energy for the support of this work through contract numbers DE-AC22-86PC91221 and DE-AC22-86PC91274.

REFERENCES

1 E.H. Rose, Controversial art of flotation, Trans. AIME, 169 (1946) 240.
2 G.S. Dobby and J.A. Finch, 114th Annual Meeting of SME-AIME, New York, Preprint No. 85-124, (1985) 10 pp.
3 R.H. Yoon and G.H. Luttrell, Frothing in Flotation: The Jan Leja Volume (J. Laskowski, ed.), Gordon and Breach, New York, (1988) in preparation.
4 G.H. Luttrell, J. Yordan and R.H. Yoon, Induction time measurements for a coal flotation system, Proc. Symp. on Interfacial Phenomena in Biotechnology and Materials Processing, Boston, MA, August 3-7, (1987).
5 R.H. Yoon and G.H. Luttrell, The effect of bubble size on fine coal flotation, Coal Preparation, 2 (1986) 179.
6 A.J.R. Bennett, W.R. Chapman and C.C. Dell, Studies in froth flotation of coal, Proc. Third Int. Coal Prep. Congr., Brussels-Liege, June, Paper E2 (1958).
7 R.H. Yoon, Flotation of coal using micro-bubbles and inorganic salts, Mining Congress J., 68(12) (1982) 76.
8 M.J. Mankosa, G.T. Adel, G.H. Luttrell and R.H. Yoon, Model-based design of column flotation, Proc., The Mathematical Modeling of Metals Processing Operations, Extractive and Process Metallurgy Fall Meeting, AIME, Palm Springs, CA, Nov. 29 - Dec. 2, (1987).
9 G.H. Luttrell, G.T. Adel and R.H. Yoon, Modeling of column flotation, SME Annual Meeting, Denver, Colorado, February, Preprint No. 87-130 (1987).
10 R.H. Yoon and T. Hilderbrand, Purification of kaolin clay by froth flotation using hydroxamate collectors, U.S. Patent No. 4,629,556 (1986).

Interfacial Phenomena in Biotechnology and Materials Processing,
edited by Y.A. Attia, B.M. Moudgil and S. Chander
Elsevier Science Publishers B.V., Amsterdam, 1988 — Printed in The Netherlands

EFFECT OF WATER RECYCLING ON SELECTIVE FLOCCULATION OF COAL

H. SOTO, P. DAUPHIN, G. BARBERY

Dept. Mines et Metallurgie. Universite Laval
Quebec City, Quebec G1K 7P4, (Canada)

SUMMARY

Selective flocculation of coal from coal-clay mixtures is studied considering
the effect of water recycling and ionic content of water. The effects of
agitation intensity, pulp density, pH, and molecular weight and ionic charge
of the polymer are also considered. The supernatant clay suspension remaining
after coal separation, was subsequently flocculated with a non ionic polymer
to eliminate the clay and allow recycling of the water. It was observed that
direct recycling of water decreased markedly the selectivity of coal
flocculation, probably due to the presence of small amounts of the non-ionic
polymer remaining in solution after flocculation of the clay suspension. This
problem was eliminated by allowing a period of time after the separation of
the clay flocs. In this case, degradation of the residual polymer was likely
to occur and results similar to those obtained in fresh tap water were
achieved i.e. ash reduction from 44% to less than 20% in a single stage with
coal recovery better than 90%. By carrying out the coal flocculation step at
very low shear rates, it was possible to increase pulp density up to about
10% solids without a noticeable decrease in selectivity.

INTRODUCTION

Polyacrylamides (PAA) have been used in selective flocculation studies of
coal with a variable degree of success (1,2,3,4,5). Hucko (1) reported that
selectivity can be obtained at alkaline pH values with non ionic PAA of low
molecular weight. Attia and Fuerstenau indicated that anionic PAA are more
selective than non ionic ones (2). A similar finding has been made by Barbery
and Dauphin (5) and by Brookes et al.(3). Although these studies show the
potential of selective flocculation for the treatment of fine coal there are
some important issues that have not been considered. For example the problem
of water consumption and related aspects of water quality and particularly
water recycling have not been given the attention they deserve. On the other
hand, the importance of mixing in coal flocculation has been stressed in
studies by Hogg and coworkers (6,7) and also recently by Kogan et al. (8) and
Waters (9). From these works it is clear that size, strength and structure of
the flocs and other characteristics of the flocculation process are largely
determined by the mixing conditions prevailing during and after polymer
addition. In the present study it is attempted to use a controlled shear rate
during flocculation in order to increase selectivity. It is hoped that an
increased selectivity will make it possible to increase pulp density during
flocculation and therefore reduce water consumption. The problems of water
quality and water recycling are studied using coal-kaolinite mixtures as a
model system. Finally the role played by molecular weight and ionic character
of the polymers is further explored.

METHODS

Materials

Kaolinite of average particle size 1 micron was obtained from Georgia Kaolin through a local dealer. Medium volatile bituminous coal 1.5 inches to 28 mesh was obtained from Cape Breton Development Co., Nova Scotia. This sample was ground to minus 100 micrometers using first a roll crusher and then a pulverizer. Average particle size of the ground coal was about 55 microns. Unless otherwise indicated, 1:1 mixtures of these minerals analyzing 44 % ash were used throughout the experimental work. It is expected that this mixture will be a good model of fine coal rejects that range from 20% to 80% coal and typically contain illitic and kaolinitic clays as the main gangue minerals. The median particle size of these rejects is reported to be 100 microns for coal and 5 microns for the mineral matter (6).
Polymers used in this work are commercially available products provided by Union Carbide (PEO) and by Cyanamid (PAA). Characteristics of these polymers, as provided by the manufacturers, are listed in Table 1.

TABLE 1. Characteristics of polymers

TRADE NAME	TYPE	CHARGE	MW (millions)
Polyox WSR coagulant	Polyethylene oxide	Nonionic	5
Superfloc 127	Polyacrylamide	Nonionic	>10
Superfloc A-120	Polyacrylamide	30% anionic	>10
Superfloc A-130	Polyacrylamide	40% anionic	>10
Superfloc A-137	Polyacrylamide	47% anionic	>10
Superfloc 204	Polyacrylamide	40% anionic	6-10

Sodium metaphosphate (SMP) was used as dispersing agent and, unless otherwise indicated, tap water averaging 24-28 ppm calcium was used in all tests.

Procedures

The bulk of the flocculation tests reported here were performed with 200 ml pulp samples. Different sets of experiments were also performed with pulp volumes of 400 and 2000 ml, but no noticeable difference was observed when the volume of the system was increased. The pulp was prepared by dispersing the required amount of mineral mixture in water containing the dispersing agent. The system was energetically stirred with a 4.6 cm diameter propeller-type impeller at 1200 rpm in a baffled beaker during 15 minutes at a constant pH value. After dispersion, the suspension was mildly agitated with a paddle-type impeller (10) at just 80 rpm in the same beaker but without baffles. Polymer was added in the form of a single instantaneous dosage of 20 ml of a diluted solution and the pulp was stirred for 20 seconds before stopping the stirrer. After 5 minutes settling, the supernatant suspension was siphoned out and the flocs were dried, weighed and analyzed for ash content.

In the water recycling tests, the mineral suspension remaining after separation of the coal flocs, was flocculated once more; this time the pulp was stirred with the propeller-type impeller at 600 rpm in the baffled beaker. After adding the polymer (Superfloc 127) the pulp was stirred for 30 seconds and then immediately filtered (or decanted after 48 hours) and reused for a new test. A scheme of this procedure can be seen in figure 1. Zeta potential tests were performed in the usual manner with a Rank Brothers Mark

II micro-electrophoresis apparatus (5).

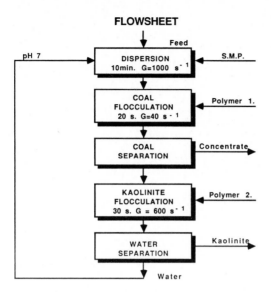

Fig. 1. Flowsheet of selective flocculation of coal-kaolinite mixtures with total water recycling.

RESULTS AND DISCUSSION

Mixing Effects

In the flocculation process mixing plays a key role because it promotes particle-particle collisions, homogenizes polymer concentration in the pulp

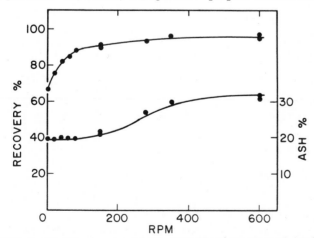

Fig. 2. Effect of stirring speed on flocculation of coal-kaolinite mixtures with 35 g/t Superfloc A-120 and 100 mg/l sodium metaphosphate at pH 7.4 and 2% solids.

and controls flocs growth by breakage of large flocs or by attrition (6-9). In the present research work, some preliminary tests showed that large, dark and fast sedimenting flocs were obtained when PAA was added to a mildly stirred pulp. Upon sedimentation a white supernatant apparently containing most of the kaolinite was obtained. On the contrary, strong stirring yielded smaller flocs and a dark supernatant.

These results suggest that coal recovery and perhaps selectivity would be strongly dependent on the shear stress during flocculation. A more detailed study of the effect of agitation intensity was therefore carried out. Some results of this study are shown in Figure 2 and Table 2. It is evident in this Figure that an extremely gentle mixing was enough to induce flocculation. The mixing obtained by just pouring the 20 ml polymer solution into the pulp allowed a 66% recovery of coal. Mild agitation increased recovery to 90% at about 80 rpm. Stronger mixing conditions did not improve noticeably the recovery of coal, on the contrary, at agitation speeds higher than those shown here, recovery actually decreased.

TABLE 2. Flocculation of pure kaolinite at different agitation speeds under conditions similar to those of figure 1.

A. With 35 g/ton polymer

POLYMER	RPM	% FLOCCULATED
NONE	80	12.5
NONE	600	12.7
A-120	80	13.3
A-120	600	15.7
127	80	16.3
127	600	20.6

B. With 50 g/ton Superfloc 127

RPM	% FLOCCULATED
60	28
100	28
200	36
400	38
600	33

On the other hand, selectivity remained practically constant from 0 to about 120 rpm. Beyond this speed, ash content of the flocs increased slowly but steadily from 20% to more than 30% at 600 rpm. For the impeller used in the low speed tests (0-100 rpm) the mean velocity gradient (G) at 60-80 rpm, according to data provided in reference 9 was estimated to be 42-45 seg. For higher agitation speeds a propeller-type impeller was used. For this impeller G was estimated to be 550 seg at 600 rpm according to data in reference 10. It is clear from this data that a shear rate higher than about 45 seg induces the incorporation of kaolinite to the flocs and therefore is detrimental to the selectivity of the process. Data in Table II also shows a higher flocculation of kaolinite at increased shear rates with a maximum recovery in the vicinity of 400 rpm. As a conclusion an agitation speed of 80 rpm was chosen for all subsequent coal flocculation tests while kaolinite flocculation was carried out at 600 rpm.

Effect of pulp density

In Figure 3 it is shown that increasing pulp density up to about 12 % solids did not affect selectivity. On the other hand, recovery increased at high pulp densities. These results indicate that under the appropriate mixing and reagent conditions, high   selectivity and recoveries would be possible at percent solids even higher than those normally used in coal flotation. In order to obtain these results, it was necessary to add sodium hydroxide to keep the pH at neutral value during dispersion; pH tended to decrease due to the larger amount of kaolinite added. It was also necessary to use more concentrated dispersant solutions. However, the increment in solution concentration was smaller than the rate of increase of solids, therefore dispersant consumption in terms of grams per ton of solids actually decreased at high percent solids. At pulp densities higher than about 14% solids, a steady increase of the ash content of the flocs was observed. This was at least partially due to the slower settling velocity of the flocs which made it necessary to increase the settling time. The lower settling velocity can be explained by the higher density and higher viscosity of the pulp. In these conditions, an interesting approach would be to attempt the separation of the flocs by floating them under an extremely mild agitation regime.

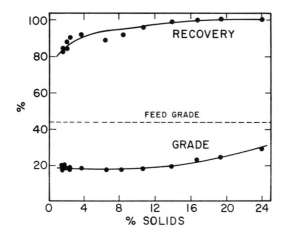

Fig. 3. Effect of pulp density on flocculation of coal-kaolinite mixtures with 35 g/t Superfloc A-120 at pH 8.0. (dispersant concentration was steadily increased from 100 mg/l to 600 mg/l to account for the larger amount of solids to be dispersed. However, in terms of g/ton of solids, dispersantdosage actually decreased from 10 kg/ton at 2% solids to 1.8 kg/ton at 24% solids).

Effect of feed grade

In Figure 4 it is shown that the feed grade in the range 23% to 58 % ash has a minor effect on the flocculation behavior. In quantitative terms, the ash reduction increased with the ash content of the feed at the cost of a lower recovery. With all feeds tested, ashes in the concentrate were reduced to less than one half the ash content of the feed.

380

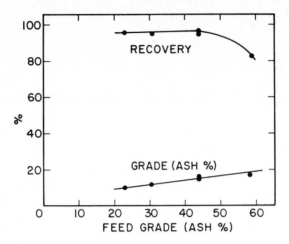

Fig. 4. Effect of feed grade on flocculation of coal-kaolinite mixtures with 35 g/ton Superfloc 204 at pH 7.0. Dispersant 350 mg/l and pulp density 9% solids.

Effect of pH and polymer characteristics

A series of flocculation tests was performed with a number of different non ionic and anionic polymers as a function of the solution pH. Results are presented in Figures 5A and 5B.

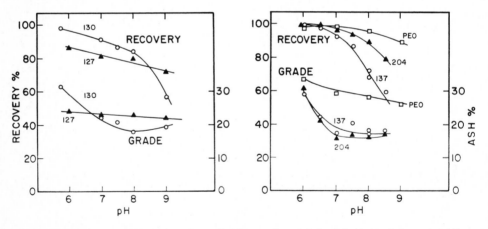

Fig. 5. Effect of pH on flocculation of coal-kaolinite mixtures with different polymers at a constant dosage of 35 g/ton. Dispersant 350 mg/l and pulp density 9% solids. See description of polymers in Table 1.

Several interesting conclusions can be drawn from these figures. It appears clear that the non ionic polymers (PEO and Superfloc 127) are less selective than the anionic PAA. This is in agreement with results reported previously (2,5). Also evident in these figures is the importance of pH on both recovery and grade when anionic polymers are used. In all cases, recovery decreases when pH increases, while grade reaches a maximum in the range of pH 6.5 to 8.0 depending on the degree of charge of the polymer. This behavior can be correlated with the electrokinetic potential of the two minerals shown in Figure 6.

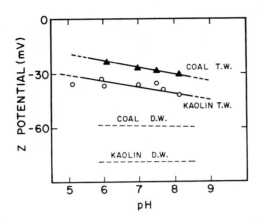

Fig. 6. Zeta potential of coal and kaolinite as a function of pH in tap water (TW) and distilled water (DW).

For both minerals zeta potential is negative and decreases at alkaline pH values. This should increase the repulsive forces between the mineral surface and the negatively charged polymers. This results in lower flocculation and eventually in lower grades, due to the reduced amount of coal settling in a given time (while there is a rather constant amount of gangue naturally settling). However, since flocculation is observed, it is evident that other adsorption forces in addition to electrostatics must be acting during the adsorption of anionic polymers on coal The effect of pH is much less important in the case of non ionic polymers where polymer adsorption is much less likely to be affected by electrostatic interactions. This behavior is also in agreement with results of PEO adsorption on graphite and anthracite reported by Gochin et al. (11).

Both the degree of anionic charge and the molecular weight of the PAA had a small yet potentially important effect on recovery and grade of the flocs. The best results were obtained with the smaller polymer tested (SF 204) which had a molecular weight of near 10 million. This suggests that smaller anionic PAA might be even more selective. A relatively high anionic charge (30-40%) seems necessary to guarantee selectivity; however, an excessive charge (more than 40% anionic) seems to reduce recovery and this could also eventually result in lower selectivity for the reason discussed above.

Water quality

The effect of dissolved ions on selective flocculation of coal is shown in Figure 7 and Table 3.
In order to isolate the effect of the specific ions studied, these tests were performed in distilled water. As shown in Figure 7, the presence of sodium chloride had a definite effect on flocculation. Recovery increased to reach a maximum with concentrations of the order of 0.5 to 1.0 g/l while ash content of the flocs increased slowly for concentrations higher than about 0.5 g/l. It is apparent that good flocculation results can be expected in waters

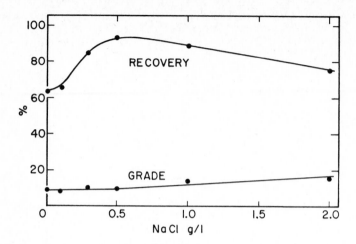

Fig. 7. Effect of NaCl concentration on flocculation of coal-kaolinite mixtures with 60 g/ton Superfloc A-120 in distilled water at pH 7.4. Dispersant 100mg/1.

TABLE 3. Effect of calcium on selective flocculation of coal-kaolinite mixtures in distilled water at pH 7.4; polymer 35 g/ton A-120; dispersant 100 mg/1. Pulp density 2.0% solids

Ca (mg/l)	Coal recovery (%)	Grade (ash %)
0	70	10.5
10	83	17
20	95	20
24-28 *	85	18
40	98	37

* Tap water with no additionnal calcium.

containing moderate amounts of sodium chloride. As seen in Table 3 the effect of calcium ions was more drastic, the presence of small amounts of this ion significantly reduces selectivity. This points to a potential problem if hard waters are used for selective flocculation of coal with PAA. It also stresses the fact that water quality considerations are an important factor in selective flocculation of coal. However, results presented here show that a high degree of selectivity and recovery are still posible in water containing up to 28 ppm calcium and 1.0 g/1 sodium chloride. This means that a moderate alkalinity level can be tolerated without undue loss of selectivity or recovery.

Water recycling

Water recycling tests were performed batchwise following the scheme presented in Figure 1. Right after coal separation, the kaolinite suspension was flocculated with a non ionic PAA (Superfloc 127) and then, either filtered and immediately reused to disperse a new mineral mixture, or allowed to stand for 48 hours before being decanted and recycled. An important consideration in the selection of a pH value to carry out the selective flocculation of coal is the effect of pH on the clay flocculation stage required for water recycling purposes. Obviously using the same pH value in both flocculation steps would present a number of advantages. Kaolinite flocculation at alkaline pH values is difficult and requires rather high polymer dosages; as pH decreases to neutral or slightly acidic, destabilization of the suspension becomes easier. Earlier water recycling tests were not successful because coal flocculation was carried out at pH 8.0, and, therefore kaolinite flocculation required excessive amounts of polymer. This could likely result in a high residual concentration of polymer in the recycled water. It was

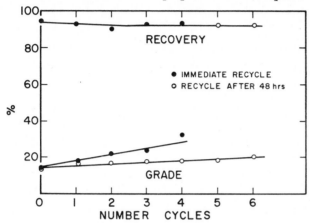

Fig. 8. Effect of water recycling on flocculation of coal-kaolinite mixtures with 35 g/ton Superfloc 204 at pH 7.0. Dispersant 350 mg/l and pulp density 9% solids.

demonstrated that the addition of very small amounts of superfloc 127 during the coal flocculation stage drastically reduced selectivity. Lowering the pH value of the solution after coal flocculation was not considered because this would result in build up of salt by neutralization of the pH regulating agents and also in additional reagent cost. A possible solution to the problem is to use, in the selective flocculation stage, those PAAS that show the best coal flocculation performance at a lower pH value i.e. at neutral or sligthly acidic pH. However, as shown in Figure 8, even when both flocculation stages were performed at neutral pH, selectivity still decreased after a few cycles of direct recycling of the filtered water. It was determined that the calcium content of the recycled solution tended to decrease after each cycle, while the dispersant concentration tended to increase. This made it increasingly difficult to flocculate the kaolinite suspension and therefore, it was necessary to increase the superfloc 127 dosage after the second cycle. It is likely then, that the reduced selectivity after each cycle, was due to a larger concentration of the non ionic polymer in the recycled water. This hypothesis seems to be confirmed by the results presented in Figure 8 which show that by allowing a period of 48 hours after kaolinite flocculation, and before starting a new cycle, the problem was practically eliminated. As shown in this figure, with this procedure essentially the same coal flocculation results were obtained even after six stages of total water recycling. It is likely that during this period of time the residual polymer was rendered inactive by natural degradation. This would not be an unrealistic solution to the problem of water recycling. In practice it would mean just the impounding of the tailings for a short period of time before water reclamation.

384

CONCLUSIONS

1. Selective flocculation of coal from kaolinite using anionic PAM is feasible. Selectivity depends on pH, ion content of the solution and molecular weight and degree of anionic charge of the polymer.

2. Selectivity can be achieved in water containing moderate amounts of sodium and calcium ions.

3. A low agitation intensity (a G value of 40 sec -1) improves selectivity.

4. Water consumption can be drastically reduced by increasing the pulp density and by recycling of water after flocculation of kaolinite using a nonionic polymer. It is convenient to carry out both flocculation steps at the same pH value.

5. Residual activity of the nonionic polymer in the recycled water reduces selectivity. This activity is likely to depend on the type and concentration of polymer used in the flocculation of kaolinite. However, this effect can be controlled by allowing a period of time after kaolinite flocculation to asure the destruction of the polymer remaining in the recycled water.

REFERENCES

1   R.E. Hucko, Beneficiation of Coal by Selective Flocculation, U.S.B.M. Int. Report 8234 (1977).
2   Y.A. Attia and D.W. Fuerstenau, Feasibility of Cleaning High Sulfur Coal Fines by Selective Flocculation, XIV Int. Miner. Proc. Congress, Toronto, Paper VII-4, (1982).
3   G.F. Brookes, M.J. Littlefair, L. Spencer, Selective Flocculation of Coal/Shale Mixtures Using Commercial and Modified Polyacrylamide Polymers, XIV Int. Miner. Proc. Congress, Toronto, Paper VII-7 (1982).
4   Y.A. Attia, S. Yu, S. Vecci, Selective Flocculation Cleaning of Upper Freeport Coal with a Totally Hydrophobic Polymer Flocculant, in: Y.A. Attia (Ed.) Flocculation in Biotechnology and Separation Systems, Elsevier, Amsterdam, 1987, pp. 547-564.
5   G. Barbery, P. Dauphin, Selective Flocculation of Coal Fines, in: Y.A. Attia (Ed.), Flocculation in Biotechnology and Separation Systems, Elsevier, Amsterdam, 1987, pp. 547-564.
6   R. Hogg, Flocculation Problems in the Coal Industry, in: P. Somasundaran (Ed.), Fine Particle Processing, Vol. 2, A.I.M.E., New York, 1980, p. 990.
7   R. Hogg, R.C. Klimpel, D.T. Rey, Agglomerate Structure in Flocculated Suspensions and its Effects on Sedimentation and Dewatering, SME AIME Annual Meeting, New Orleans, 1986.
8   V. Kogan, K. Driscoll, Y.A. Attia, Polymer Mixing, Selective Flocculation, in: Y.A. Attia (Ed.), Flocculation in Biotechnology and Separation Systems, Elsevier, Amsterdam, 1987, pp. 321-334.
9   K.J. Ives, Solid Liquid Separation, Chapter 4, part II, in: L. Svarovski (Ed.) Butterworths, (1981) 2nd Edition.
10 G.E. Joosten, J.G.M. Schilder, A.M. Broere, The Suspension of Floating Solids in Stirred Vessels, Trans. Inst. Chem. Eng., 55 (1977) 220-222.
11 R.J. Gochin, M. Lekili, H.L. Shergold, The Mechanism of Flocculation of Coal Particles by Polyethylene Oxide, Coal Preparation 2 (1985) 19 and 33.

Interfacial Phenomena in Biotechnology and Materials Processing,
edited by Y.A. Attia, B.M. Moudgil and S. Chander
Elsevier Science Publishers B.V., Amsterdam, 1988 — Printed in The Netherlands

385

SURFACTANT ADSORPTION AND WETTING BEHAVIOR OF FRESHLY GROUND AND AGED COAL

B. R. Mohal and S. Chander

Mineral Processing Section, Department of Mineral Engineering, College of Earth and Mineral Sciences, The Pennsylvania State University, University Park, PA 16802

SUMMARY
    The adsorption and wetting behavior of freshly ground and aged coals has been determined in the presence of a nonionic surfactant. The amount of surfactant adsorption changes with the concentration and "age" of the coal. The freshly ground coal has smaller amounts of adsorption at lower surfactant concentrations than the "aged" coal whereas at large surfactant concentrations, the order is reversed. In the latter case, more surfactant is adsorbed on fresh coal than on the "aged" coal. It is proposed that at low concentrations, the surfactant adsorbs through both hydrophobic and hydrophilic interactions with the molecule lying parallel to the surface, whereas at higher concentrations surface micelles form at the hydrophobic sites. The wetting behavior correlates well with the adsorption behavior.

INTRODUCTION

    Nonionic surfactants have many applications in the coal and mineral processing industry. They can be used as dispersing agents to modulate flotation, flocculation and filtration of particulates (ref. 1). They can also be used to increase the wettability of hydrophobic solids such as coal and can hence be used as dust suppressants (ref. 2,3). Although the adsorption of representative nonionic surfactants such as Triton N-101 (an ethoxylated nonylphenol with 9.5 moles of ethylene oxide and HLB equal to 13.4) have been studied extensively on substrates such as calcium carbonate (ref. 4), silica (ref. 5) carbon black (ref. 6) and graphite (ref. 7), very few studies are available for coals. Coals differ in structure from other substrates with respect to two important characteristics: 1) coal are porous and 2) coals consist of a patchwork of hydrophobic and hydrophilic sites (ref. 8). Since the adsorption characteristics of a surfactant are influenced significantly by the nature of the substrate the characteristics of the coals must be taken into consideration.

    Kinetics of surfactant adsorption is likely to have a major influence on the wetting of coal, since the coal particles are wetted in a few seconds or less (ref. 9). Equilibrium adsorption measurements are normally carried out for a period ranging from of 2 to 18 hours to achieve equilibrium (ref. 4,5). For porous solids such as coal, the surfactant molecules can diffuse into the pores of the coal thereby showing a much greater adsorption than would be possible on the external surface only. Therefore for wetting studies in which the coal particles are wetted in a few seconds, the wetting test data might not be directly relatable to equilibrium adsorption measurements. In addition, it is well known that coal particles undergo disintegration when being agitated during the adsorption tests. The disintegration and dissolution of dissolved species (ref. 10) is a function of time and would therefore have an influence on the equilibrium adsorption results. The influence of the dissolution of species from coal probably does not exist during the time frame of a few seconds during which wetting occurs. Thus the rapid adsorption studies become even more important to understand the mechanisms of surfactant-coal interactions during rapid processes such as wetting.

In this investigation, rapid adsorption measurements have been made to study the influence of aging on wetting behavior of coals in the presence of surfactants. It is well known that oxidation or aging of coal increases the relative proportion of the hydrophilic sites of coal and effect their floatability (ref. 11). These studies have been conducted to determine the effect of the relative amounts of hydrophobic to hydrophilic sites on coal on the mechanism of surfactant adsorption.

EXPERIMENTAL MATERIALS AND METHODS
Coal and Reagents

The two coal samples used in this investigation were: a HVA bituminous coal from the Upper Freeport seam (PSOC 1361p*) and a sub-bituminous A coal from Colorado (PSOC 1382p*), both were obtained from the Penn State Coal Data Bank. The analysis of the coals are given in Table 1. The coal samples were crushed and screened to obtain a 250 x 150 μm fraction used for the tests. For the aging tests coal samples were aged in ambient air for a period of one week.

The surfactant used in this investigation was an ethoxylated nonylphenol (Triton N-101) obtained from Rohm and Haas. Conductivity water (18 mohm cm^{-1}) obtained from a Millipore Milli-Q water system was used for the adsorption experiments. Distilled water from a tin lined Barnstead still equipped with a Q baffle system was used for the wetting experiments.

Surfactant Adsorption

For adsorption measurements a 10 g sample of 250 x 150 μm coal was deslimed over a 150 μm screen in distilled water. The coal was then placed in an agitated vessel and enough water added to make up the volume to 300 ml. A

---

*(The PSOC numbers refer to the Penn State coal Data Base Catalog numbers.)

TABLE 1

Analysis of the coal samples

	HVA-bituminous	Sub-bituminous
H_2O	3.2 %	13.03 %
Proximate analysis (Dry basis %)		
Ash	8.47	5.65
VM	38.16	34.20
FC	53.37	47.12
Ultimate analysis (DAF basis %)		
C	81.94	75.19
H	5.7	5.04
N	1.54	2.01
S	2.12	0.40
O (Diff)	8.7	17.37
Surface area (250 μm x 150 μm fraction)		
N_2 (BET) cm^2/g	3,300.0	2530.0*
Blaine cm^2/g	618.0	1942.0

* Estimated from Esposito (ref. 13)

liquid sample was continuously withdrawn through a 10 μm filter with the help of a peristaltic pump and transferred to the spectrophotometer. A schematic of the apparatus is shown in Figure 1. At the start of the experiment the solution in the flow cell was referenced as the blank using an HP 8451A Diode Array Spectrophotometer. The UV spectra in the range 190-300 nm was then monitored every 1.2 seconds. A known quantity of surfactant (1-5 ml depending on the desired concentration) was injected 10 seconds after the solution was referenced. The adsorption test was carried out typically for about 6-7 hours. The coal was then filtered, dried and weighed. The surfactant Triton N-101 has two absorbance peaks at 224 nm and 276 nm due to the phenol. In the concentration range investigated the 224 nm peak was linear and had a stronger absorbance than the 276 nm peak. Due to greater sensitivity of the former peak it was used for the adsorption isotherms. Preliminary experimentation has shown that the agitation rate had no effect on the absorbance measurements.

Wetting Rate of Powders by the Modified Walker Method

In the Modified Walker Test a 30 mg sample of 250 x 150 μm coal was dropped on the surface of a surfactant solution and the amount of coal wetted

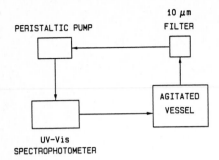

Fig. 1. Schematic of the setup for adsorptions measurements.

by the liquid was measured by placing the pan of a Cahn electrobalance beneath
the liquid surface. The initial wetting rates were determined from measurements of the amount of coal wetted as a function of time. The rate of wetting
was found to be independent of the amount of the coal sample in the range
20-100 mg. The data reported for the wetting rates is an average of at least
four replicate measurements in which the coefficient of variation was about
0.08. Additional details of the measurements are given elsewhere (ref. 9,12).

RESULTS AND DISCUSSION

Adsorption Kinetics

The absorbance at a wavelength of 224 nm of a solution containing
3×10^{-5} M (initial concentration) Triton N-101 is shown in Figure 2 with and
without the addition of a HVA bituminous coal sample. The solution (without
coal) reached 90 % of the equilibrium value in about 40 seconds and 100% in
about 5 minutes after the addition of the surfactant. A constant absorbance
thereafter showed negligible adsorption of the surfactant on the sides of the
tubes or the agitated vessel.

In the experiment in which coal was present in the solution the absorbance
first increased rapidly after the addition of the surfactant and then it
started to decrease as the surfactant adsorbed on coal. The maximum absorbance
was lower than the case where no coal was present. The plot of the absorbance
vs. $\sqrt{t}$ after the initial rapid adsorption is linear for a certain time period
that ranged from about three to seven hours depending on the surfactant
concentration. The mechanism of adsorption in this time period is considered
to be that of slow diffusion into the pores of coal. Extrapolation of the
linear diffusion curve to short times gives the amount of surfactant adsorbed
at the external surface as shown in Figure 2. The absorbance becomes constant
at longer times indicating equilibrium adsorption at the inner surface of the

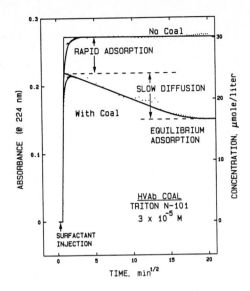

Fig. 2. Absorbance of a solution containing Triton N-101 with and without the presence of a HVA bituminous coal.

pores. The results of this aspect of the investigation will be reported elsewhere. Since different amounts of surfactant may be adsorbed depending upon the time for which adsorption is allowed to occur the amount of surfactant adsorbed during "rapid" adsorption on the external surface was chosen as the parameter for comparison. The external area was measured using the Blaine surface area measurement device, so that the adsorption density at the external surface could be determined.

<u>Disintegration and Dissolution of Coal During Surfactant Adsorption</u>

It is well known that various soluble species are leached out from coal in an aqueous environment (ref. 10). These soluble species have an absorbance in the UV range and therefore can interfere in the determination of surfactant concentration using UV spectrophotometery. In order to determine the effect of the soluble leached species from coal on the absorbance measurements the spectra of the solution containing an aged sub-bituminous coal was measured after equilibration for 2.5 hours and is shown as curve A in Figure 3. In addition, some fine particles may also be detached from coal particles (by disintegration or deaggregation processes) and may contribute to the background absorbance. To test this hypothesis the spectra of the same solution after centrifugation at 15,000 RPM for 0.5 hr is shown as curve B in Figure 3. The decrease in the absorbance of the solution after centrifugation indicates that some fines are also being produced but they are not the sole cause for the increase in background absorbance. The amount of fines produced and the

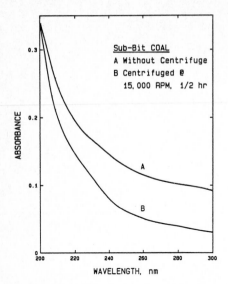

Fig. 3. Spectrum of a solution containing an aged sub-bituminous coal
A) before centrifugation and B) after centrifugation at 15,000 RPM for 1/2
hour.

species dissolved depend upon the surfactant concentration as shown in Figure
4. In this figure the absorbance at 298 nm is plotted as a function of Triton
N-101 concentration at various times of equilibration. Similar plots would
result for other wavelengths also. The results at 298 nm were selected to

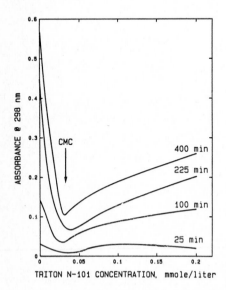

Fig. 4. Background absorbance at 298 nm for an aged sub-bituminous coal as a
function of the Triton N-101 concentration.

minimize the effect of Triton N-101 absorbance on the spectra. At concentrations below the Critical Micelle Concentrations (CMC) the background absorbance decreases with increase in concentration whereas at concentrations above CMC the background absorbance increases.

For concentrations lower than the CMC the greater quantities of the monomer present in the solution tend to block the pores and slow the dissolution and disintegration of coal. For concentrations greater than the CMC the micelles can solublize the leached species and lead to an increase in the extent of dissolution and disintegration. The leaching is time dependant and therefore to minimize the interference from the leached species the "rapid" adsorption on the external surface was chosen for subsequent experimentation.

Adsorption Isotherms

The rapid adsorption isotherms for Triton N-101 are given in Figure 5 for the fresh and aged sub-bituminous coal and in Figure 6 for the HVA bituminous coal. The corresponding Langmuir plots are given in Figure 7 and 8, respectively. Compared to the freshly ground sample the aged sub-bituminous coal adsorbs a greater amount of surfactant at concentrations less than the CMC and adsorbs a lesser amount at concentrations greater than the CMC. The break in the Langmuir plot near the CMC indicates two modes of adsorption. For concentrations less than the CMC the adsorption is due to monomers, whereas for concentrations greater than the CMC the adsorption is due to "surface micelles". At longer times the adsorption continues to occur by diffusion of monomers into the pores of the coal as stated in a previous section.

At concentrations lower than the CMC the surfactant can adsorb on both the hydrophobic and hydrophilic sites of coal. The schematic of a surfactant molecule adsorbing on a hydrophobic substrate is shown in Figure 9. The adsorption on the hydrophobic sites is due to the hydrocarbon chain and the adsorption on the hydrophilic sites is due to hydrogen bonding between the substrate and the ethylene oxide part of the molecule. The free energy of adsorption for the aged coal surface is greater than the free energy of adsorption for the fresh coal surface for concentrations lower than the CMC as shown in Table 2. Since both types of adsorption (hydrocarbon chain interaction and hydrogen bonding) may take place simultaneously at low surfactant concentrations, the total adsorption would depend upon the relative proportion of hydrophobic and hydrophilic sites. The results show that the aged coal with a greater proportion of hydrophilic sites adsorbs a greater amount of surfactant as compared to the freshly ground coal.

At concentrations greater than the CMC, the relatively small area per molecule (11.4 $\overset{o}{A}^2$/molecule for the surfactant on the HVA bituminous coal and 20 $\overset{o}{A}^2$/molecule for the sub-bituminous coal) as shown in Table 2 is much greater

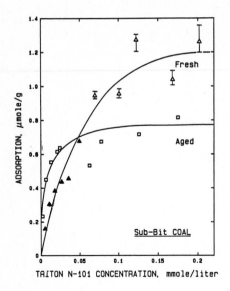

Fig. 5. Adsorption of Triton N-101 versus equilibrium concentration for a (△) freshly ground and (□) aged sub-bituminous coal. (For filled symbols error bars are less than symbol size.)

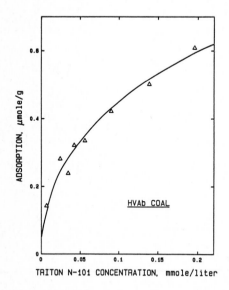

Fig. 6. Adsorption of Triton N-101 versus equilibrium concentration for a HVA bituminous coal.

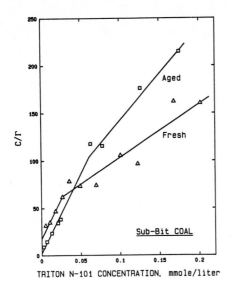

Fig. 7. Langmuir plot for adsorption of Triton N-101 on (△) freshly ground and (□) aged sub-bituminous coal.

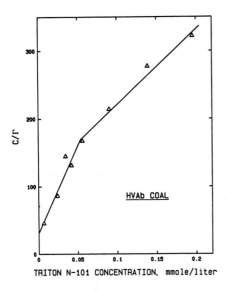

Fig. 8. Langmuir plot for adsorption of Triton N-101 on a HVA bituminous coal.

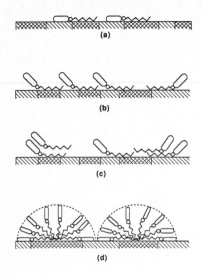

Fig. 9. A) Surfactant adsorption on a patchwork of hydrophobic and hydrophilic sites. B) Surfactant adsorption on a predominantly hydrophobic surface. C) Growth of Surface Micelles on hydrophobic sites. D) Schematic of a surface micelle.

TABLE 2

Free Energy and Area Per Molecule for Adsorption of Ethoxylated Nonyl Phenol

Concentration Range	ΔG^0 Kcal/mole	Area Per Molecule, $\AA^2$ --------basis--------------		
		CO_2 (DP)	N_2 (BET)	Blaine
Sub-bituminous				
Fresh				
Conc. < CMC	-9.07	6,400	70	53.7
Conc. > CMC	-8.07	2,400	26	20.0
Aged				
Conc. < CMC	-10.03	60,000	66	50.6
Conc. > CMC	-8.07	35,700	39	30.2
HVA bituminous				
Fresh				
Conc. < CMC	-9.19	105,000	151	28.2
Conc. > CMC	-7.83	42,000	61	11.4

than what can be accounted for by a monolayer adsorption. The cross sectional area for Triton N-101 is about 56 Å^2 (ref. 14) for the vertical orientation. Adsorption in multilayers might be considered but the shape of the adsorption isotherm is Langmuirian. Alternatively, one may consider that the micelles present in solution might attach through colloidal interaction but this mode is considered less likely because the coal is hydrophobic whereas the micelle is hydrophilic. Therefore the concept of a "surface micelle" has been proposed in this investigation. A surface micelle is formed by a mechanism in which the surfactant adsorbs on the hydrophobic coal surface by hydrocarbon chain interaction in the horizontal orientation. More molecules can adsorb on this first layer due to hydrocarbon chain interaction by a mechanism very similar to the formation of a micelle in the bulk solution as shown in Figure 9b-d. The surface micelle can be considered to be equivalent of half a micelle with the hydrophilic portion (ethoxylated part) exposed to water. Coal particles are wetted very easily at these surfactant concentrations as discussed in the paragraphs that follow. The interaction energies for the surfactant at concentrations greater than the CMC with the three substrates: aged sub-bituminous coal, freshly ground sub-bituminous coal and a HVA bituminous coal are very similar as shown in Table 2 which is consistent with the hypothesis that surface micelles are formed at energetically, similar sites which we consider to be the hydrophobic sites. Since the aged coal has a smaller number of hydrophobic sites the adsorption on it is less as shown in Figure 5 (at concentrations greater than CMC).

From Figures 7 and 8 it can be seen that the break in the Langmuir plots does not occur exactly at the CMC. This shift may in fact be real. It is well known that certain substances increase while others might decrease the CMC of a surfactant depending on their nature (ref. 15). From Figure 3 it is clear that certain species are dissolving from the coal and these may cause the CMC shift. Further experimentation is needed to establish the nature of the dissolved species.

Wetting Rates

The initial wetting rates are shown in Figure 10 for fresh and aged sub-bituminous coals and in Figure 11 for a fresh and aged HVA bituminous coal. The results show that there is an increased wetting of aged coal at low concentrations of the surfactant for both the sub-bituminous and the HVA bituminous coals. Although the changes in wetting rate are small, the results correlate well with the adsorption results, shown in Figure 5. Wetting of the coal in this low concentration range implies the adsorption of the surfactant onto the hydrophobic coal surface by the hydrophobic chain such that the hydrophilic portion of the surfactant extends outwards, as represented in Figures 9b and c.

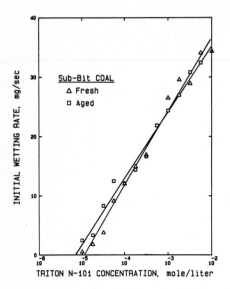

Fig. 10.   Initial wetting rate vs. concentration for a (△) freshly ground and
(▢) aged sub-bituminous coal.

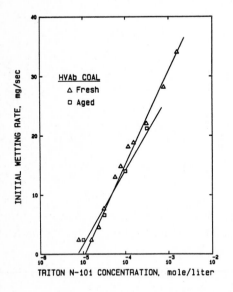

Fig. 11.   Initial wetting rate vs. concentration for a (△) freshly ground and
(▢) aged HVA bituminous coal.

At higher surfactant concentrations, the wetting rate of aged coal is less than that of the freshly ground sample for both the coals.

With the aged sub-bituminous coal at surfactant concentrations greater than the CMC the area per molecule is 30.2Å^2 as compared to 20.0Å^2 for the freshly ground coal, in the adsorption experiments (Table 2). The small differences in the wetting rate at higher concentrations is not entirely unexpected since the time for coal surfactant interaction in wetting is much lower (approximately a few milliseconds, ref. 9) than in the case of adsorption experiments. As stated above the aged coal has a smaller density of surface micelles which gives rise to a smaller rate of wetting.

CONCLUSIONS

The adsorption of ethoxylated nonyl phenol on coal may be summarized as follows: rapid adsorption of the surfactant on the external surface; slow equilibration by diffusion of the surfactant into pores and adsorption on the internal surface. For coals investigated in this study the pores accessible by nitrogen in surface area determinations are available for surfactant adsorption also. The break in the Langmuir plot near the CMC indicates two different modes of adsorption. For concentrations below the CMC the adsorption is due to surfactant monomers. For concentrations above the CMC the adsorption is through the formation of surface micelles at the external hydrophobic surface and as monomers at the internal surface.

This investigation shows that the presence of the surfactant affects the disintegration and dissolution of species from coals. Significantly greater amount of fine particles are generated and soluble species leach out from the aged sub-bituminous coal as compared to the freshly ground coal. The amount of substances released from coal depends upon the surfactant concentration with a minima at the CMC. For concentrations below the CMC, an increase in concentration of the monomer prevents the dissolution of the soluble species whereas for concentrations above the CMC the surfactant micelles can solublize the dissolving species and thus promote dissolution.

As compared to the freshly ground coal the aged coal adsorbs a greater amount of surfactant at concentrations less than the CMC and adsorbs a lesser amount at concentrations above the CMC. The effect of aging on surfactant adsorption can be explained if it is considered that the monomers adsorb (at conc. < CMC) on both hydrophilic and hydrophobic sites whereas the surface micelles form (at conc. > CMC) only at the hydrophobic sites. The wetting rates correlate very well with the adsorption results for both the HVA bituminous and the sub-bituminous coals.

ACKNOWLEDGEMENTS

The authors acknowledge the support from the Mineral Institutes Program under Grant No. G1135142 from the Bureau of Mines, U.S. Department of the Interior, as part of the Generic Mineral Technology Center for Respirable Dust, The Pennsylvania State University.

REFERENCES

1 G.D. Parfitt, Dispersion of Powders in Liquids, Applied Science Publishers, Third Ed., 1982.
2 R. Harold, Coal Age 84, No. 6 (June 1979) 102-105.
3 Anon., Rock Products, 65 No. 8 (August 1962) 92-96.
4 H. Kuno and R. Abe, Kolloid Zt., 177 (1961) 40-50.
5 D.N. Furlong and J.R. Aston, Colloids and Surfaces, 4 (1982) 121-129.
6 H. Kuno, R. Abe and S. Tahara, Kolloid Zt., 198 (1964) 77-81.
7 V.N. Moraru, F.D. Ovcharenko, L.I. Kobylinskya and T.V. Karmazina, Kolloid Zh., 46 (1984) 1148-1153.
8 J.A. Gutierrez-Rodriguez, R.J. Purcell Jr. and F.F. Aplan, Colloids and Surfaces, 21 (1984) 1-25.
9 B.R. Mohal and S. Chander, Colloids and Surfaces, 21 (1986) 193-203.
10 D.W. Fuerstenau and Pradip, Colloids and Surfaces, 4 (1982) 229-243.
11 J.A. Gutierrez-Rodriguez and F.F. Aplan, Colloids and Surfaces, 21 (1984) 27-51.
12 S. Chander, B.R. Mohal and F.F. Aplan, Colloids and Surfaces, 26(1987) 205-216.
13 C. Esposito , M.S. Thesis, The Pennsylvania State University, 1986.
14 L. Hsiao, H.H. Dunning and P.B. Lorenz, J. Phys. Chem., 60 (1956) 657-660.
15 M.J. Rosen, Micelle Formation by Surfactants, Surfactants and Interfacial Phenomena, John Wiley & Sons, New York, 1978.

Interfacial Phenomena in Biotechnology and Materials Processing,
edited by Y.A. Attia, B.M. Moudgil and S. Chander
Elsevier Science Publishers B.V., Amsterdam, 1988 — Printed in The Netherlands

INTERFACIAL AND COLLOIDAL EFFECTS IN LIQUID-LIQUID SEPARATION OF
ULTRAFINE COAL

T.C. HSU and S. CHANDER

Mineral Processing Section, Department of Mineral Engineering, College
of Earth and Mineral Sciences, The Pennsylvania State University
University Park, PA 16802

SUMMARY

Partitioning and separation of ultrafine particles of coal and ash-
forming minerals in liquid-liquid separation process depends on several kinds
of interfacial and colloidal interactions between dispersed "particles". The
type of these interactions are analyzed for the coal-mineral matter-water-oil
system. The analysis is used to explain the kinetics of transfer of coal and
mineral particles from the aqueous to the organic or the emulsion phase. The
results show that initially oil-in-water emulsions are produced which invert
to water-in-oil emulsion in the presence of hydrophobic particles. The type
of emulsions formed depend upon the hydrodynamic and solution conditions.
After long periods of mixing, 'bridged' emulsions are produced if the speed of
mixing is high. The stability of these emulsions strongly effects the performance of the process in terms of both the grade and the recovery.

INTRODUCTION

Addition of oil to improve the flotation of fine coal or oxidized coal is
a common practice. Oil is known to improve the separation performance by
enhancing the hydrophobicity of coal. Although the use of oil for beneficia-
ting naturally hydrophobic graphite was patented more than a century ago in
Germany (German Patent 42, Class 22, ref. 1), the use of oil in coal cleaning
has increased in recent years and several oil based processes, other than
flotation, have emerged. Two of these are: spherical agglomeration and
liquid-liquid separation. The processes in which two immiscible liquids are
present as distinct phases are sometimes referred to as selective coalescence
processes, a term which connotates coalition of dispersed particles (liquid
droplets or particles). Some examples of these processes include the
Puddington spherical agglomeration process in which oil is used as a bridging
liquid to selectively agglomerate hydrophobic particles (ref. 2), the
so-called Otisca-T process (ref. 3) in which a hydrocarbon (pentane or butane)
is used as the liquid immiscible with water or the LICARDO process in which
liquid carbon dioxide is used as the immiscible phase (ref. 4).

The amount of oil used is an important parameter in each of these processes. The role of oil in a particular process, depending upon its amount,
can be a surface modifying agent (dissolved oil), an agglomerating medium

(liquid bridges), or a collecting phase (oil droplet). When oil is present in sufficient quantities so as to form a separate phase, the performance of the process might depend upon a number of interfacial and colloidal interactions between the dispersed phases. The nature of these interactions have been analyzed in this paper for the coal-water-hydrocarbon oil as the principal phases. The analysis is used to explain the transfer of coal particles from aqueous phase to the emulsion.

THEORETICAL BASIS

Hierarchy of the Liquid-Liquid Separation System

The liquid-liquid separation process can be conveniently analyzed as a hierarchy system, shown in Table 1, in which the upper levels control the behavior of the lower levels. The parameters at the same level might be considered independent, though parallel interactions can exist. Parameters at Level II are the variables related to interfacial properties and the formation of colloids. The interactions between dispersed "particles" in a medium, among coal-water-oil phases, are considered in the colloidal level (Level III) which controls the recovery and grade of product through the characteristics of respective colloids. Separation performance can be estimated from the information obtained in Level IV. Some details of the various levels are discussed in the paragraphs that follow.

TABLE I

A Hierarchy System Proposed for the Liquid-Liquid Separation Process

LEVEL	
I	PHASE PROPERTIES/HYDRODYNAMICS: concentration, particle size, agitation intensity, etc.
II	PHASE INTERACTIONS: S/W, O/W, S/O interfacial properties, multi-phase contact
III	COLLOIDAL INTERACTIONS: dispersion, emulsion aggregation, agglomeration
IV	MACROSCOPIC BEHAVIOR: recovery, grade separation efficiency

Level I - Bulk Phase Properties/Hydrodynamics

In Level I, the parameters included are the physical and chemical properties of the phases in contact, such as density, viscosity, and size and shape
of particles. The variations in chemical conditions such as the concentration
of a salt or a surfactant, pH, and ionic strength are also considered as Level
I parameters. In addition, the hydrodynamic conditions determined by such
factors as geometry of the agitation system, stirrer speed, etc., are considered in this level.

Level II - Phase Interactions

In the multi-phase system considered in this paper, namely, coal-mineral
matter-oil-water, many kinds of interfaces are present. Level II variables
include various interfacial tensions (or energies) which determine the conditions for three phase contact among the interacting phases. For example, the
equilibrium position of particles at the oil/water interface is determined by
interactions at this level. Level II interactions also include the effect of
hydrodynamics on the dispersion of immiscible liquids in an agitated system.
The properties of an emulsion droplet are influenced by parameters in this
level.

Level III - Colloidal Interactions

Level III interactions include colloidal interactions between dispersed
"particles", which can be oil or water droplets, and coal or ash forming
mineral particles. As a result of these interactions, complex colloids such
as stabilized particle emulsions, coal-oil-water agglomerates, and two
dimensional aggregates of particles (or a solid film of particles) at the
oil-water interface, might form during the separation process. The characteristics of these particle-containing colloids will determine the grade and
recovery of the process.

Level IV - Macroscopical Properties

Level IV parameters are mainly the recovery and grade of the desired
product (clean coal), which are direct manifestations of colloidal interactions. In fact, the grade of product depends on the relative recovery of the
clean coal and gangue mineral. One may consider the recovery of solids as the
only independent variable in this level, since the recovery at any instant
will be defined by the entire history of the "particles" in the shear field.
The macroscopic picture of the system can be represented in the form of a
kinetic curve, or the recovery versus time plot. Any change in the upper
levels of the hierarchy can affect the kinetic curve by changing the shape of

the curve or shifting the curve in the time domain as discussed in a later
section.

Thermodynamics of the Collection of Fine Particles by Emulsion Droplets

Equilibrium Position of Particles at the Water/Oil Interface. The
equilibrium position of fine particles at the water/oil interface, according
to von Reinders (ref. 5), is determined by the relative magnitude of the
interfacial energies:

a) if $\gamma_{so} > \gamma_{wo} + \gamma_{sw}$ solid will remain dispersed in the
aqueous phase.

b) if $\gamma_{sw} > \gamma_{wo} + \gamma_{so}$ solid will be dispersed in the oil phase.

c) if $\gamma_{wo} > \gamma_{so} + \gamma_{sw}$ or if none of these three interfacial
tensions is greater than the sum of the
other two, the solid will concentrate at
the oil-water interface.

Case c) is actually the condition that

$$\gamma_{so} - \gamma_{sw} < \gamma_{wo}$$

where three phase contact becomes possible. The thermodynamic equilibrium can
be represented by Young's equation:

$$\gamma_{so} - \gamma_{sw} = \gamma_{wo} \cdot \cos \theta$$

or $\cos \theta = (\gamma_{so} - \gamma_{sw})/\gamma_{wo}$

If $\gamma_{so} > \gamma_{sw}$, the equilibrium three phase contact angle will be smaller than
90° and the solid is wetted preferentially by the aqueous phase. If $\gamma_{so} <
\gamma_{sw}$, the contact angle will be larger than 90° and the major part of the solid
is surrounded by the oil phase. A change in the contact angle at the
oil-water-solid interface will change the equilibrium position of the solid
and determine the type of emulsion formed. Schulman and Leja (ref. 6)
concluded that (particle-stabilized) oil-in-water emulsions formed when the
three phase contact angle was less than 90° (measured through water) while
water-in-oil emulsions formed when the contact angle was larger than 90°. The
contact angle of coal for the air-coal-water system is generally less than 90°
and varies with rank. Replacement of air by the oil phase would increase the
contact angle. Some typical values, measured as a part of this investigation,
are given in Table 2 for a bituminous coal.

Colloidal Interactions in the Coal-Oil-Water System

Several colloidal (Level III) interactions, which are likely to be present in the coal-oil-water system are listed in Table 3. The free energy of interaction between various "dispersed" particles under some typical conditions was calculated by the DLVO theory of colloid stability and the values for the primary maximum (V_m, h_m) and the secondary minimum (V_s, h_s) are also given in Table 3. The free energies, V's are given in kT units and the position of the free energy maximum, h_m or minimum, h_s is given in nm. The values of some of the parameters used in the calculations are given in Table 3, additional details may be found elsewhere (ref. 7). These calculations show that coalescence of oil droplets in water might occur through coagulation in the secondary minimum i.e. the coalescence may be a non-activated phenomena. The coalescence of water droplets in oil must occur by crossing the energy barrier of magnitude V_m at h_m, which therefore becomes a process involving activation energy. The attachment of oil droplets to coal particles could occur by both the coagulation in secondary minimum (non-activated) or by coagulation in primary minimum (activated) depending upon the pH and ionic strength of the solution. Although the interactions between coal particles are dominated by their electrical double layers, under certain conditions coagulation in the secondary minimum is possible.

METHODS AND MATERIALS

A baffled agitated tank of standard geometry was used (ref. 8). The tank was modified at the bottom so that the contents could be discharged by opening a valve. Most of the tests were conducted in Tank B of 420 ml capacity. Some preliminary studies were made in Tank A of 300 ml capacity. A six-blade turbine impeller was used as the agitator. In a typical test, a predetermined amount of coal sample was added to the aqueous solution of the desired pH and ionic strength. Unless stated otherwise, a slurry concentration of 1% solids by weight was maintained throughout the investigation. The slurry was first conditioned for 30 minutes at 2000 rpm to ensure dispersion and chemical equilibrium. No variation in separation performance was observed if the conditioning period was increased to 2 hours. After the conditioning period,

TABLE 2

Contact Angles for Some Coals in Coal-Air-Water and Coal-Oil-Water Systems

COAL	θ_a	θ_o
HVAB	72	124
SUB-B	44	110

the stirrer speed was set to the desired value and oil was added through a funnel within a period of 5 seconds. The volume of oil used was 15% of the total liquid volume. After a preset time for transfer of particles, the stirrer was stopped and the phases were allowed to separate for 5 minutes. The aqueous phase containing the ash-forming minerals was drained from the bottom of the vessel. It was necessary to wash the emulsion phase to remove the entrained ash-forming minerals. The material in the two aqueous phases and the emulsion phase was analyzed separately but for the purpose of calculating recoveries, the two aqueous phase samples were combined.

The ash tests were performed in a commercial ashing oven using ASTM procedure. The coal samples for analysis were predried at 110°C for one hour and then heated slowly to 750°C for at least two hours. The ash content was calculated from the weight loss. The total sulfur content was analyzed by a Leco Sulfur Analyzer.

The coal sample used in these tests was from Pittsburgh seam, Robena Mine, PA. Raw coal with a top size of ¼ inch analyzed 14.0% ash, 2.7% total sulfur (of which 1.15% was pyritic and the remainder was organic) and less than 1% moisture. The raw coal sample was ground to 80% -400 mesh in a laboratory coffee grinder and was further ground in a pulverizer (Buehler Rotary mill). The -400 mesh fraction was classified in a Donaldson Acucut air classifier. Unless otherwise specified, a $-10\mu m$ size fraction was used in the tests. The coal sample analyzed 12.4% ash and 2.5% sulfur.

Distilled water and reagent grade chemicals were used in all the experiments.

TABLE 3

Interaction Energies for Various Colloidal Interactions

Particle-Medium-Particle	V_m, kT	h_m, nm	V_s, kT	h_s, nm
oil-water-oil	>20,000	0.3	-50	30
water-oil-water	∿100	5	∿-1	>1000
coal-water-oil	>100	2	>-20 to-50	∿20
coal-water-coal	>1000	∿1	∿-1 to-20	∿20

Zeta Potential of oil droplets = - 50 mV

Zeta Potential of water droplets = - 10 mV

Ionic strength = 0.01 $\underline{M}$

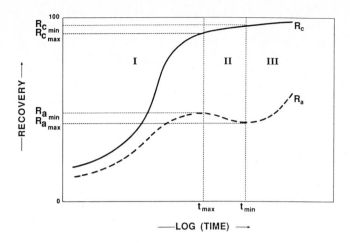

Fig. 1. A schematic representation of the kinetics of liquid-liquid separation process showing characteristic regions I, II, and III. Rc and Ra are the combustible matter and ash-forming mineral recoveries, respectively.

RESULTS AND DISCUSSION

<u>Kinetics</u>

The kinetic curves for combustible matter recovery (R_c) and ash mineral recovery (R_a) are schematically shown in Fig. 1. The R_c increases gradually with time of separation but R_a first increases, then decreases and finally increases again. On the basis of the ash recovery curve, the separation process is divided into three stages shown in the figure. Stage I is designated as the region from the start of the process to where ash recovery reaches its local maximum, $R_{a,max}$ at time, t_{max}. The period from t_{max} to t_{min} where the ash recovery reaches local minimum is designated as Stage II, and at times larger than t_{min} the system is considered to be in Stage III. The behavior of the system in various stages is discussed in the paragraphs that follow.

Stage I. Both R_c and R_a increase with time in this stage. Two types of dispersions can form when oil and water are mixed, namely, oil-in-water (O/W) and water-in-oil (W/O) emulsions. The type of dispersion formed under a given set of conditions depends on such factors as volume ratio of the liquid phases, type and intensity of agitation, position of agitator, etc. The fluid flow pattern for a turbine impeller in a baffled tank is represented in Figure 2a. The stress components at point P are schematically shown in Figure 2b.

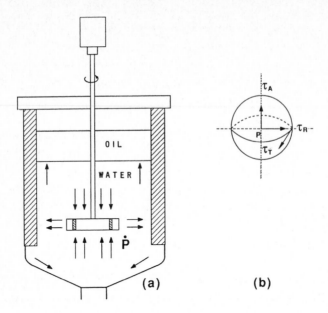

Fig. 2. (a) A representation of fluid flow pattern in a baffled tank.
(b) Various stress components at point P in the tank.

The axial stress component τ_A may be responsible for the initial dispersion of
the fluids while radial and tangential components, τ_R and τ_T, respectively,
might control the droplet size by fragmentation and coalescence processes.

The position of the agitator and the liquid to volume ratio used in the
present investigation favor the formation of an O/W emulsion. At the onset of
agitation sufficient amount of oil is not dispersed and therefore the interfacial area available for particle capture is small. The system is considered
to be under dispersion-control at this initial period. Since thermodynamic
conditions favor the capture of particles with a non-zero contact angle,
particles with small hydrophobicity are also captured by the droplets. In
addition, hydrophilic particles are known to stabilize O/W emulsions. These
conditions favor transfer of both hydrophobic and hydrophilic particles to the
emulsion phase resulting in poor selectivity or low efficiency of separation.

After a certain period of time in Stage I, the R_c increases rapidly. At
this time the oil droplet size approaches its steady state value, that is, the
available interfacial area reaches its maximum value. The amount of coal
recovered is determined by the rate at which particles are collected by oil
droplets. Under these conditions the system is considerd to be under
particle-capture control. The transition from dispersion-control to
particle-capture control is associated with the decrease in rate of the

combustible matter recovery. The decrease in rate of recovery is also due to the decrease in the quantity of coal available for capture. For the purpose of this discussion, the time at which slope of the recovery versus time curve is maximum is taken as the transition time, t_{tr}. At times less than t_{tr}, the system is under dispersion control whereas at times greater than t_{tr} the system is under particle-capture control.

Shortening the duration of the dispersion sub-process, by methods such as increasing agitation speed or pre-emulsification of the oil phase, can greatly shorten the time needed to achieve a certain recovery. This increase in rate can be represented in Fig. 1 by shifting the kinetic curve horizontally to the left or to shorter times. Higher recovery at a given time may also be obtained if coal is more hydrophobic. Such an increase in flotation recovery will shift the kinetic curve upwards on the recovery axis.

Stage II. The on-set of Stage II is marked by a decrease in the ash mineral recovery. At the same time, the combustible matter recovery continues to increase though at considerably lower rate than in Stage I. At the beginning of Stage II at time t_{max}, sufficient hydrophobic particles are collected at the surface of oil droplets to favor W/O emulsion and induce phase inversion. The inherent instability of the system under these conditions provides sufficient energy for readjustment of the particles at the interface and possible crowding-out of the particles with lower contact angles. Such a rejection of ash minerals, prompted by the phase inversion process improves the separation efficiency. An increase in pyrite rejection in flotation of coal has been observed by Aplan and associates (refs. 9,10). In most cases, about 10% increase in separation efficiency was observed in this stage. The relatively high value of $R_{a, min}$ is attributed to the presence of locked particles or inherent ash in coal matrix. Especially large value of inherent ash in this coal was confirmed by the washability techniques developed by Dumm and Hogg (ref. 11). The clean coal fraction (1.4 specific gravity) was 2.76% ash with a yield of only 36%, for example.

Stage III. The stage III is marked by an increase in the ash-minerals pick-up. As a result of the inversion of O/W to W/O emulsion in Stage II, particles-coated-water-droplets (PCWD's) are dominant in Stage III. Entrainment of ash particles within the aqueous droplets is very likely. Entrainment phenomena was confirmed by adding kaolin particles to the coal at the beginning of the separation tests. When the emulsion droplets were broken after the separation tests, they contained milky white suspension of the kaolin particles.

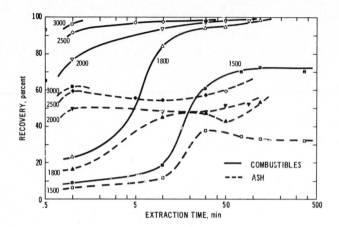

Fig. 3. The effect of agitator speed (given as rpm on respective curves) on the combustible matter and ash recoveries.

The coal particles coating the water droplets prevent the coalesence of the emulsion. These particles may promote aggregation of the droplets, however. The product obtained in this stage is a paste-like emulsion which could neither be dispersed by oil nor by water. The paste is considered to be a three-phase network of emulsion droplets and particles which is termed as a "bridged emulsion". The particles are held together by liquid bridges. Such bridged emulsions may be the precursors for the agglomerate formation in the spherical agglomeration process. Since considerable amounts of entrainment and entrappment of particles is likely to occur, the formation of bridged emulsion is considered detrimental to the separation process. The separation efficiency decreases rapidly with increasing time, once bridged emulsion has formed.

Effect of Hydrodynamics

The effect of stirrer speed on the combustible matter recovery and the ash-mineral recovery is shown in Fig. 3. These results show that the general mechanism of transfer is the same for all the speeds tested as can be seen from the similar shape of the kinetic curves. The kinetic curve shifts to the left (shorter times) and upward (high recovery) with increase in stirrer speed. With the exception of the lowest stirrer speed of 1500 rpm, the combustible matter recovery (Rc) reaches a value of 100% at sufficiently long times. At 1500, rpm, Rc decreased slightly on prolonged agitation.

The values of various kinetic parameters, defined in the previous section, are listed in Table 4 as a function of the speed of agitation. The following observations may be made with regard to the effect of hydrodynamics on the separation process:

a) Both the t_{max} and t_{min} decrease significantly as the stirrer speed increases.

b) Both $R_{a,max}$ and $R_{a,min}$ increase with increase in stirrer speed.

c) Except for the 1500 rpm, $R_{c,max}$ is about 90% of the total recoverable combustible matter.

d) The highest value of separation efficiency is observed at a moderate speed of 1800 rpm.

Based on the mechanism proposed in the previous section, the decrease in t_{max} and t_{min} can be recognized as simply due to faster dispersion of oil at higher stirrer speeds. Higher stirrer speeds increase the available interfacial area with a resultant increase in recovery of both the combustible matter and the ash-minerals at a given time on the kinetic curve. Although the stirrer speed is likely to affect the rate of collisions between particles and emulsion droplets also, the above observations can be simply interpretted in terms of the increase in interfacial area available for capture of particles.

The slight decrease in R_c in Stage III was observed at 1500 rpm. Such a decrease was observed in the presence of different salts also. An examination of the emulsion type revealed that W/O/W complex emulsion was formed instead of the bridged emulsion in stage III when the stirrer speed was 1500 rpm. That is, the formation of bridged emulsion might be a process which requires passing a certain energy barrier that can be overcome only when the stirrer speed exceeds a minimum value.

TABLE 4

Parameters Related to Onset of the Inversion Process

SPEED, rpm	t_{max},min	Ra_{max},%	Rc_{max},%	Es,%
1500	35	40	65	25
1800	15	48	90	42
2000	2	50	85	35
2500	1 ?	60	92	32

Note: Es = Rc − Ra

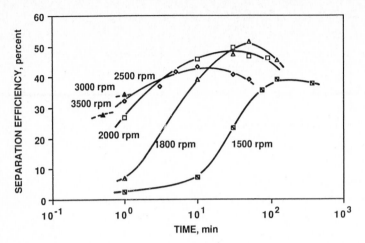

Fig. 4. The effect of agitator speed on the separation efficiency for separation of ash-forming minerals from coal.

The separation efficiency, E_s defined as the difference between the combustible matter recovery and the ash recovery (R_c-R_a), is plotted in Fig. 4 for several speeds. For each stirrer speed the E_s reaches a maximum value at t_{min}. The E_s is highest for the stirrer speed of 1800 rpm.

A logical step to describe the effect of hydrodynamics may be to find some shift factors to relate variation in stirrer speed with time, or to find some rate parameter to account for both the time and stirrer speed. Finding of both the horizontal (time) and vertical (recovery) shift factors is not possible from the present data, however, because a very accurate speed control is needed between 1500 to 1800 rpm. Attempts to construct a universal kinetic curve (recovery vs. some rate parameter) based on the total energy input were unsuccessful. The energy input into an agitated system after time t, is proportioned to $(N^3.t)$, where N is the stirrer speed. It is possible that the recovery depends upon the available interfacial area ehich depends on the parameter $(N^m.t^n)$ where m and n are constants. Typically, m = 0.72-3.5 and n < 0.5 (Ref. 7).

CONCLUSIONS

The mechanism of liquid-liquid separation of ultrafine particles (-10μm) of coal from mineral matter has been investigated in this study. A theoretical analysis of the process is presented in the form of a hierarchical system which takes into consideration the interfacial and colloidal interactions in an agitated system of two immiscible liquids in the presence of hydrophobic and hydrophilic particles. In the initial stages of mixing oil-in-water

emulsions form which invert to water-in-oil emulsions in the presence of
hydrophobic particles. The rejection of mineral matter is maximum under
incipient conditions of phase inversion. At prolonged agitation water-in-oil-
in-water complex emulsions form which increase mineral matter pick-up through
entrainment. Intense agitation results in bridged emulsions formed through
oil acting as the liquid bridge. Bridged emulsions contain substantial
quantities of entrapped mineral matter which lowers the grade of the product.

ACKNOWLEDGEMENTS

The authors acknowledge the financial assistance from the Coal Research
Section and the Ben Franklin Program of the Pennsylvania State University.

REFERENCES

1   Crabtree and Vincent, Historical outline of major flotation
    developments, D. W. Fuerstenau in: Froth Flotation, (Ed.), AIME, New
    York, 1962, pp. 39-53.
2   H.M. Smith, and I.E. Puddington, Sperical agglomeration of barium
    sulfate, Canadian J. Chemistry, 38 (1960) 1911.
3   F.J. Simmons and D. V. Keller, Two ton-per-day production of otisca
    T-process ultra-clean coal/water slurry, 10th International Coal
    Preparation Congress, CIM, 1986, pp. 1-8.
4   G. Araujo, D.X. Xe, S.M. Chi, B.I. Morsi, G.E. Klinzing and S. H. Chiang,
    Use of a liquid Co_2 for fine coal cleaning, Energy Progress, 7 (1987) 72.
5   W. von Reinders, Die verteilung eines suspendierten pulvers oder eines
    kolloid gelosten stoffes zwischen zwei losunggmittein," Kolloid Z, 13
    (1913) 235-241.
6   J.H. Schulman, J. Leja, Control of contact angles at the oil-water-solid
    interfaces, Trans. Faraday Soc., 50 (1984) 593-604.
7   T.C. Hsu, Liquid-Liquid Separation of Ultrafine Coal, M.S. Thesis, The
    Pennsylvania State University, 1987.
8   F.A. Holland and F.S. Chapman, Liquid Mixing and Processing in Stirred
    Tanks, Reinhold, New York, 1966.
9   C.M. Bonner and F.F. Aplan, The Influence of Oil on the Flotation of
    Pyrite during Coal Flotation, manuscript in preparation.
10 T.J. Olson and F.F. Aplan, The effect of frothing and collecting agents
    on the flotation of coal, pyrite and locked particles in a coal flotation
    System, Processing and Utilization of High Sulfur Coals II, Y.P. Chugh
    and R.D. Caudle, eds., Elsevier, NY, (1987) pp 71-82.
11 T.F. Dumm and R. Hogg, Distribution of sulfur and ash in ultrafine coal,
    Processing and Utilization of High Sulfur Coals II, Y.P. Chugh and R.D.
    Caudle, eds., Elsevier, NY, (1987) pp.23-32.

Interfacial Phenomena in Biotechnology and Materials Processing,
edited by Y.A. Attia, B.M. Moudgil and S. Chander
Elsevier Science Publishers B.V., Amsterdam, 1988 — Printed in The Netherlands

SURFACTANT ENHANCED ELECTRO-OSMOTIC DEWATERING IN MINERAL PROCESSING

C.S. GRANT[1] and E.J. CLAYFIELD[2]

[1,2]School of Chemical Engineering, Georgia Institute of Technology Atlanta, Georgia 30332 (U.S.A)

SUMMARY

Electro-osmotic dewatering studies were conducted on iron oxide (ochre) mineral ultrafines as an alternative to hydraulic dewatering techniques which are limited by the presence of small particles. The absence of a significant electro-osmotic effect on the slurry indicated the ochre particles were uncharged in their natural state. The hydrophobic adsorption of cetyl trimethyl ammonium bromide (CTAB), a cationic surfactant, onto the partially hydrophobic particles significantly increased the rate and extent of dewatering. Subsequent samples were hydrophilic as evidenced by the lack of surfactant adsorption. Base conditioner (NaOH) was added to the hydrophilic slurry to facilitate a change in potential-determining ions. The NaOH-alone slurry showed a slight increase in electro-osmotic dewatering. A marked synergistic improvement in electro-osmotic dewatering was produced with the addition of NaOH and CTAB. Models of surfactant adsorption based on experimental behavior are postulated. The effect of NaOH and CTAB on particle flocculation and slurry filterability is also reported, with supporting evidence provided by capillary suction time (CST) and sediment volume (SV) studies.

INTRODUCTION

The effective dewatering and separation of ultrafine particle dispersions is of paramount importance in the mineral processing industry. Presently, fine particle dispersions are difficult to dewater based on the economic and physical constraints of current prevailing dewatering techniques. In mineral processing, as ore grade decreases, there is the inevitable production of mineral ultrafines. Dispersions that exist in the form of fine particulates in wastewater treatment, industrial waste fines disposal, and biological systems also present problems in dewatering.

The feasibility of a given dewatering technique is dependent on (1) the final desired moisture content, (2) cost per unit slurry dewatered, and (3) critically on particle size that governs the bed pore size. The factor critically limiting conventional dewatering methods is the reduction in hydraulic permeability when ultrafines are present.

This paper addresses the development of an electro-osmotic method to improve the overall dewatering of mineral ultrafines. Electro-osmotic dewatering, in theory, is not limited by the presence of ultrafine material; thus electro-osmotic dewatering is an attractive concept provided the surface electrical properties of the particles can be utilized to generate an electro-osmotic effect (refs. 1-3).

With the ochre (iron oxide) system involved in this work, there is no significant natural charge on the particles. To achieve the desired electro-osmotic effect, it is necessary to condition the ochre surface by the adsorption of an ionogenic surfactant. Mechanisms are proposed for the enhanced electro-osmotic dewatering which are supported by experimental results.

THEORY

Limitations of Conventional Dewatering Methods

Two of the most widely used techniques of dewatering are vacuum/pressure filtration and thermal drying. Hydraulic dewatering methods (e.g. vacuum and pressure filtration) are severely retarded in the presence of fine particles. These ultrafine particles lead to a considerable reduction in hydraulic permeability, which is a measure of the dewaterability of a system under the influence of a pressure gradient.

The concept of hydraulic permeability reduction is predicted by the following equation:

$$\Delta P = \frac{2\gamma\cos\theta}{r_i} \tag{1}$$

which expresses the capillary pressure drop (ΔP) across interstices between particles in the filtercake as a function of the surface tension (γ), particle/liquid contact angle (θ), and finally the interstice radius (r_i).

In spite of the progress made with regard to the application of an adequate pressure gradient, the volumetric rate of water removal is highly dependent on particle size. The volumetric filtrate flow through an interstice is represented by the Poiseuille equation for fluid flow through a packed bed as follows:

$$V' = \frac{r_i^4 \, \Delta P \, \pi}{8 \, \mu_f \, l} \tag{2}$$

where μ_f is the viscosity of the liquid medium, l is the length of the interstice, and r_i is the interstice radius. Although thermal drying is independent of particle size, further examination indicates not only an incompatibility with heat sensitive materials but also a high energy requirement. This illustrates the importance of developing an improved dewatering technique for ultrafine suspensions. Research has shown that as a result of the electrical surface charge on particles, electro-osmosis may be utilized to increase the rate and extent of dewatering.

Electrokinetic theory

The theoretical treatments of electrokinetic phenomena combines the theory of the electric double layer of charge surrounding a particle with that of liquid flow, to obtain relationships of electro-osmotic displaced liquid as a function of zeta potential. Smoluchowski (ref. 4) developed an equation to describe the rate of water removed in an electro-osmotic system. The volumetric flow of liquid through a capillary or porous plug (V') by electro-osmosis can be represented by:

$$V' = \frac{\epsilon \, I \, \zeta}{\mu_m \, k_o} \tag{3}$$

where ϵ is the medium permittivity, I is the current across the system, ζ represents the zeta potential, k_o is the liquid conductivity and μ_m is the viscosity of the medium.

EXPERIMENTAL PROCEDURE

Ochre Slurry

The ultrafine material utilized in this study was ochre, which is an inorganic pigment. Ochre, a naturally occuring iron oxide, is represented by a general chemical formula, $Fe_2O_3 \cdot xH_2O$, with associated water. A complex iron ore, ochre is comprised in varying amounts of the following minerals: goethite-$HFeO_3$; limonite-$HFeO_3 \cdot nH_2O$, $Fe_2O_3 \cdot nH_2O$, and hydrous iron oxides; and hematite-Fe_2O_3.

The ochre slurry samples were obtained from the New Riverside Ochre Company, which is a full scale processing plant in which the rate determining step is the vacuum filtration. These slurry samples contained no additives or flocculants and had a moisture content of approximately 60-62%. Earlier analysis of the ochre indicated an approximate particle size range of 2-10 μm.

Electro-osmotic Dewatering Cell

The electro-osmotic dewatering cell (Figure 1) is comprised of a hydrophobic polyurethane buchner funnel, filter paper, and a plastic ring to secure the filter paper. Based on its resistance to rust and the relative ease of soldering, bronze screening was used to fabricate the circular electrodes for electro-osmosis. The filtrate collection reservoir was simply a 100 ml Pyrex glass graduated cylinder.

Experimental Run

A 150 ml sample of the ochre slurry was charged into the dewatering cell and allowed to drain under a constant vacuum to partially consolidate the ochre. When the hydraulic filtration rate approached a minimum, the upper electrode was positioned on top of the filter cake and a potential applied. A series of runs were performed alternating the bottom electrode between a cathode and an anode, in order to determine the direction of the electro-

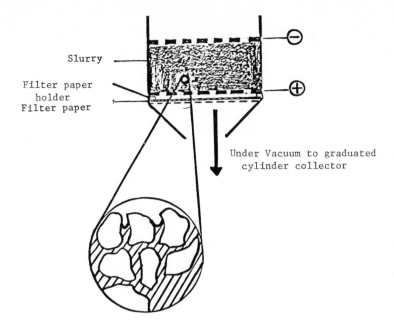

Slurry

Filter paper
holder
Filter paper

Under Vacuum to graduated
cylinder collector

Fig. 1. Electro-osmotic dewatering cell.

osmotic effect. When the drainage had ceased, the vacuum and the potential
were removed; the resulting filter cake was weighed and oven dried to
determine the final dried moisture content.

Surfactant and Base Additives

The cationic surfactant used in all experiments was cetyl trimethyl
ammonium bromide (CTAB). An aqueous solution of the surfactant was combined
with the ochre slurry to ensure adequate surfactant dispersal and adsorption
onto the ochre particles.

Subsequent runs included the addition of aqueous sodium hydroxide to
modify the surface charge of the particles and the pH of the dispersion. A
final series of runs were conducted with the addition of the base conditioner
followed by surfactant.

RESULTS AND DISCUSSION

<u>Electro-osmotic</u> <u>Results</u> <u>and</u> <u>Model</u>

Initial vacuum and electro-osmotic dewatering experiments were performed on Batch 1 slurry in the absence of chemical additives. Batch 1 showed no significant electro-osmotic effect with the bottom electrode positive (B +) or negative (B -), indicating a lack of natural surface charge on the ochre ultrafines.

With the addition of small amounts of CTAB (Figure 2), there is an electro-osmotic enhancement of dewatering; this occurs when the bottom electrode is positive. Convincing evidence that this is an electro-osmotic phenomenon is found when electrode polarity reversal causes a decrease in the original dewatering rate.

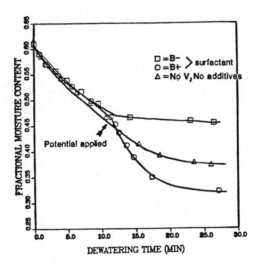

Fig. 2. Electro-osmotic effect with the addition of 10^{-3} M CTAB surfactant (Batch 1), 35 volts applied potential.

The generation of an electro-osmotic dewatering effect, with the bottom electrode positive, indicates CTAB adsorbs to give a positive charge on the particles, with an associated double layer of negative counterions. This suggests that the CTAB adsorbs by hydrophobic bonding of its hydrocarbon tail with hydrophobic patches in this ochre slurry (ref. 5). A likely cause for the partial hydrophobicity of the particle surface is the presence of adsorbed humic acids on this natural pigment material.

The alternative "reverse orientation" model would be adsorption of the CTAB molecule with its cationic headgroup associating with the oppositely charged particle surface (ref. 5). The reverse orientation model is ruled out because of the demonstrated uncharged non-ionic nature of this ochre surface.

In contrast, subsequent ochre material (Batch 2), although similiarly charged, exhibited a wholly hydrophilic behavior. Experimental results obtained showed for example, (1) the untreated ochre showed no electro-osmotic effect with applied potential and (2) CTAB did not adsorb on the particle to give an electro-osmotic effect.

For effective CTAB adsorption to occur, appropriate conditioning of the particle surfaces is necessary. For such oxide surfaces, research has shown that this can be readily accomplished by the appropriate addition of acids and bases to control the H^+ and OH^- potential-determining ions (ref. 6). In flotation work a cationic surfactant conditioner is often adsorbed rendering the particles hydrophobic. The air bubbles used in the subsequent flotation then attach to the hydrophobic particles removing the particles from the bulk liquid phase. Jaycock and Ottewill (ref. 7) have extensively studied the correlations between percent flotation, degree of surfactant adsorption, and electrokinetic properties of dispersions with cationic surfactants.

By changing the pH of the aqueous phase, the adsorbing surface charge (potential determining ions) and the degree of surfactant ionization is altered (ref. 8). Flotation work performed by Iwasaki et. al. (ref. 9) investigates how the adsorption of dodecyl ammonium chloride (RNH_3Cl), a cationic surfactant, onto the iron oxide goethite, is influenced by pH (Figure 3). As the pH is increased, there is a dramatic increase in the flotation recovery, because the adsorbed surfactant renders the particles hydrophobic.

The original slurry has a pH of approximately 6.3, which is very close to

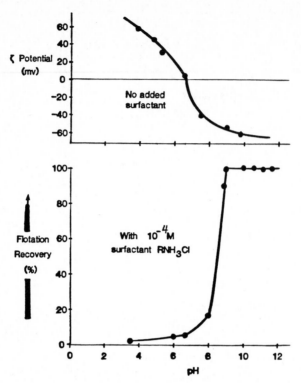

Fig. 3. Flotation of goethite with collector addition (ref. 9).

the point of zero charge in the goethite (iron oxide) system (Figure 3). This slurry neutrality is supported by the observed absence of electro-osmotic effect in the original ochre slurry. By imparting a negative charge to the ochre, the point of zero charge is surpassed making the surface more susceptible to surfactant adsorption.

In this study, the pH of the ochre slurry was altered by adding a small amount of the aqueous sodium hydroxide (NaOH). Evidence to support hydroxide ion adsorption came from the large difference in the calculated and the actual slurry molarity.

The base conditioned slurry was electro-osmotically dewatered without any added surfactant. There was a slight enhancement of the rate of dewatering over that of the untreated slurry (Figure 4). Data obtained in this system

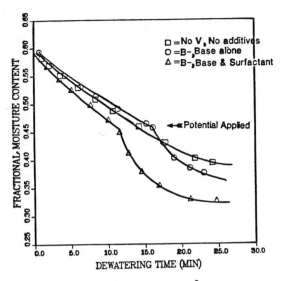

Fig. 4. Electro-osmotic dewatering with 10^{-3} M NaOH addition, synergistic improvement with NaOH and CTAB addition with the bottom electrode negative, 35 volts applied potential (Batch 2).

indicate a positive electro-osmotic effect when the bottom electrode is negative, due to the shearing off of the positive counter ion double layer associated with the OH⁻ potential determining ions.

After modifying the pH of the slurry, CTAB was added and subsequent dewatering studies performed. Figure 4 illustrates the significant increase in the electro-osmotic dewatering rate with NaOH/CTAB over that of the NaOH alone system with the bottom electrode negative. The extent and synergistic nature of this improvement in dewatering performance with the addition of base and surfactant is clearly shown by the results.

Flocculation

A comparison of the dewatering curves for the original slurry and NaOH alone samples (Figure 4) indicates an initial reduction in water removal rate in the NaOH curve with vacuum filtration alone. This reduction can be attributed to the induced repulsive field between the negatively charged particles, resulting in a more dispersed suspension (Figure 5). In a dispersed

422

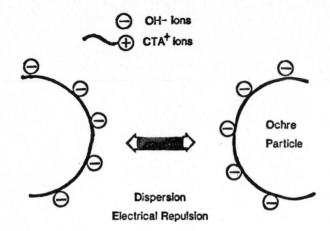

Fig. 5. Electrical repulsion and dispersion of ochre particles upon adsorption of potential determining OH ions.

system, with minimal particle/particle interaction, individual particles can form a close packed filter cake that results in a high specific resistance to filtration. In contrast, the addition of NaOH and CTAB causes an initial increase in the vacuum filtration rate. This result implies the presence of a more flocculated ochre system, an open structured cake with improved filterability due to the interaction of the hydrophobic CTAB tails.

Capillary Suction Time and Sedimentation Volume

The specific resistance to filtration can be easily measured (ref. 10) by using a capillary suction time (CST) apparatus (Figure 6). Commonly used in the area of environmental sludges, the CST operates on the principle of capillary suction. Experiments were conducted by pouring the ochre slurry into a metal collar reservoir that is positioned on a piece of filter paper. The capillary action of the paper draws the filtrate out of the slurry and consequently saturates a growing filter paper area. The time required for the advancing liquid front to travel between two metal probes contacting the filter paper, is refered to as the capillary suction time. A low CST is indicative of a low specific resistance to filtration, and a relatively

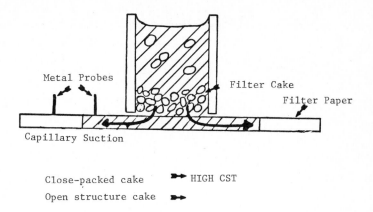

Fig. 6. Schematic of capillary suction time apparatus.

flocculated open structured filter cake.

Further evidence of the degree of flocculation was quantified by measuring another relatively easy technique, the sediment volume (SV), (ref. 11). Dispersed and flocculated systems yield a low and high sediment volume respectively, due to the extent of floc formations and packing.

Results from the CST and SV measurements supported the above flocculation theory in both the NaOH and the NaOH/CTAB runs (Figures 7,8). Using the original slurry values as a reference, NaOH alone caused a significant increase in CST, with an accompanying decrease in the measured SV. The system containing the base and surfactant responded predictably with (1) a decrease in CST and (2) a corresponding increase in SV. The CTAB/ NaOH treated slurry appears to be slightly more flocculated that the original slurry due to the increase in the affinity of the particles for each other.

CONCLUSIONS

For the cost effective dewatering of ultrafine systems, surfactant enhanced electro-osmotic dewatering of ochre is a promising alternative to conventional dewatering techniques. This preliminary work indicates a

424

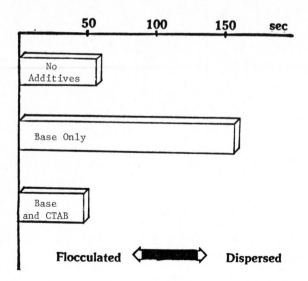

Fig. 7. Capillary suction time values in flocculated and dispersed ochre systems (Batch 2).

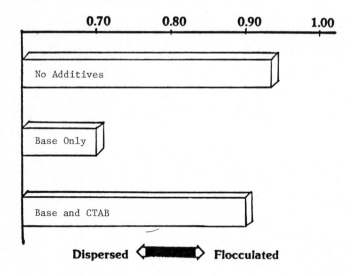

Fig. 8. Sediment volume values for flocculated and dispersed systems (Batch 2).

synergistic improvement in the rate and extent of dewatering with base and surfactant additives. Capillary suction time and sediment volume findings confirm the influence of chemical additives on the flocculation properties of the ochre slurry. The electro-osmotic dewatering results agree with the proposed model for base and surfactant adsorption, based on hydrophobic and electrostatic particle interactions.

REFERENCES

1   H. Yoshida, T. Shinaka, and H. Yukawa, Water Content and Electric Potential Distributions in Gelatinous Bentonite Sludge with Electro-osmotic Dewatering, **J. Chem. Eng. Japan,**18(4) (1985) 337.

2   N.C Lockhart, Electro-osmotic Dewatering of Clays, **Colloids and Surfaces** 6 (1983) 229.

3   C.S. Grant, Electro-osmotic Dewatering of Mineral Ultrafines, Masters thesis, Georgia Institute of Technology, Atlanta, GA., 1986.

4   D.J. Shaw, **Electrophoresis,** Academic, New York, 1969.

5   J.B. Kayes, The Effect of Surface Active Agents on the Microelectro-phoretic Properties of a Polystyrene Latex Dispersion, **Interface Sci.,** 53(3) (1976) 426.

6   G.A. Parks, and P.L. De Bruyn, The Zero Point of Charge of Oxides, **J. Phys. Chem.,** 66 (1962) 967.

7   M.J. Jaycock, and R.H. Ottewill, Cationic Surfactant Dispersions, **Bull. Inst. Mining Met.,** 677 (1963) 497.

8   M.E. Ginn, Adsorption of Cationic Surfactants on Mineral Substrates, in E. Jungerman (ed.), **Cationic Surfactants,** Marcel Dekker Inc., New York, 1970.

9   I. Iwasaki, S.R.B. Cooke,and A.F. Columbo, Flotation of Goethite, U.S. **Bur. Mines, Rep. Invest.,** no. 5593 (1960).

10 R.C. Baskerville, and R.S. Gale, A Simple Automatic Instrument for Determining the Filtrability of Sewage Sludges, **Wat. Pollut. Control,** 67 (1968) 233.

11 D. Attwood,and A.T. Florence, **Surfactant Systems - Their Chemistry, Pharmacy and Biology,** Chapman and Hall, London, 1983.

PART 4

COLLOID FORMATION AND CHARACTERIZATION

Interfacial Phenomena in Biotechnology and Materials Processing,
edited by Y.A. Attia, B.M. Moudgil and S. Chander
Elsevier Science Publishers B.V., Amsterdam, 1988 — Printed in The Netherlands

ASSESSMENT OF FILM FLOTATION EFFICACY FOR THE CHARACTERIZATION OF SOLID
SURFACES

D. W. Fuerstenau, J. L. Diao, K. S. Narayanan and R. Herrera-Urbina
Dept. of Materials Science and Mineral Engineering
University of California, Berkeley, CA 94720

SUMMARY
     The efficacy of film flotation for characterizing the lyophobic/lyophilic
nature of solid particles has been established through experimentation with
particles of different densities and shapes that have homogeneous hydrophobic
surfaces.   Such materials as sulfur, silanated glass beads and quartz, and
paraffin wax-coated particles of coal, graphite, calcite, magnesite and pyrite
were film floated and their critical wetting surface tensions determined. In
separate tests, it was also confirmed that the technique is nearly independent
of particle shape, size, and density over a fairly wide range. These results
strongly suggest that the process involved in film flotation is predominantly
controlled by interfacial forces and that the effect of gravitational forces
is essentially negligible.

INTRODUCTION
     Characterization of solid surfaces in terms of their wettability or
surface energy is important for understanding the behavior of solid
particulates in surface-based processes such as froth flotation, flocculation,
agglomeration, detergency, etc.   The wetting behavior of solids has been
traditionally assessed through measurement of contact angles and immersion
time, determination of the heat of immersion, and even by engulfment in a
freezing front. Although these techniques provide information for comparing
the wetting behavior of various solids in a given liquid or of several liquids
for a given solid, each of these methods has its specific difficulties either
in carrying out the experiments or in interpretation of the results.
Furthermore, unless combined with other techniques, most of these methods can
only assess average wetting properties of massive or particulate samples.

     In response to the need for a methodology that allows characterization of
the lyophobicity/lyophilicity of an assembly of particulate materials, a
simple film flotation technique to characterize the wetting behavior of a mass
of particles has been recently devised in our laboratories (1,2).   The
application of film flotation for the characterization of coal powders has
proven to be extremely useful because it gives information on the surface
energy, wetting characteristics and the degree of surface heterogeneity of the
coal particles, and it offers a means for correlating surface properties with
particle behavior in surface-based processes (3-5).

The present investigation was undertaken to assess the efficacy of film flotation for characterizing the lyophobic/lyophilic nature of an assembly of particles that have a distribution of surface energies. The effect of such variables as particle size, particle density, particle shape and surface homogeneity on film flotation was delineated.

Surface Parameters Obtained from Film Flotation

Whereas conventional methods to assess the wetting behavior of powders provide, irrespective of the surface heterogeneity of the solid, a single value of the critical wetting surface tension of the material, film flotation results can be used to determine several surface-related parameters. Fuerstenau and Williams (5,6) have defined the critical wetting surface tension of the most lyophobic particles in the assembly as γ_c^{min}, and the critical wetting surface tension of the most lyophilic particles in the powder as γ_c^{max}. The difference between these two parameters is a rough indicator of the surface heterogeneity of the material. That is, for a fully homogeneous surface, γ_c^{min} must be equal to γ_c^{max}.

The mean critical wetting surface tension of each distribution, $\bar{\gamma}_c$, can be calculated from the film flotation frequency histogram using the equation:

$$\bar{\gamma}_c = \int \gamma_c \, f(\gamma_c) \, d\gamma_c \tag{1}$$

where γ_c is the critical wetting surface tension of particles and $f(\gamma_c)$ is the density function of the critical wetting surface tension of particles.

The standard deviation of the density function of the critical wetting surface tension, σ_{γ_c}, which reflects the heterogeneity of the surface, is given by:

$$\sigma_{\gamma_c} = [\int (\gamma_c - \bar{\gamma}_c)^2 f(\gamma_c) \, d\gamma_c]^{1/2} \tag{2}$$

High $\sigma_{\gamma c}$ values correspond to heterogeneous materials.

EXPERIMENTAL MATERIALS AND METHODS

Materials used in this study included Cambria #33 bituminous and anthracite coals, Ceylon graphite (99.0% carbon), sulfur (Nevada), quartz (Brazil) and Superbrite glass beads (manufactured by Minnesota Mining and Manufacturing Co., Reflective Products Division). Monosized fractions of coal, graphite, sulfur and quartz were prepared by grinding the as-received materials in a small ceramic mill to avoid iron contamination, followed by sizing in the √2 Tyler series of sieves.

Glass beads and quartz particles were cleaned with hot concentrated hydrochloric acid in order to remove impurities, specifically ferric ions. The leached samples were rinsed repeatedly with triply-distilled water until the conductivity of the supernatant remained constant. These materials were then dried in an oven at 40°C for 36 hours and finally stored in air-tight glass bottles.

Particles with homogeneous hydrophobic surfaces were prepared by coating different substrates with trimethylchlorosilane (TMCS) or with paraffin wax. TMCS-coated materials were prepared by mixing about 20 grams of either glass beads or quartz particles with 20 ml of pure TMCS, and then stirring the suspension for 2 hours at room temperature. Subsequently, the methylated particles were washed three times with pure cyclohexane and dried in an oven at 110°C for 24 hours. The methylated samples were stored in a desiccator over silica gel to prevent moisture adsorption. Paraffin wax-coated samples were prepared by a simple vapor deposition procedure. To accomplish this, the vapor generated by heating paraffin wax was passed through a bed of particles (1.5 grams) by applying vacuum. To ensure a uniform wax coating on all particles, the bed was tumbled every fifteen minutes. In order to prepare surfaces having various degrees of hydrophobicity, the coating process was carried out for 0.5, 3 and 8 hours, after which the wax-coated samples were stored in air-tight glass bottles. The chemicals used in this study were all of reagent grade. Triply distilled water was used throughout the experiments.

Film flotation experiments were conducted following the procedure reported elsewhere (1,2). Briefly, sufficient particles were placed on the surface of the test liquid to make a monolayer. Liquids used for this purpose were usually aqueous methanol solutions of a range of compositions so that the surface tension could be varied from 22.5 to 72.8 mN/m. For each solution, the fraction of particles that float due to their lyophobicity was determined. Unless otherwise specified, all the film flotation experiments were carried out in a 20°C constant temperature room using aqueous methanol solutions as the wetting liquid.

RESULTS AND DISCUSSION

For a given assembly of particles having fully homogeneous surfaces, a single critical wetting surface tension should characterize the lyophilicity. In terms of film flotation parameters, this means that $\bar{\gamma}_c = \gamma_c^{min} = \gamma_c^{max}$, and $\sigma_{\gamma c} = 0$. Under these conditions, the film flotation partition curve must be virtually vertical. That is, below $\bar{\gamma}_c$ all the particles are imbibed by the liquid and above $\bar{\gamma}_c$ all the particles reside at the liquid/gas interface.

To test the validity of film flotation for characterizing the wetting behavior of solid surfaces, coal particles were coated with a uniform film of

432

paraffin wax in order to obtain homogeneous hydrophobic particles of known critical wetting surface tension. The purpose of using coal particles was to have material of the same shape and density as that used in film flotation studies on coal (5,6). The film flotation partition curves of both as-received and paraffin wax-coated Cambria #33 bituminous and Pennsylvania anthracite coal samples (48 x 65 mesh particle size) are shown in Figure 1. From these results one can observe a distinct shifting and a pronounced steepening of the partition curves after the samples had been wax-coated. In addition, the shifts are similar irrespective of the coal and exhibit a sharp inflection near γ_c^{min}. For the as-received coals, $\bar{\gamma}_c$ = 46.8 and 69 mN/m, while γ_c^{min} = 35.5 and 50 mN/m for Cambria #33 and anthracite coal, respectively. When these same samples of coal are coated with paraffin wax, their γ_c^{min} was found to be 22.5 mN/m in both cases. The values of $\bar{\gamma}_c$ were found to be 26.9 and 27.9 mN/m for wax-coated samples of Cambria #33 and anthracite coals, respectively.

The critical wetting surface tension (γ_c) of homogeneous surfaces such as those of paraffin wax, Teflon, polystyrene, etc., has been estimated by earlier researchers through contact angle (θ) measurements. Zisman (7), who introduced the concept of critical wetting surface tension, measured the contact angle on the solid both with a series of liquids of different surface tensions and with a series of solutions. From contact angle data, a plot of

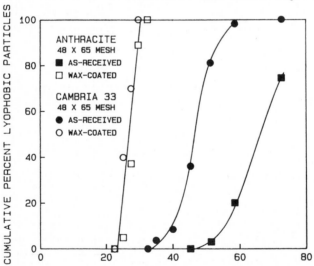

Fig. 1 - Cumulative distribution of as-received and wax-coated Cambria #33 and anthracite coal particles as a function of their critical wetting surface tension determined from film flotation with aqueous methanol solutions.

cosine θ versus the liquid surface tension can be constructed (the well-known Zisman plot). The critical wetting surface tension is defined as that surface tension which corresponds to zero contact angle, that is the condition where cosine θ = 1. Since γ_c values reported for paraffin wax range from 21 to 25 mN/m (7-12), the value of γ_c^{min} obtained from film flotation of paraffin wax-coated coal samples, namely 22.5 mN/m, corresponds to the critical wetting surface tension of paraffin.

In another series of experiments, graphite particles were coated with paraffin wax for periods of 0.5, 3 and 8 hours. In this way, particles having different degrees of hydrophobicity and homogeneity were obtained. Thus, $\bar{\gamma}_c$ and $\sigma_{\gamma c}$ of wax-coated graphite particles should decrease with increasing coating time since the longer the particles are coated with paraffin the more hydrophobic and homogeneous they will be. The film flotation distribution curves of wax-coated graphite are given in Figure 2. From these results one can observe that the longer the coating time, the more pronounced is the shift and steepness of the curves. These distribution curves tend to approach the limiting case corresponding to a vertical line at the critical wetting surface tension of paraffin wax.

The four wetting parameters for different degrees of surface coverage of wax-coated graphite are given in Table 1. As expected, these values decrease

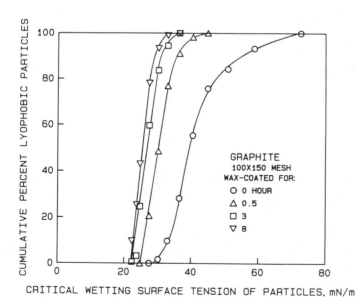

Fig. 2 - Cumulative distribution of as-received and wax-coated graphite particles as a function of their critical wetting surface tension
obtained by film flotation using aqueous methanol solutions.

Table 1 - Film flotation wetting parameters of as-received and wax-coated 100x150 mesh graphite particles.

Treatment	γ_c^{min}	γ_c^{max}	$\bar{\gamma}_c$	σ_{γ_c}
		mN/m		
As-received	28.0	72.0	41.7	8.00
Wax-coated for 0.5 hour	24.8	42.7	30.7	3.91
Wax-coated for 3 hours	22.0	36.3	27.2	3.16
Wax-coated for 8 hours	22.0	33.7	25.8	2.60

with increasing coating times. At higher wax coverages the particles exhibit great difficulty towards being imbibed by the liquid and their γ_c's obtained from film flotation are lower. This indicates that the higher the surface coverage by paraffin wax the lower the surface energy and the higher the contact angle. Since the most lyophobic particles are expected to be completely covered by wax, γ_c^{min} of wax-coated samples should be equal to γ_c of paraffin wax. The value of γ_c^{min} for wax-coated graphite was also found to be in good agreement with the critical wetting surface tension of paraffin reported in the literature (7-12).

Separate film flotation experiments were conducted using spherical glass beads and quartz particles coated with TMCS. As expected, the acid-cleaned glass beads and quartz particles exhibit lyophilic surfaces, that is they sink completely in triply-distilled water. Upon methylation, glass and quartz surfaces are covered with a homogeneous layer of hydrophobic methyl groups (13), and their cumulative distribution curves should be nearly vertical. These plots are shown in Figure 3. The three critical wetting surface tension parameters for this homogeneous hydrophobic surface have almost the same value, namely $\gamma_c^{min} \approx \bar{\gamma}_c \approx \gamma_c^{max} = 23$ mN/m. For a substrate covered with a monolayer of methyl groups, a value of 22-24 mN/m has been reported for γ_c (7). Another important conclusion drawn from these results is that particle shape does not affect the wetting behavior of the particles in film flotation experiments. Even though the shape of methylated glass beads and quartz particles is different, their wetting tension distribution diagrams are identical. Figure 3 also shows that uncoated clean quartz particles are hydrophilic in all solutions tested.

The next step in ascertaining the efficacy of film flotation for characterizing the wetting behavior of particles involved film flotation of naturally hydrophobic and homogeneous sulfur particles. The cumulative critical wetting surface tension distribution obtained from these experiments is shown in Figure 4. The values of the four wetting parameters (in mN/m) are: $\gamma_c^{min} = 28.0$, $\gamma_c^{max} = 34.5$, $\bar{\gamma}_c = 31.3$ and $\sigma_{\gamma c} = 1.55$. Since $\bar{\gamma}_c$ and $\sigma_{\gamma c}$ of

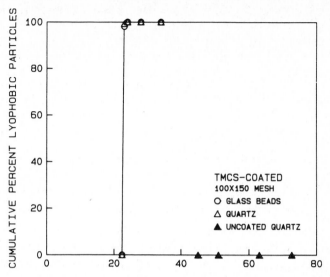

Fig. 3 - Cumulative distribution of TMCS-coated spherical glass
beads and quartz particles as a function of their
critical wetting surface tension determined from film
flotation with aqueous methanol solutions.

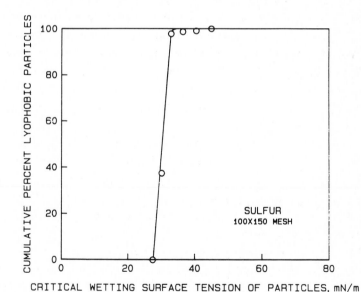

Fig. 4 - Cumulative distribution of sulfur particles as a function
of their critical wetting surface tension determined from
film flotation with aqueous methanol solutions.

436

sulfur are much lower than $\bar{\gamma}_c$ and σ_{γ_c} of graphite, sulfur is more lyophobic and homogeneous than graphite. However, these materials have the same γ_c^{min}, namely 28.0 mN/m, indicating that the most lyophobic particles in the two samples have the same hydrophobicity. The reported γ_c values for sulfur range from 27.5 to 31.5 mN/m (14-16), while those for graphite are between 28 and 30 mN/m (17,18).

Effect of Particle Density

Subsequently, the effect of particle density on film flotation was investigated. To have particles with different densities but the same hydrophobicity, particles of graphite, quartz, calcite, magnesite and pyrite were coated with paraffin wax. Figure 5 presents the results of film flotation experiments with these wax-coated minerals. The shape and the steepness of these curves are similar to each other even though the density of wax-coated particles varies from 2.2 to 5.0 g/cm^3. The wettability paramaters and the density of all the wax-coated samples are summarized in Table 2. As received quartz, calcite and magnesite samples, being extremely hydrophilic, are totally engulfed into the liquid phase when triply distilled water is used as the wetting medium. These results confirm the prediction made by our energy analysis of particles that transfer from the gas phase to the liquid phase in film flotation [19]. Even for particles with a density of 6.0 g/cm^3,

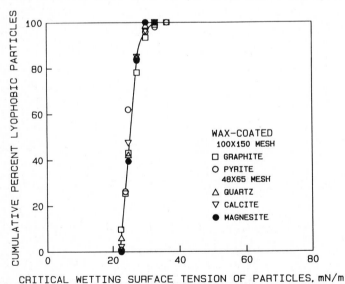

Fig. 5 - The effect of particle density on the cumulative distribution of wax-coated graphite, pyrite, quartz, calcite and magnesite as a function of their critical wetting surface tension determined from film flotation with aqueous methanol solutions.

Table 2 - Critical Wetting Surface Tension of Wax-Coated Particles.

Material	Specific Gravity	γ_c^{min}	γ_c^{max} mN/m	$\bar{\gamma}_c$
Coal	1.1-1.6	22.5	27-35	26-31
Graphite	2.2	22.5	31.4	27.1
Quartz	2.6	22.0	27.3	26.3
Calcite	2.7	22.5	30.0	26.5
Magnesite	3.0	22.5	30.0	26.5
Pyrite	5.0	22.0	33.7	25.8

the contributions of interfacial energy and gravitational energy are 81% and 19%, respectively. Therefore, the surface energy of particles controls the film flotation process, and the density of the particle has little effect on practical, quiescent film flotation.

Effect of Particle Size

The wetting characteristics of graphite samples of different particle sizes (425 x 300 µm, 300 x 212 µm, 212 x 150 µm, 150 x 106 µm, 106 x 75 µm and 75 x 53 µm) were also assessed by film flotation to delineate the effect of particle size on this technique. For the various graphite particles listed above, the values of $\bar{\gamma}_c$ were found to be 39.8, 40.5, 41.1, 41.7, 43.6 and 44.1 mN/m, respectively. These results clearly indicate that $\bar{\gamma}_c$ does not significantly depend on particle size and that gravitational forces can be considered negligible in the size range tested. Because graphite has two distinct and different crystal surfaces, the surface energy of each surface varies. The surface energy of an edge is higher than that of a face, the slight increase in $\bar{\gamma}_c$ with decreasing particle size is most likely due to an increase in the edge/face ratio for small graphite particles. As shown in Figure 5, the distribution curves of wax-coated particles are nearly identical for two size fractions (106 x 150 µm and 212 x 300 µm).

Effect of Wetting Liquid Temperature

Since the surface tension of the liquid is a strong function of temperature, film flotation was expected to be affected by the temperature of the solution. Figure 6 presents the results of film flotation experiments with graphite using 40 volume percent methanol solutions at different temperatures As can be seen from this figure, the surface tension of the solution decreases linearly with increasing temperature, in accordance with the Eotvos equation

$$\frac{d\gamma_{LV}}{dT} = -K \tag{3}$$

438

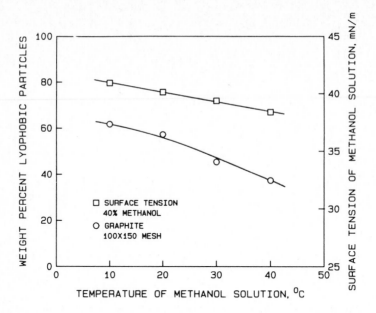

Fig. 6 - The effect of solution temperature on the film flotation response
of graphite particles using aqueous methanol solutions.

with the constant K being equal to 0.085. The surface tension decreases from
40.9 mN/m at $10^{\circ}C$ to 38.4 mN/m at $40^{\circ}C$, while the fraction of the particles
that are lyophobic decreases from 62 to 37 over the same temperature range.
When film flotation experiments were carried out at these same surface
tensions by varying the methanol concentration at $20^{\circ}C$, the percentage of
material exhibiting lyophobic surfaces was 60 and 35, respectively. Again,
this confirms our conjecture that the process is controlled mainly by
interfacial forces.

Effect of Nature of Wetting Liquid

In analyzing the wetting behavior of partially lyophobic solids in aqueous
solutions of polar organic reagents, one is concerned with effects due to
preferential adsorption of organic species. To delineate the possibility of
adsorption and its effects on film flotation, graphite particles were film
floated using aqueous solutions made up of methanol, ethanol, propanol,
tertiary butyl alcohol, and acetone. The surface tension of these solutions
was taken from the literature (19). Film flotation results obtained from
these experiments are presented in Figure 7. Since the experimental data fall
reasonably well onto a single curve, preferential adsorption of organic
species appears to be negligible in these systems. The values of $\bar{\gamma}_c$ (in mN/m)
determined from these curves are 41.7, 39.4, 39.0, 37.6, and 39.3 for aqueous

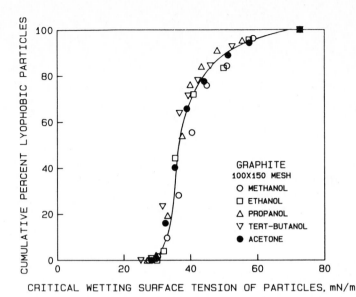

Fig. 7 - Cumulative distribution of graphite particles as a function of their critical wetting surface tension determined from film flotation with various polar organic solutions.

solutions of methanol, ethanol, propanol, tertiary butanol, and acetone, respectively. These findings are similar to those obtained from the film flotation of coal in different aqueous polar organic solutions (6).

As part of this study, an attempt was also made to estimate the wettability parameters of coal using pure organic liquids of different surface tensions. Many investigators have used pure organic liquids to determine the critical wetting surface tension of low energy solids (7). Liquids tested in this investigation include o-xylene, carbon disulfide, monobromobenzene and aniline. However, when film flotation was performed using these liquids, a discoloration of the liquid was noticed. Such a phenomenon indicates an interaction between the coal and the liquid. Therefore, reproducible and reliable partition curves could not be obtained. The important point to note here is that chemical interaction between the solid and the liquid should be absent for the success of this film flotation technique.

SUMMARY

The wetting characteristics of particles having homogeneous hydrophobic surfaces were assessed through film flotation experiments in order to establish the efficacy of this technique for the characterization of solid surfaces. Such hydrophobic materials as sulfur, silanated glass beads and quartz, and wax-coated particles of coal, graphite, quartz, calcite, magnesite

440

and pyrite were used in this investigation. The minimum critical wetting surface tension obtained from film flotation of wax-coated materials agrees well with the γ_c of paraffin wax determined by various researchers from contact angle measurements. Similarly, the γ_c^{min} for sulfur and TMCS-coated particles was found to agree with the γ_c values available in the literature. Film flotation results using TMCS-coated glass beads and quartz indicate that film flotation is independent of particle shape. From this work, it is concluded that, provided there is no chemical interaction between the liquid medium and the particle surface, film flotation is predominantly controlled by interfacial forces. The effects of such factors as particle size, particle density, and particle shape on film flotation are negligible, and this technique is sensitive only to the surface energy and heterogeneity of the particles.

ACKNOWLEDGEMENT

The authors wish to acknowledge the U.S. Department of Energy, Pittsburgh Energy Technology Center, Grant No. DE-FG22-84PC70776, and the U.S. Bureau of Mines Grant No. G1164106 for the support of this research.

REFERENCES

1   M.C. Williams and D.W. Fuerstenau, A Simple Flotation Method for Rapidly Assessing the Hydrophobicity of Coal Particles, Int. J. Miner. Process., 20 (1987) 153-157.
2   D.W. Fuerstenau and M.C. Williams, Characterization of the Lyophobicity of Particles by Film Flotation, Colloids and Surfaces, 22 (1987) 87-91.
3   D.W. Fuerstenau, M.C. Williams and R.H. Urbina, The Physical and Chemical Properties of Coal and Their Role in Coal Beneficiation, Proceedings, Int. Conf. on Coal Science, Sydney, Australia, 1985, pp. 517-520.
4   D.W. Fuerstenau, K.S. Narayanan, R. Herrera-Urbina and J.L. Diao, Assessing the Effect of Oxidation on the Wettability and Flotation Response of Coal, Proceedings, Int. Conf. on Coal Science, Eurohal, The Netherlands, 1987.
5   D.W. Fuerstenau, M.C. Williams, K.S. Narayanan, J.L. Diao and R.H. Urbina, Assessing the Wettability and Degree of Oxidation of Coal by Film Flotation, submitted to Energy and Fuels, 1987.
6   D.W. Fuerstenau and M.C. Williams, A New Method for Characterization of the Surface Energy of Hydrophobic Particles, Part. Charact.,4 (1987) 7-13.
7   W.A. Zisman, Relation of the Equilibrium Contact Angle to Liquid and Solid Constitution, in: R.F. Gould (Ed.), Contact Angle, Wettability and Adhesion, ACS, Adv. in Chem. Series No. 43, 1964, pp. 1-51.
8   J.R. Dann, Forces Involved in the Adhesion Process, I. Critical Surface Tensions of Polymeric Solids as Determined with Polar Liquids, J. Colloid Interface Sci., 32 (1970) 302-320.
9   W.R. Good, A Comparision of Contact Angle Interpretations, J. Colloid Interface Sci., 44 (1973) 63-71.
10 F.E Bartell and H.H. Zuidema, Wetting Characteristics of Solids of Low Surface Tension such as Talc, Waxes and Resins, J. Am. Chem. Soc., 58 (1936) 1449-1454.

11   H.W. Fox and W.A. Zisman, The Spreading of Liquids on Low-Energy
      Surfaces. III. Hydrocarbon Surfaces, J. Colloid Sci., 7 (1952) 428-442.
12   Von E. Wolfram, New Data on the Wettability of Polymers, Kolloid-Zeit.,
      Zeit. Polymere, 211 (1966) 84-94.
13   J. Laskowski and J.A. Kitchener, The Lyophobic-Lyophilic Transition of
      Silica, J. Colloid and Interface Sci., 29 (1969) 67-679.
14   B. Yarar and J. Kaoma, Estimating of the Critical Surface Tension of
      Wetting of Hydrophobic Solids by Flotation, Colloids and Surfaces,
      11 (1984) 429-436.
15   S. Kelebek and G.W. Smith, Selective Flotation of Inherently Hydrophobic
      Minerals by Controlling the Air/Solution Interfacial Tensions, Int. J.
      Miner. Proces., 14 (1985) 275-289.
16   D.A. Olsen, R.W. Moravec and A.J. Osteraas, The Critical Surface Tension
      Values of Group VIA Elements, J. Phys. Chem., 7 (1967) 4464-4466.
17   S. Kelebek, Surface Properties and Selective Flotation of Inherently
      Hydrophobic Minerals, Ph. D. Thesis, Mc Gill University, Montreal, Canada,
      1984.
18   F.M. Fowkes and W.D. Harkins, The State of Monolayers Adsorbed at the
      Interface Solid-Aqueous Solution, J. Amer. Chem. Soc., 61 (1940)
      3377-3393.
19   J.L. Diao, Characterization of Wettability of Solid Particles by Film
      Flotation, M.S. Thesis, University of California, Berkeley, 1987.

Interfacial Phenomena in Biotechnology and Materials Processing,
edited by Y.A. Attia, B.M. Moudgil and S. Chander
Elsevier Science Publishers B.V., Amsterdam, 1988 — Printed in The Netherlands

FORMATION AND CHARACTERIZATION OF ULTRAFINE SILVER HALIDE PARTICLES

M.J. HOU and D.O. SHAH

Center for Surface Science and Engineering, University of Florida, Gainesville,
Florida 32611

SUMMARY

Ultrafine silver chloride particles were precipitated by the reaction of silver nitrate and sodium chloride in microemulsion medium. The formation mechanism, aggregation state and colloidal stability of these ultrafine particles were studied by stopped-flow photometry, quasi-elastic light scattering, spectrophotometry, and electron microscopy. The ultrafine particles were found to be 50 to 100 A in diameter with very uniform size distribution. The dispersions of these particles were very stable as a result of the steric stabilization provided by the surfactant layer. The colloidal stability of such dispersions was found to increase as one decreased the chain length of alkane, decreased the concentration of alcohol, increased the chain length of alcohol, or added the nonionic surfactant Arlacel-20. The observation was explained from both equilibrium and dynamic points of view.

INTRODUCTION

Preparation of monodispersed, ultrafine metallic particles has been one of the most pursued goals in many industries. It is, however, difficult to obtain small and nearly monodispersed particles by classical methods, particularly for particle sizes smaller than 10 nm. This difficulty stems from the fact that a lot of metallic particles sinter easily during the reactions of their precursor species (1). In practice, the precipitation of metallic particles has been done sometimes in the presence of protective colloids such as gelatins and surfactants to control the particle size. In this respect, water-in-oil (W/O) microemulsions have been shown to be ideal media for the precipitation of metallic particles (2-8).

Microemulsions are thermodynamically stable, isotropic dispersions consisting of microdomains of oil and/or water stabilized by an interfacial film of surface active molecules (9). The microstructure of microemulsions has been described as spherical droplets of 10-60 nm dispersed in the continuous medium (10-19), although it has also been reported that microemulsions may also exist as other structures such as bicontinuous (18,20-21) or cubic structure (22). Several experimental studies (23-25) have shown that the reacting species solubilized in the water pools of microemulsion droplets can exchange rapidly among different droplets so that reaction took place in the "cages" constituted by

the microemulsion droplets. This is made possible due to the energetic collisions among microemulsion droplets, which lead to the temporary fusion and
split of droplets. Utilizing the dynamic nature, small dimension and uniform
size distribution of microemulsions, several researchers (2-8) have successfully prepared, by precipitation in microemulsions, small and nearly monodispersed particles of various materials. However, there is little discussion
on the mechanism of the growth of these ultrafine particles and the colloidal
stability of such ultrafine particle dispersions. The importance of this
knowledge cannot be over emphasized in the preparation of ultrafine particles
in microemulsions. Furthermore, it is less known with regard to the effects of
the microemulsion components on the precipitation in microemulsions, which
would be extremely vital in designing the optimal formulation to precipitate
ultrafine particles. This paper reports the results of our study on the precipitation of silver chlorides in microemulsions consisting of AOT (sodium di-
2-ethylhexyl sulphosuccinate), alkane and aqueous solutions of metal salts.
Particular attention has been focused on the discussion of the sequence of
particle growth, the effects of microemulsion components on precipitation and
the colloidal stability of dispersions.

EXPERIMENTAL

Materials

Silver nitrate of analytical reagent grade was purchased from J.T. Baker
Chemical Co.. Sodium chloride and potassium chromate of certified reagent
grade were purchased from Fisher. AOT of 99% sodium purity was from Sigma.
Alkanes of 99 mole % purity, including n-hexane, n-heptane, n-octane, n-nonane,
n-decane and n-dodecane, were purchased from Fisher. Alcohols of 99 mole %
purity, including n-propanol, n-butanol and n-pentanol, were purchased from
Sigma. Arlacel-20 was obtained from ICI Americas. All reagents were used
without further purification.

Methods

Sample preparation. Microemulsions containing different metal salts were
prepared by solubilizing the salt aqueous solutions into AOT/alkane solutions
under vigorous stirring. The precipitation of silver chlorides was done by mixing a microemulsion containing silver nitrate with the one containing sodium
chloride.

Stopped-flow photometry. A stopped-flow apparatus can provide rapid and
homogeneous mixing in 1 to 2 ms. In this study, the microemulsion containing
silver nitrate was mixed, by stopped-flow method, with the one containing
sodium chloride and the indicator, potassium chromate. Meanwhile, an attached

photometer was used to monitor the transmittance of the mixture at 420 nm (the characteristic absorption of silver chromate). The stopped-flow photometer used was purchased from Durrum Instrument Co. (Durrum Rapid Kinetics Systems Series D-100).

Quasi-elastic Light Scattering. Quasi-elastic light scattering (QELS) was used to monitor the variation of the hydrodynamic radius of the dispersions after mixing of microemulsions containing different metal salts. The optical source of the light scattering apparatus was a Spectra Physics Argon ion laser (model 2000) operating at 514.5 nm. The scattered light was collected at an angle of 90 degree by a laser light scattering goniometer (model 200SM) and was focused on a photomultiplier cathode (EMI-9865), both from Brookhaven Instrument. The time dependent correlation function of the scattered intensity was recorded by using a fast clipped real time correlator (model BI 2030) with 72 channels and a time resolution of 100 ns. During the measurement, the data were rejected to avoid any error caused by dust contamination if the difference between the calculated and measured baselines was greater than 1%. The normalized correlation function of the scattered electric field, $g^{(1)}(t)$, was analyzed by using the method of cumulant (26-27), in which the logarithm of the normalized $g^{(1)}(t)$ was fitted to a polynomial equation using a nonlinear least-square fitting routine. The fitting equation is described as follows:

$$\ln[(c_1 b)^{1/2} g^{(1)}(t)] = \ln(c_1 b)^{1/2} - \Gamma t + \mu_2(t^2/2!) \qquad [1]$$

where c_1 and b are baseline and instrument constant. The first cumulant, Γ, of linewidth distribution is equal to Dq^2, where D is the z-averaged diffusion coefficient. The magnitude of scattering vector, q, is defined as $q = (4\pi n/\lambda_0)\sin(\theta/2)$, where n is the refractive index of continuous medium, λ_0 the wavelength of light in vacuum and θ the scattering angle. The second cumulant μ_2 provides the polydispersity of the linewidth distribution. The polydispersity, Q, is defined as $Q = \mu_2/\Gamma^2 = (\bar{D^2} - \bar{D}^2)/\bar{D}^2$. The hydrodynamic radius, R_H, was calculated by Stokes-Einstein equation $D = kT/(6\pi\eta R_H)$, where k is the Boltzmann constant, T the absolute temperature and η the viscosity of the continuous medium.

Transmittance measurements. The transmittance of mixture was measured as the function of time at 550 nm after mixing the microemulsion containing silver nitrate with the one containing sodium chloride. The spectrophotometer used was from Perkin-Elmer (Model 575 UV-VIS Spectrophotometer).

Electron Microscopy. Electron microscopy study was done at The Major Analytical Instrumentation Center (MAIC) at The University of Florida. All electron micrographs were taken by a JEOL 200CX transmission electron microscope. The samples were prepared by directly dropping a very small amount of silver chloride dispersion on the polyvinyl fomal film stretched on a copper mesh. Electron diffraction patterns of AgCl microcrystals were also taken by the same instrument. In the latter case, the voltage used is 200 kV and the camera length is 82 cm.

RESULTS

Completion of Reaction between Silver Nitrate and Sodium Chloride

Dvolaitzky et al. (5) have shown that the rapid exchange of reagents between water pools can be inferred from the instantaneous formation of a red silver chromate precipitate after mixing the microemulsion containing silver nitrate with the one containing sodium chloride and an indicator, potassium chromate. In this procedure, known as Mohr method (28), the chlorides were titrated by silver nitrate solution, and the indicator formed a red precipitate with an excess of silver nitrate solution after all chlorides have been precipitated as silver chlorides. Thus, the precipitation rate of silver chromate is negligibly small in the presence of Cl^-, and one would expect this rate to increase after the concentration of Cl^- has decreased to a negligible level. We have used a stopped-flow photometer to monitor the precipitation rate of silver chromate through the decrease of the transmittance of the mixture at 420 nm, the characteristic absorption wavelength of silver chromate.

Figure 1.a shows a typical decay of the transmittance at 420 nm after mixing the microemulsion containing potassium chromate with the one containing silver nitrate. Note that the initial rise of the curve is due to the replacement of old sample (previous determination) with the fresh mixture. Immediately after that, the curve showed a quick and exponential decay. Figure 1.b shows the variation of transmittance at 420 nm after mixing the microemulsion containing silver nitrate with the one containing both sodium chloride and potassium chromate. We observed that the decay of transmittance was very slow initially due to the presence of chlorides. After approximately 60 ms, we found that the transmittance of the mixture suddenly dropped, and the transmittance decayed exponentially, indicating that the chloride concentration have been decreased to a negligible level.

QELS Study During the Growth of AgCl Particles

We have used QELS to monitor the variation of the time dependent correlation function of the scattered light intensity after mixing the microemulsions containing different reactants. The system studied is AOT/n-heptane/metal salts aqueous solution, with [AOT] = 0.14 M (in heptane), $n_w/n_{AOT} = 8$, and $[AgNO_3] = [NaCl] = 0.4$ M (in water). Note that n_w and n_{AOT} are the number of water and AOT molecules. This sample did not show any variation in transmittance during

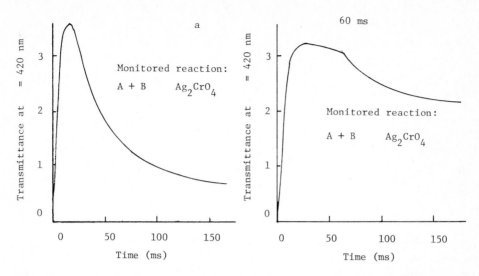

Figure 1. Decrease of transmittance at = 420 nm due to the formation of red silver chromate precipitate. (a) Controlled observation in the absence of NaCl, A: 15 mL AOT/n-heptane(o.15M) + 0.4 mL $AgNO_3$(0.4M), B: 15 mL AOT/n-heptane(0.15 M) + o.1 mL K_2CrO_4(0.077M) + 0.3 mL distilled water. (b) Inhibited precipitation in the presence of NaCl, A: same as in (a), B: same as in (a) except replacing distilled water with 0.4 M NaCl aqueous solution.

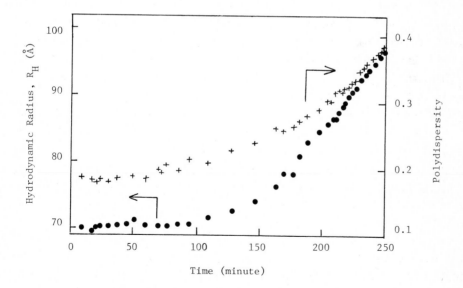

Figure 2. Effective hydrodynamic radius and polydispersity of particle dispersion as a function of time after mixing.

the QELS study and remained visually clear for weeks after mixing. By fitting the time dependent correlation function with equation [1], we obtained two important parameters, the first cumulant of linewidth, Γ, and the second cumulant, μ_2. The z-averaged hydrodynamic diameter calculated from Γ and the polydispersity were shown in Figure 2. It was found that both hydrodynamic diameter and polydispersity remained almonst constant initially and started to increase after approximately 100 minutes. We also noted that the microemulsions containing metal salts also showed strong time dependent correlation function in the scattered intensity before their mixing. Thus, the contributions of both AgCl microcrystals and microemulsion droplets have to be considered in the interpretation of Figure 2, although AgCl microcrystal are generally better scatters. The constancy of the hydrodynamic diameter and the low polydispersity suggested that we only observed a single mode of translational diffusion in the initial period of observation. This is consistent with the picture that in the initial stage the AgCl microcrystals are very small and restricted inside the microemulsion droplets. Thus, in this stage, AgCl microcrystals have to grow through the constant collision, fusion and split of microemulsion droplets. The increase in both polydispersity and mean hydrodynamic diameter inferred a more significant contribution of AgCl microcrystals to the observed correlation function of intensity, as a result of the growth or aggregation of microcrystals. The sudden increase in the polydispersity also suggested that the correlation time of the translational diffusion of microcrystals started to deviate from that of microemulsion droplets. Thus, the motion of microcrystal at this stage probably is not restricted in the "cages" of microemulsion droplets.

Effects of the Variation in Microemulsion Components

It is well known that the turbidity of the aqueous solution increases rapidly upon mixing silver nitrate and alkali halide in the absence of protective colloids. Matijevic and Ottewill (29-30) have attributed such turbidity increase to the fast coagulation and precipitation of silver halide sols. When precipitation occured in microemulsions, the increase of turbidity was drastically reduced or even negligible. Such slow increase in turbidity has been shown to be the slow flocculation of primary particles in dispersions by X-ray diffraction study (5). In this study, the variation of transmittance of the mixture has been monitored after mixing the microemulsion containing $AgNO_3$ with the one containing NaCl. The effects of microemulsion components on the variation of transmittance was studied by systematically changing various components of microemulsions. Basically, transmittance T is related to turbidity τ by (ref. 29):

$$T = I_t/I_o = \exp(-\tau l) \qquad\qquad [2]$$

for a scattering but non-absorbing sample, where I_t and I_o are respectively the intensities of transmitted and incident beams and l the length of optical path. For small particles ($r < \lambda/20$), τ is further related to the number of particles per unit volume, N_p, and their individual volume, v_p, by Rayleigh equation

$$\tau = AN_p v_p^2 = A\phi_p v_p \qquad\qquad [3]$$

where ϕ_p is the volume fraction of particles and A is an optical constant defined by

$$A = \frac{24\,\pi^3 n_o^4}{\lambda^4}\,\frac{n^2 - n_o^2}{n^2 + 2n_o} \qquad\qquad [4]$$

In the above expression, n and n_o are the refractive index of particles and solvent, while λ is the wavelength of light used (in vacuo). Therefore, one expects the transmittance to decrease as the mean aggregated volume increases.

Figure 3.a shows the effect of alkane on the growth of AgCl microcrystal aggregates. In general, the growth rate increased as the chain length of alkane used was increased. It is noted that the microcrystal dispersions were relatively stable for the systems with n-hexane, n-heptane or n-octane, while only moderate stability (but still better than in aqueous medium) was observed in systems with n-nonane or n-decane. The effect of the addition of alcohols is shown in Figure 3.b. When n-pentanol was added to relatively stable n-octane system, the growth rate of microcrystal aggregates was greatly enhanced, and this enhancement increased with the amount of n-pentanol added. Figure 3.c shows the effect of the chain length of the alcohols used. The enhancement of the growth rate of aggregates was found to increase with the decrease of the chain length of the alcohol used. Thus, the addition of shorter-chain alcohol further destabilized the microcrystal dispersions. Figure 3.d shows the effect of the addition of long chain nonionic surfactant, Arlacel-20. For relatively unstable n-decane system, we found that the addition of Arlacel-20 improved the colloidal stability of microcrystal dispersions so that the growth rate of aggregates was greatly slowed down.

Electron Microscopy Study

We have used transmission electron microscopy to study the morphology and the aggregation state of AgCl microcrystals prepared from various microemulsions. In this study, the following variables were kept constant for all samples, i.e., [AOT] = 0.14 M (in alkane), [NaCl] = [AgNO$_3$] = 0.4 M (in aqueous solution), and n_w/n_{AOT} = 8. Figures 4.a-b show the electron micrographs of

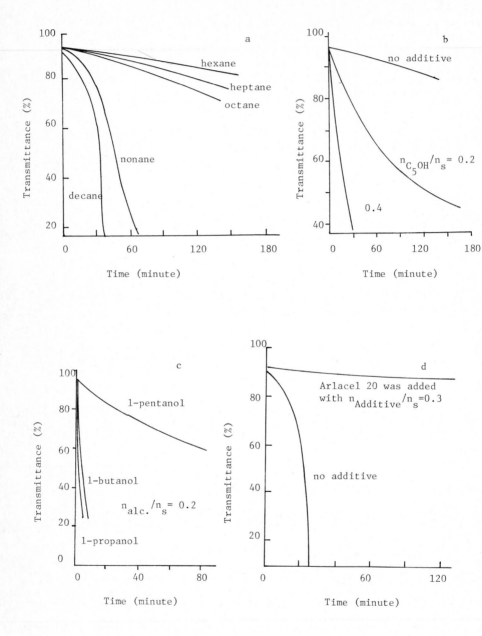

Figure 3. Variation of transmittance as a function of time. (a) Effect of chain length of alkane. (b) Effect of the addition of 1-pentanol. (c) Effect of the chain length of alcohol (d) Effect of nonionic surfactant, Arlacel 20.

AgCl microcrystals precipitated in n-heptane system at different time corresponding to 1 day and 1 week after mixing. As shown in these micrographs, no significant aggregation of AgCl microcrystals was observed even after 1 week, indicating this system is a very stable dispersion. We also noticed that this dispersion remained visually clear and did not show any sign of sedimentation. From these electron micrographs, the shape of AgCl microcrystals appeared to be very close to sphere and the size of AgCl particles was estimated to be 50 to 100 A. Figures 4.c-e show the electron micrographs of AgCl microcrystals precipitated in n-dodecane system. For this system at the above specified concentration, no significant aggregation was observed 4 hours after mixing. Note that the concentrations of surfactant used in TEM study are much smaller than in turbidity measurement. This may explain the observation of relative dispersed particles. Significant aggregation of AgCl particles was observed after 5 days. However, such aggregation of microcrystals seemed easy to be dispersed by sonication, as shown in Figure 4.e. The shape and the size of AgCl particles observed in this system are similar to that found in n-heptane system. The diffraction pattern of these microcrystals was also obtained by focusing electron beam on the selected area shown in Figures 4.d and 4.e. Four d-planes of AgCl crystal have been identified from these diffraction patterns, namely, (1,1,1), (2,0,0), (2,2,0) and (4,2,0). From the electron micrographs, it is easy to see that the big aggregates in Figure 4.d are consisted of many small microcrystals. We also noticed that the size of microcrystals did not show significant difference when precipitated in different systems.

DISCUSSIONS

A Possible Sequence of the Events during Precipitation

From the results presented in the preceeding sections, we can roughly sketch a sequence of the events occurred during the precipitation of of silver chloride in microemulsions. First of all, we found that the reaction between Ag^+ and Cl^- completed in a very short period of time (approximately 60 ms for the case studied). This observation is not surprising at all considering the time scale of the dynamic processes occurred in microemulsions. The time scale of the exchange of reagents in water pools has been reported to be less than 1 microsecond (31-35), although the presence of high ionic strength in the systems studied may slow down this exchange process (5, 24) due to the

452

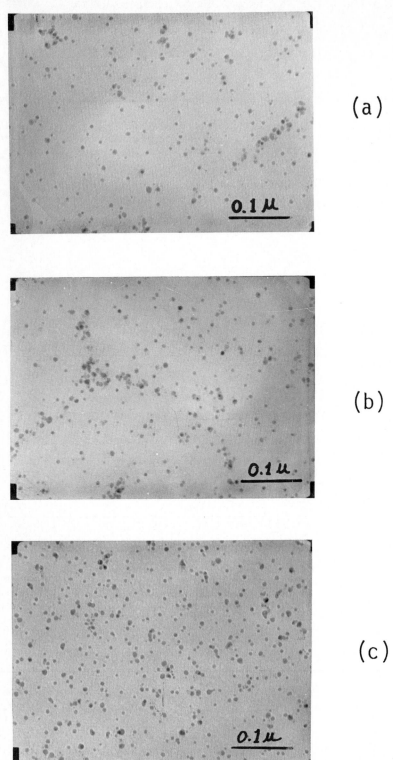

(a)

0.1 μ

(b)

0.1 μ

(c)

0.1 μ

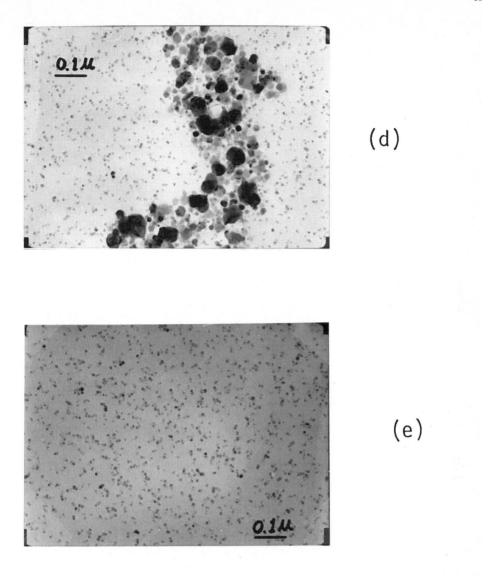

(d)

(e)

Figure 4. Transmission electron micrographs of AgCl particles precipitated in microemulsions; the composition is as follows: [AOT] = 0.14 M in alkane, [NaCl] = [AgNO$_3$] = 0.4 M in aqueous solution, n_w/n_s = 8. (a) Picture taken 1 day after reaction with n-heptane as solvent. (b) Picture taken 1 week after reaction with n-heptane as solvent. (c)Picture taken 4 hours after reaction with n-dodecane as solvent. (d) Picture taken 5 days after reaction with n-dodecane as solvent. (e) Picture taken 5 days after reaction with n-dodecane as solvent; the sample was sonicated before being examined by TEM.

454

condensation of interfacial film. After the reaction was completed, the nucleation and the initial growth of microcrystals were expected to take place inside the microemulsion droplets through the collision, fusion and split of droplets. Such a realization is very important in explaining the ability of microemulsions to precipitate ultrafine particles by providing the steric barrier (interfacial film) to prevent these nucleus or microcrystals from rapid sintering. Considering the salt concentration we used (0.4 M in aqueous solution), the mass of AgCl in each droplet is only 1/200 of the amount needed to grow to a 50 Å AgCl particle, thus numerous of fusions have to occur during the growth of microcrystals either by the deposition of ions on growing nucleus and microcrystals or by the coagulation of these microcrystals. One would expect the former process to be the dominant one before the depletion of ions. The latter process, however, is the only way, if it is possible at all, for microcrystals to grow further after all ions have been depleted. One also expects more difficulty for larger microcrystals to diffuse through the channel opened during the lifetime of the transient dimer of droplets. On the other hand, there is not much difficulty for water to redistribute among the droplets. Once all these processes are over, the AgCl microcrystals probably are left with a layer of surfactants adsorbed on particle surface. At this stage, we can consider the following possibilities. On one hand, AgCl microcrystals can grow bigger through the coagulation of existing microcrystals. In this case, these AgCl particles have to overcome the energy barrier provided by the surfactanr film and stick long enough to sinter. The experimental results showed that most of the AgCl particles obtained were about 50 to 100 Å in diameter. Thus, this process must stop, if it exists at all, when microcrystals grow to this size. On the other hand, these microcrystals can present as a fairly stable dispersion or undergo slow flocculation (aggregation of microcrystals). This picture seemed to be consistent with the experimental results of this study and Dvolaitzky et al. (5).

Stability of the Ultrafine AgCl Particles Formed in Microemulsions

For dispersions of fine particles in liquid media, encounters between particles frequently occur due to Brownian motion and gravitation. Whether such encounters result in permanent contact or the particles rebound and remain free is determined by the inter-particle forces between them. In the dilute dispersions such as the case we considered, it is sufficient to consider only interaction between pairs of particles. Thus, the work, V_t, required to bring the particle surface from infinity to an inter-particle distance, h, is given by $V_t(h) = F(h) - F(\infty)$, where $F(\infty)$ represented the total free energy contributed from all interparticle forces at infinite separation of two particles. Therefore, we can write

$$V_t = V_A + V_S + V_E \qquad\qquad [5]$$

where V_A, V_S and V_E are the potential energies due to van der Waals attraction, steric interaction and electrostatic repulsion.

Our theoretical calculation (36) revealed that V_S and V_A are relatively large and comparable in magnitude, while the magnitude of V_E is negligibly small, indicating that the electrostatic repulsion is not significant compared to either van der Waals attraction or steric repulsion. Furthermore, the results showed that the magnitude of the primary maximum decreased when the chain length of alkane used increased o that the solvancy became poor. This may partly explain the observation of the increase in the growth rate of micro-crystal aggregates with alkane chain length. Nevertheless, it is disputable whether the increase of alkane chain length from C_6 to C_{12} could induce so drastic change in the solvency that the instability of the dispersion was induced. Thus, other factors may also need to be considered to fully explain the observed enhancement of the flocculation of AgCl particles. The effect of surfactant adsorption was also analyzed by the same calculation. For a marginally good solvent, the decrease of volume fraction from 0.3 to 0.1 eventually eliminated the primary maximum and totally destabilized the particle dispersions. This result may partly explain the profound effects of the addition of short chain alcohols or long chain nonionic surfactant Arlacel-20.

Dynamic Point of View

It is important to notice that the precipitation of AgCl is in statu nascendi when using microemulsion as reaction media. One would expect the dynamic processes of microemulsions have significant influences on reaction, nucleation and crystal growth. Thus, it is pertinent to discuss, from the dynamic point of view, the effects of the variation in microemulsion components on the observed transmittance variation. If we assumed the commonly used second order flocculation kinetics (37-38), i.e.,

$$\frac{dN_p}{dt} = -\frac{k_o}{W}N_p^2 \qquad\qquad [6]$$

where $k_o = 4kT/(3\eta)$ and

$$W = 2a \int_{2a}^{\infty} \frac{\exp[V(R)/(kT)]}{R^2} \, dR \qquad\qquad [7]$$

Thus, the flocculation rate is also dependent on the concentration of particles. One would expect that the number of the particles precipitated in the initial stage depends on the effectiveness of the collision and fusion of the microemulsion droplets. This effectiveness of mass exchange has been shown (33-34) to strongly depend on the fluidity of the interface of microemulsion droplets. A more fluid interface not only facilitates the exchange of the content of water pools during the collisions, but also drastically increases the attractive interaction between microemulsion droplets, which may further increases the frequency of sticky collisions. From many studies on the dynamics and the inter-droplet interaction of microemulsions, it has been shown that (39-41) the droplet interface became more fluid and the inter-droplet interaction became more attractive as one increased the chain length of alkane, or increased the concentration of alcohols, or decreased the chain length of alcohols. or decreased the concentration of the nonionic surfactant Arlacel-20 (42). Therefore, one would expect the increase in the number of particles due to more effective collisions when the microemulsion components were changed as described above. Our observation of the effects of the microemulsion components on the variation of transmittance thus can also be accounted for by the above explanations from the dynamic point of view.

CONCLUSIONS

We have studied the precipitation of AgCl in Microemulsions by uing stopped-flow photometry, quasi-elastic light scattering, spectrophotometry and transmission electron microscopy. We also analyzed the colloidal stability of the dispersions of ultrafine AgCl particles formed in microemulsions. The following conclusions were made based on the experimental results and foregoing discussions.

1. The reaction between $AgNO_3$ and NaCl in microemulsion media was completed in a very short period of time (approximately 60 ms for the studied case), as would be expected from the dynamic nature of microemulsions.

2. Ultrafine AgCl particles were obtained by precipitation in microemulsions. From the electron micrographs, the particle size was found to be 50 to 100 Å and the distribution of particle size appeared to be very uniform. From the analysis of electron diffraction patterns, the face-centered cubic lattice structrue of AgCl was identified.

3. The dispersions of these AgCl microcrystals were found to be very stable. Some dispersions (such as n-heptane system) can stay visually clear for weeks. From the TEM study, such system did not show any significant aggregation of particles even after 1 week.

4. From the transmittance and TEM studies, some dispersions were found to undergo very slow but reversible flocculation. The aggregates observed appeared to be consisted of many small AgCl particles with the same size as the dispersed particles. No significant difference in the size of primary particles was found from system to system.

5. The colloidal stability of the dispersions of AgCl ultrafine particles was found to increase as one decreased the chain length of alkanes, decreased the concentration of short chain alcohols, increased the chain length of the alcohols, or added the nonionic surfactant Arlacel-20.

6. The analysis of the colloidal stability of the dispersions revealed that the most important stabilization factor is the steric barrier provided by the surfactant film adsorbed on the particle surface. On the other hand, the electrostatic repulsion is negligibly small.

ACKNOWLEDGEMENT

We are grateful to the National Science Foundation (Grant No. NSF-CPE 8005851), American Chemical Society Petroleum Research Fund (Grant No. PRF-14718-ACS) and ALCOA Foundation for supporting this research.

REFERENCES

1   H. Topsoe, J.A. Dumesic, E.G. Derouane, B.S. Clausen, S. Morup, J. Villadsen and N. Topsoe, Stud. Surf. Sci. Catal., 3 (1979) 365.
2   K. Kon-no, M. Gobe, K. Kandoni and A. Kitahara, International Conference on Surface and Colloid Science, Jerusalem, 1981.
3   M. Boutonnet, J. Kizling, P. Stenius, G. Maire, Colloid and Surfaces, 5 (1982) 209.
4   N. Lufimpadio, J.B. Nagy, and E.G. Deroune, in: K.L. Mittal and B. Lindman, (Eds.), Surfactants in Solution, 3, Plenum Press, New York, 1984, 1483.
5   M. Dvolaitzky, R. Ober, C. Taupin, R. Anthore, X. Auvray, C. Petipas and C. William, J. Dis. Sci. Technol, 4 (1983) 29.
6   M.J. Hou, D.O. Shah, 60th National Symposium of Colloid and Surface Science, Atlanta, Georgia, 1986.
7   K. Kurihara, J. Kizling, P. Stenius and J.H. Fendler, J. Am. Chem. Soc., 105 (1983) 2574.
8   M. Gobe, K. Kon-no, K. Kandoni and A. Kitahara, J. Colloid Interface Sci., 93 (1983) 293.
9   R. Leung, M.J. Hou, C. Manohar, D.O. Shah and P.W. Chun in: D.O. Shah (Ed.) ACS Symposium Series, 272 (1985) 325.
10 J.H. Schulman, D.P. Riley, J. Colloid Sci., 3 (1948) 383.
11 J.H. Schulman, W. Stoeckenius and L.M. Prince, J. Phys. Chem., 63, (1959) 1677.
12 K. Shinoda and S. Friberg, Adv. Colloid Interface Sci, 4 (1975) 281.
13 J.H. Schulman and J.A. Friend, J. Colloid Interface Sci. 4 (1949) 497.
14 C.A. Miller, R.N. Hwan, W.J. Benton and T. Fort, Jr., J. Colloid Interface Sci., 61 (1977) 547.

15   M. Dvolaitzky, M. Guyot, M. Lagues, J.P. Le Pesant, R. Ober, C. Sauterey
      and C. Taupin, J. Chem. Phys., 69 (1978) 3279.
16   S.I. Chou and D.O. Shah, J. Phys. Chem., 85 (1981) 1480.
17   D.J. Cebula, R.H. Ottewill, J. Ralston and P.N. Pusey, J. Chem. Soc.
      Faraday Trans., 77 (1981) 2585.
18   E.W. Kaler, K.E. Bennet, H.T. Davis and L.E. Scriven, J. Chem. Phys. 79
      (1983) 5873.
19   E. Gulari, B. Bedwell and S. Alkhafaji, J. Colloid Interface Sci., 77
      (1980) 202.
20   L.E. Scriven, Nature, 263 (1976) 123.
21   P. Guering, B. Lindman, Langmuir, 1 (1985) 464.
22   J. Tabony, Nature, 319 (1986) 400.
23   F. Menger, J. Donahue, R. Willium, J. Am. Chem. Soc., 95 (1973) 286.
24   J.F. Eicke, J.C. Shepherd, and A. Steinemann, J. Colloid Interface Sci.,
      56 (1976) 168.
25   H.F. Eicke, P.E. Zinsli, J. Colloid Interface Sci., 65 (1978) 131.
26   J.C. Brown, P.N. Pusey, R. Dietz, J. Chem. Phys., 62 (1975) 1136.
27   D. Koppel, J. Chem. Phys., 57 (1971) 4814.
28   A.I. Vogel, Quantitative Inorganic Analysis, Chap. 1, London, 1951.
29   E. Matijevic and R.H. Ottewill, J. Colloid Interface Sci., 13 (1958)
      242.
30   R.H. Ottewill and A. Watanabe, Trans. Faraday Soc., 56 (1960) 855.
31   B.H. Robinson and P.D. Fletcher, Ber. Bunsenges. Phys. Chem., 85 (1981)
      863.
32   P.D. I., Fletcher, B.H. Robinson, F. Bermejo-Barrera and D.G. Oakenfull,
      J.C. Dove and D.C. Steytler, in: I.D. Robb (Ed.) Microemulsions, Plenum
      Press, New York, 1982, p. 221.
33   S. Atik and J.K. Thomas, Chem. Phys. Lett., 79 (1981) 351.
34   S. Atik and J.K. Thomas, J. Am. Chem. Soc., 103 (1981) 3543.
35   C. Tondre and R. Zana, J. Disp. Sci. Technol., 1 (1980) 179.
36   M.J. Hou and D.O. Shah, submitted to J. Colloid Interface Sci.
37   M. Von Smoluchowski, Ark. Phys. Chem., 92 (1917) 129.
38   N. Fuchs, Z. Phys., 89 (1934) 736.
39   A.M. Cazabat and D. Langevin, J. Chem. Phys., 74 (1981) 3148.
40   S. Brunetti, D. Roux, A.M. Bellocq and P. Bothorel, J. Phys. Chem., 87
      (1983) 1028.
41   B. Bedwell and E. Gulari, J. Colloid Interface Sci., 102 (1984) 88.
42   M.J. Hou and D.O. Shah, J. Colloid Interface Sci., in press.

Interfacial Phenomena in Biotechnology and Materials Processing, 459
edited by Y.A. Attia, B.M. Moudgil and S. Chander
Elsevier Science Publishers B.V., Amsterdam, 1988 — Printed in The Netherlands

NEW DEVELOPMENTS IN KNOWLEDGE OF ALUMINUM COLLOIDS

J.Y. BOTTERO[1], D. TCHOUBAR[2], J.M. CASES[1], J.J. FRIPIAT[3], F. FIESSINGER[4]

[1]Equipe de Recherche sur la Coagulation-Floculation et le Traitement des Eaux, CNRS, U.A. 235, rue du Doyen Marcel Roubault, 54501 Vandoeuvre Cedex (France)

[2]Laboratoire de Cristallographie, CNRS, U.A. 810, Université d'Orleans, U.F.R. de Sciences, rue de Chartres, 45067 Orleans Cedex 2 (France)

[3]University of Wisconsin, Department of Chemistry, P.O.B. 413, Milwaukee – Wisconsin 53201 (USA)

[4]Societé Lyonnaise des Eaux, 38, rue du President Wilson, 78230 Le Pecq (France)

SUMMARY

The hydrolysis products formed from aluminum chloride solutions progressively neutralized by caustic soda have been studied by small-angle X-ray scattering, ^{27}Al high resolution liquid and solid state nuclear magnetic resonance, microelectrophoresis and potentiometric titration. The polymer formed during hydrolysis is essentially the "Al_{13}" species. From R = OH/Al $\geq$ 2.3, the Al_{13} ions begin to aggregate. At R = 2.5 the aggregates have a fractal geometry with a fractal dimension D_f = 1.43. At R = 2.6 the fractal aggregates are bigger and denser ($D_f \sim 1.8$). This is characteristic of a cluster-cluster aggregation mechanism. This last mechanism does not depend on the pH of the solution. The driving force of the aggregation is the charge (σ) of the Al_{13} surface during the hydrolysis. Between pH 4 and pH 6, σ decreases from 1.5 to 0.45 $(H^+)/nm^2$. The Al_{13} aggregates are more and more dense (D_f increases from 1.42 to 1.8). For pH$_{13}$ 6, decreases but the structure of the aggregates does not change very much.

INTRODUCTION

The removal of mineral colloids, as well as some organics from drinking water, is achieved by coagulation-flocculation using aluminum salts. Addition of these salts is generally made in the pH range 6.5 to 8.5, and their final concentrations lie between 10^{-3} and 10^{-4} mol L^{-1}. Alum is commonly used to flocculate particles and to adsorb organics. Large aluminum hydroxide particles are generated by alum additions which remove "pollutants" by adsorption through van der Waals forces. There are, however, problems with the use of alum, e.g. it does not operate efficiently at all water temperatures.

In order to adapt the aluminum flocculant to the raw water "quality" we have synthesized a series of aluminum species, obtained from hydrolysis of aluminum chloride, and studied their chemistry, structure and surface properties in aqueous solution. This paper deals with a new class of aluminum hydroxide colloids used for flocculation and adsorption in water treatment, which are present in the original salt solutions and after pH adjustment and salt addition.

MATERIALS

Starting with a 0.5 M stock solution of aluminum chloride (obtained from $AlCl_3 \cdot 6H_2O$ (Merck reference no. 1084) and with a NaOH solution (Titrisol, Merck) we prepared solutions with a final 0.1 M Al concentration." In this preparation, 50 ml of the 0.5 M $AlCl_3$ solution were added to a 300-mL reactor which was thermostatically regulated at $20^{\circ}C$ under nitrogen. 200 mL of a solution containing the required amount of sodium hydroxide were added, while stirring at high shear using a peristaltic pump, to obtain the molar ratio r = $(NaOH)/(AlCl_3)$ = $(OH)/(Al)$ desired. The time required for this addition was fixed at 1 hr. The characteristic parameters are the total concentration of aluminum (Al_T) = 10^{-1} M, r = $(OH)/(Al)$, and the aging time t.

In certain stock solutions or after pH and dilution, aluminum hydroxide colloids were separated at various aging times by centrifugation at 5 x 10^4 g for 30 min. The solid residue in the bottom of the centrifugation tube was then immediately lyophilized and kept in a dessicator under vacuum ($\simeq 10^{-2}$ Torr). Table 1 shows the different samples investigated at different states.

TABLE 1

Samples investigated in the present work at various molar ratios, r, and aging times, t.

r	0 to 2	2.5	2.6
t			
10 min	liquid	liquid solid	liquid solid
- 1 hr	liquid	liquid solid	liquid solid
1 day to 10 days	liquid	liquid solid	liquid solid

METHODS
Liquid sample

^{27}Al NMR spectra have been obtained at 23.45 MHz in the Fourier
transform mode with a Brucker HX 90 interface to a Nicolet 1080 computer
(ref.1). 10 mm o.d. tubes were employed, the substance used as intensity
reference being dissolved in deuterium oxide (whose deuterium signal placed
the field frequency lock), in a Wilmad coaxial cell. A 3-kHz spectral width
was selected. The free induction decays were accumulated using 2K words,
while the Fourier transform was performed on a 16K array, by using the zero
filling technique; 90° pulses were used; the time elapsing between two
pulses was sufficient to allow a complete return of the magnetization to
equilibrium. In order to find the amount of aluminum bound in the different
species, experiments were made by using reference samples of the ion
$Al(OH)_4^-$ at 10^{-1}, 5×10^{-2}, and 2×10^{-2} M, placed in the capillary tube of
the coaxial cell. The peak of the reference $Al(OH)_4^-$ is located at 79.9 ppm
from the signal of $Al(H_2O)_6^{3+}$. Thus by simple integration of the peaks we
can calculate the amount of aluminum bound in each detectable species.

S.A.X.S. experiments. For this work, we made use of the synchrotron
beam from the DCI storage ring at L.U.R.E. (University of Orsay); this
source has particular advantages, in beam symmetry and energy, which make
it possible to study weakly scattering systems in very short times : 1) The
beam diverges very little in the vertical plan (2×10^{-3} rad.). It is
practical to obtain a point image of the source using a single-curve
monochromator which focuses the beam in a horizontal plane, thus
eliminating any instrumental deformation relating to the shape of the beam
and its spectral content; 2) Since the beam is very intense, counting times
are about 200 times shorter than when a conventional source is used. The
focal distance chose was longer than 3 M. The sample-detector distance can
be varied from a few centimeters to 2 m. We used a localization detector of
the LC-delay type. Its resolving power is 150 μm between two measuring
points. The range of the small angles investigated was obtained with the
detector in the vertical position and a sample detector distance of 2 m.
The Bragg zone analyzed was between 20 and 1500 Å and called the
observation window. The wavelength used was 1.6 Å. As a result of the
narrow beam, the only corrections necessary were those for absorption by
the sample. In order to obtain the scattering due to the aluminum species,
we subtracted the scattering by the solvent, i.e., solutions of NaCl of
suitable concentration determined by the neutralization ratio r =
$(NaOH)/(Al_T)$.

METHOD OF INTERPRETATION

Taking into account the low aluminum concentration, $[Al_T] = 0.1$ M g and the motion of the aluminum species in the solution, the samples can be considered as statistically isotropic diluted systems. Consequently, all the geometrical features of the particles can be deduced from the reduced scattering intensity $I_n(s)$

$$I_n(s) = I(s)/PO \qquad (1)$$

where $I(S)$ is the experimental spectrum, PO the total scattering power, and $s = 2\theta/\lambda$,

$$PO = \int_0^\infty 4\pi s^2 I(s)ds \qquad (2)$$

The correlation function $\gamma(r)$ in direct space is obtained by the Fourier transform of $I_n(s)$ using the relationship

$$\gamma(r) = \int_0^\infty 4\pi s^2 I_n(s)\frac{\sin 2\pi sr}{2\pi sr} dr \qquad (3)$$

Two kinds of data can be obtained from scattering curves. If the particle size is smaller than the observation window size, i.e. 1500 Å, it is possible to measure the radius of gyration (ref. 2).

$$I_n(s) = I_n(0) \exp -\frac{4\pi^2 R_g^2}{3} s^2 \qquad (4)$$

$I_n(0)$ is the value at the origin of the intensity, and Rg is the radius of gyration of the particle, which is determined, from the slope at the origin of the function $\ln(S)$ vs s^2. For a cylindrical particle of diameter $2R_c$ and thickness or height $2H_c$, we have

$$R_g^2 = R_c^2/2 + H_c^2/3 \qquad (5)$$

For a spherical particle of radius R

$$R^2 = 5/3 R_g^2 \qquad (6)$$

In the general case of dense particles dispersed in a solvent, a more interesting function is the distance distribution $P(r) = r^2\gamma(r)$ (eqn.7). Let us recall that the whole profile and the amplitude of this function depend on the geometric shape and the volume of matter within the particle.

Note also that P(r) = 0 for the value of r which is called the maximal chord.

For aggregates, the P(r) function is indicative of the internal correlation of the subunits forming the cluster. The amplitude of P(r) is directly related to the density of the whole aggregate (ref. 3-4). The scattering intensity of the aggregates is :

$$I(s) = I_o(s) \cdot G(s) \tag{7}$$

where $I_o(s)$ is the scattering by the subunit of the aggregates and $G(s)$ is the interference function between these subunits. In addition, the decrease of the scattering intensity is significant of the type of aggregation. As shown previously, the $I(s)$ function decays as s^{-D_F} where D_F is known as the fractal dimension (ref. 5). All the reported results were obtained by comparing the P(r) function with those computed for different geometric shapes and types of aggregation. The fractal dimension was determined on the $I(s)$ versus s in a log-log plot.

Solid samples

MAS NMR. The ^{27}Al high-resolution solid-state NMR spectra were obtained at 130.8 MHz at a spinning rate of 3.9 kHz. The experimental conditions were those described by Plee et al (ref. 6).

INTERPRETATION METHOD

When a spectrum presents octahedral and tetrahedral bands, because of the asymmetry and of the broadness of the peak observed near the reference frequency ($Al^3(H_2O)_6$ aq) assigned to octahedral aluminum, the ratio of the area of the line attributed to tetrahedral aluminum (near 62 ± 1 ppm) to the area of the whole signal could not be obtained from direct integration.

For ^{27}Al (I = 5/2), the m = $\pm$ 1/2 transition is independent of the first-order quadrupolar interaction but it is affected by the second-order quadrupolar effect. Large variations in the quadrupolar coupling constant (QCC) provoke distortion of the peak shape because of the influence of second-order terms. The sideband pattern is also modified, QCC is sensitive to change in the distribution of electrical charges and thus to perturbation of the symmetry of the coordination shell. When the lines assigned to Al^{IV} and Al^{VI} are narrow and well separated, such as in well-crystallized micas, straight integration of the peaks gives a reasonable estimate of the relative Al^{IV} and Al^{VI} contents, especially if the spectrum is recorded at very high field, as in this work (11.7 T). For amorphous or quasi-amorphous samples, the A^{IV} and Al^{VI} contributions are quite broad, as illustrated in Fig. 1. In that example the FWHH (full width at half height) of the Al^{IV} line is 13.4 ppm whereas that of the Al^{VI} line is 21.2 ppm. It is indeed very common that the QCC at the tetrahedral site is smaller than at the octahedral site (ref. 7). This means that the Al^{VI} signal contributes to the overall signal intensity in the domain of the Al^{IV} contribution. Then if $I_t = 100$ % is the total area of the ^{27}Al signal, I_u/I_t largely overestimates the Al^{IV} relative content (see Fig. 1).

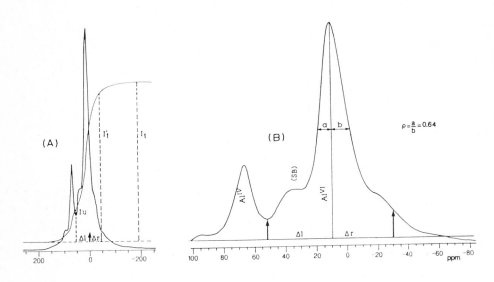

Fig. 1. (A) ^{27}Al integrated spectrum. (B) The same with extended scale. Experimental conditions : resonance frequency, 130.2 MHz ; spinning rate, 3.8. KHz ; pulse duration, 2 µs ; recycling time, 100 ms ; number of accumulations, 638 ; delay time, 1 ms ; reference, Al^{3+} aq. ; sample, amorphous aluminium hydroxide.

It is thus necessary to correct I_u in order to obtain an acceptable Al^{IV} content. In the first approximation this correction can be carried out as follows. The Al^{VI} peak is always asymmetrical, $b > a$ (Fig. 1B), and thus $\rho = a/b < 1$. The Al^{VI} line broadening toward the negative shift transforms the Al^{VI} sideband, which is well defined in the positive shift region, into a shoulder in the negative shift domain. It follows that I_u can be measured quite accurately on the Δl side of the maximum of the Al^{VI} line. If this line were narrow and symmetrical, I'_t, measured at $\Delta r = \Delta l$ would be identical to I_t. This is not the case and $I'_t/I_t < 1$. The difference $(I_t-I'_t/I_t)$ multiplied by ρ accounts for the contribution of the Al^{VI} peak in the spectral domain where the Al^{IV} line is observed, and finally the corrected Al^{IV} relative content I_c can be estimated as

$$I_c/I_t = I_u/I_t \frac{I_t - I_t'}{I_t} \rho \qquad (8)$$

In the example shown in Fig. 1, I_t = 100 %, ρ = 0.64 ($I_t - I_t'$) = 6 %, I_u = 20.9 %, and I_c = 16.6 %. Thus the Al^{IV} content is lowered by about 4 %. The correction represents 23 % of the estimated Al^{IV} relative content. The same kind of correction applied to crystalline smectites with tetrahedral substitutions, studied earlier, is much smaller and of the order of, or less than, 10 % (relative). This uncertainty may be considered within the margin of error of the MAS NMR technique applied to quadrupolar nuclei.

SURFACE CHARGE DETERMINATION

Potentiometric titration

Experiments were carried out on fresh precipitates in the presence of NaCl electrolyte. All solutions were prepared with distilled water, boiled, then re-cooled in a balloon in which nitrogen was bubbling. In a cell thermostated at 25°C, an adequate quantity of aluminum was diluted in NaCl solution under vigorous stirring in order to obtain an aluminum concentration of $2 \cdot 10^{-3}$ M. During this period the pH is maintained at 7.5 $\pm$ 0.1 and stabilized for approximately 3 min. Then, the aluminum hydroxide suspension is titrated by NaOH and HCl using an automatic instrument TITRIMAX - TACUSSEL (ref. 8). During the entire operation a nitrogen current, previously bubbled in NaOH 3N, is passed into the solution. An identical suspension is centrifuged at 80 000 x g. The supernatant constitutes the "blank". This blank is titrated under exactly the same conditions as the suspension. The surface charge $Q_o = \Gamma_{H}^+ - \Gamma_{OH}^-$ is calculated from the difference between the titration of suspension and blank solution. Titration was carried out at three initial concentrations of NaCl : $|NaCl|_{(i)}$ = 0 , 10^{-1} and 1 mol.l^{-1}. The symbol i corresponds to the concentration before the dilution of aluminum.

Electrokinetic potential

The aluminum hydroxide suspensions are prepared as above but the total aluminum concentration is 10^{-3} mol.l^{-1}. The pH is rapidly adjusted with NaOH or HCl. Electrophoresis measurement is carried out very rapidly (in less than thirty seconds), using a Laser Zee Meter Pen Kem (model 501) equipped with a video system, in order to i) minimize the flocculation, especially near the I.E.P. ii) minimize chemical and physical modifications of the flocs with time (ref. 9).

INTERFACE ELECTROCHEMICAL MODEL

For the NaCl-Aluminum hydroxide system, the surface of the solid is constituted by Ns amphoteric sites $=Al-OH$ which can ionize as follows:

$$=Al-OH_2^+ \;\rightleftarrows\; =Al-OH + H_{(d)}^+ \tag{9}$$

$$=Al-OH \;\rightleftarrows\; =Al-O^- + H_{(d)}^+ \tag{10}$$

The charged sites $=Al-OH_2^+$ and $=Al-O^-$ interact with the electrolyte ions, e.g.,

$$=Al-OH_2Cl \;\rightleftarrows\; =Al-OH_2^+ + Cl_{(d)}^- \tag{11}$$

$$=Al-ONa \;\rightleftarrows\; =Al-O^- + Na_{(d)}^+ \tag{12}$$

The uncharged sites $=Al-OH$ can also react with the electrolyte ions, e.g. :

$$=Al-OHNa^+ \;\rightleftarrows\; =Al-OH + Na_{(d)}^+ \tag{13}$$

$$=Al-OHCl^- \;\rightleftarrows\; =Al-OH + Cl_{(d)}^- \tag{14}$$

Reactions (9) to (12) are responsible for the development of electrical double layer at the interface, e.g.,

a) an internal layer of global electrical charge σ_s :

$$\sigma_{s_n} = \frac{F}{n_{AVO}} \{N\,[=Al-OH_2^+] + N\,[=Al-OHNa^+] - N\,[=Al-O^-] - N\,[=Al-OHCl^-]\} \tag{15}$$

where F and n_{AVO} are the Faraday and the Avogadro number respectively, and $N[j]$ is the density of surface site j.

b) an external diffuse layer made of $H_{(d)}^+$, $OH_{(d)}^-$, $Na_{(d)}^+$, and $Cl_{(d)}^-$. From the GOUY-CHAPMAN equation,

$$\sigma_d = -11.74 \cdot 10^{-2}\, c^{1/2}\, \sinh\,(0,0194\,\zeta)\ \text{coul} \cdot \text{m}^{-2} \quad (\text{at } 25^\circ C) \tag{16}$$

where the electrokinetic potential ζ is the slipping plane potential.

$$c = [Cl^-] + [OH^-] = [Na^+] + [H^+] \tag{17}$$

and expresses electroneutrality in the bulk solution.

Equilibrium of the double layer requires :

$$\sigma_s + \sigma_d = 0 \tag{18}$$

The Table 2 summarizes the surface reactions. In column 2, EY parameter corresponds to $\exp\,(+38.9\ 10^{-3})$ where ζ is the electrokinetic potential.

The interfacial equilibrium constants, K_j, and the active sites density N_s, were estimated by fitting titration and electrokinetic curves by using the gradient method (ref. 8).

Table 2. Surface reactions assumed in order to calculate the pK'.

No.	Reactions	Equilibrium constants	Description
1	$-AlOH_2^+ \leftrightarrow Al-OH + H$	K_1 EY	Proton adsorption to form a positive site
2	$-Al-OH \leftrightarrow -Al-O^- + H$	K_2 EY	Equivalent to OH adsorption to form a negative site
3	$-Al-OH\ Cl \leftrightarrow -Al-OH + Cl^-$	K_3 / EY	Adsorption onto
4	$-Al-ONa \leftrightarrow -Al-O^- + Na^+$	K_4 / EY	the charged sites
5	$-Al-OHNa^+ \leftrightarrow -Al-OH + Na^+$	K_5 EY	Adsorption onto the
6	$-Al-OHCl^- \leftrightarrow -Al-OH + Cl^-$	K_6 / EY	uncharged sites

RESULTS AND DISCUSSION

NMR results

Liquid state. Apart from the ion signal taken as reference $(AlOH)_4^-$, we observed a maximum of three signals whose importance varied with the value of $r = (OH)/Al_T)$.

From $r = 0$ to $r = 2.5$, a signal appears at 17 ppm from the reference, corresponding to aluminum ions in tetrahedral coordination. As we shall see later, the height and the area of this peak increase up to a value of $r = 2.2$ and then decrease (Fig. 2).

From $r = 0.5$ to $r = 2.4$ a signal appears at 79.9 ppm from the reference, corresponding to aluminum ions in octahedral coordination. This signal corresponds to aluminum ions in monomeric species of the type $Al(H_2O)_6^{3+}$, $Al(OH)(H_2O)_5^{2+}$, and $Al(OH)_2(H_2O)_4^+$. This peak does not shift. Its width at half-maximum increases with r; when r is between 2.2 and 2.3, it is ca. 3 times as wide as when $r = 0.5$. The height and the area of this peak show a corresponding decrease from $r : 0.5$ to $r = 2$, followed by an increase up to $r = 2.2$ and then a decrease.

The peak located at 17 ppm from the reference signal was attributed to the tetrahedral aluminum of the polymer $Al_{12}^{VI}(OH)_{24} Al^{IV}O_4^- (H_2O)_{12}^{7+}$. It has been described by Johansson (ref. 10) and by Akitt et al (ref. 11) (Fig. 3). The 12 other octahedral aluminum are presumed not to have a symmetrical environment. The electric field gradient at their level is therefore relatively high, and the octahedron may be distorted. All this causes so great a broadening of the peak, due to the quadrupolar relaxation, that the signal cannot be detected at high resolution.

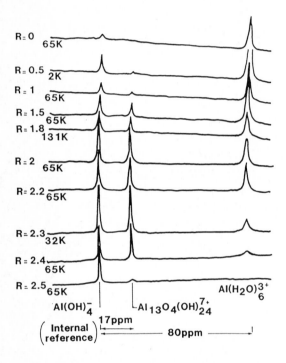

R= 0 65K
R= 0.5 2K
R = 1 65K
R= 1.5 65K
R= 1.8 131K
R = 2 65K
R= 2.2 65K
R= 2.3 32K
R = 2.4 65K
R= 2.5 65K

$Al(OH)_4^-$ $Al_{13}O_4(OH)_{24}^{7+}$ $Al(H_2O)_6^{3+}$

$\left(\begin{array}{c} \text{Internal} \\ \text{reference} \end{array}\right)$ 17ppm |← 80ppm →|

Fig. 2. ^{27}Al NMR typical spectra as a function of $R = [OH]/[Al]$.

From the areas of the peaks, it was possible to calculate the amount of aluminum bound in the monomeric, and the condensed $Al_{13}O_4(OH)_{24}(H_2O)_{12}^{7+}$ forms. Figure 4 shows the variation in the percentage of aluminum bound in the various species as a function of r and of the pH. If one allows for experimental errors, the sum of the concentrations expressed as a prcentage of the total aluminum is between 98 and 105 % when r is between 0.5 and 2.3. Beyond this last value a part of the aluminum ions are not detected by high resolution ^{27}Al NMR in liquid state. We should note that the turbidity

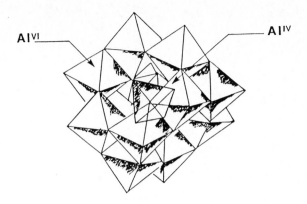

Fig. 3. Schematic structure of Al_{13} polycation.

increases when $r \geq 2.3$. For $r = 2.6$ one observes a flocculation of the colloids (Table 3).

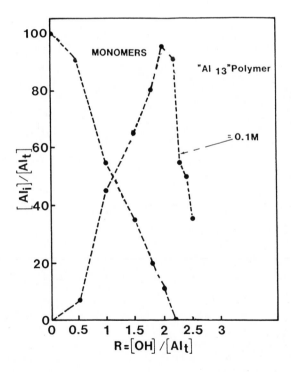

Fig. 4. Distribution of the different species, from NMR data, versus $R = [OH]/[Al]$.

Table 3. Relation between r, pH and turbidity of the concentrated solutions of aluminum.

$r = \frac{OH}{Al}$	0.5	1	1.5	1.8	2	2.2	2.3	2.4	2.5	2.6
pH	3.59	3.59	3.87	3.99	4.06	4.28	4.37	4.5	4.8 5.2	6
FTU	0.6	0.6	0.6	0.6	0.7	0.8	3	9	30 150	Floc-cula-tion thre-shold

Solid state. The ^{27}Al MAS NMR spectra are shown in Fig. 5, the experimental conditions being practically those detailed in Fig. 1. The numerical parameters are reported in Table 2. For samples 2-7, the FWHH of the Al^{VI} line is of the order of 22 $\pm$ 2 ppm. For samples 3-7 the FWHH of the Al^{IV} line is 16.5 $\pm$ 2 ppm. Sample 2 has a narrower FWHH.

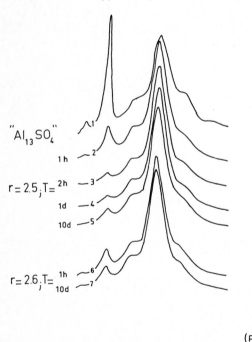

"Al$_{13}$SO$_4$"

1 h — 2

r = 2.5 ; T = 2h — 3

1d — 4

10d — 5

r = 2.6 ; T = 1h — 6
10d — 7

(ppm)

100 80 60 40 20 0 -20 -40 -60 -80

Fig. 5. ^{27}Al MAS NMR spectra obtained for the samples listed in Table 4. The experimental conditions are similar to those detailed in Fig. 2.

The reference sample for isolated Al_{13} is the sulfate salt obtained upon precipitating the Al_{13} species by the sodium sulfate from a solution with r = 2. Probably, the precipitation carried out with a large divalent anion does not modify the building unit AL_{13}. Indeed as shown in Table III the shift of the Al^{IV} line is within experimental error, the same as that observed in solution (Table 4), and the relative Al^{IV} content is the expected one. It is worth pointing out that the FWHH of the Al^{IV} line (7.4 ppm) is about the same as that measured for synthetic beidellite (ref. 6). This means that the tetrahedral symmetry is not appreciably distorted. This, according to Ghose and Tsang (ref. 12), suggests that the O-Al-O tetrahedral angle is not far from ≃ 109.5°. On the other hand the FWHH of the Al^{VI} line is more than three times larger than in beidellite (ref. 6),

Table 4. ^{27}Al chemical shifts, δ (ppm), FWHH (ppm) for tetrahedrally (Al^{IV}) and octahedrally (Al^{VI}) coordinated species.

No. spectrum	Sample	Al tetra		Al octa		% Al_{tetra}/al_t
		δ	FWHH	δ	FWHH	
1	"$Al_{13}SO_4$"	63.3	7.4	3.6	29.6	6.1
2	r=2.5 T=1h	63.1	11.5	6.3	23.1	9.9
3	r=2.5 T=2h	62.6	17.6	5.7	20.3	4.7
4	r=2.5 T=1d	62.3	17.6	5.4	20.4	4.2
5	r=2.5 T=1w	62.1	16.7	5.4	22.2	3.9
6	r=2.6 T=1h	63.7	16.7	6.8	23.1	8.5
7	r=2.6 T=1w	63.3	16.7	6.7	24.1	6.7

which would means that there is a rather broad distribution of Al-O distances within the Al octahedra. Thus the QCC of the two kinds of sites (Al^{IV} and Al^{VI}) should be very different. It is also known (ref. 13) that the shielding at octahedra aluminum increases with increasing polymerization. This means the chemical shifts of the Al^{VI} line should become more positive as the size of the octahedral layer grows. The chemical shift of the Al^{IV} line is, within experimental uncertainty, constant from spectrum 1 to spectrum 7 whereas δ Al^{VI} decreases with time for r = 2.5. The octahedral symmetry in the solids 2-7 (Table 4) is less perturbed than in the "Al_{13} sulfate" since the Al^{VI} line FWHH decreases from ≃ 30 to 21 $\pm$ 2 ppm for r increasing from 2 to 2.6. On the other hand the tetrahedral symmetry is very significantly damaged as the evolution toward the layered hydroxide proceeds. With respect to the Al MAS NMR spectral characteristics, the hydrolysis ratio seems to play a more important role thant the aging time, at least within the maximum 10 days aging time used in the present work.

The dilution and pH shock characterizing the use of aluminum salts in water treatment has been evaluated : The dilution of r = 2.0; r = 2.5 from $(Al_T) = 0.1$ mol.l^{-1} to $(Al_T) = 10^{-3}$ mol.l^{-1} at pH 6.5, 7.5 and 8.5 corresponds to the formation of large particles. MAS NMR experiments of these colloids after lyophilization at different aging time at the pH of precipitation gives the same results than in concentrated solution (acidic pH). Figure 6 shows an example of spectra of precipitates obtained from dilution of r = 2.5 and r = 2 at pH = 7.5. Following the same procedure one have calculated the $Al^{IV}/Al^{IV} + Al^{VI}$ ratio. Table 5 indicates than the Al_{13} content in the flocs decreases as pH of the dilution increases. Note

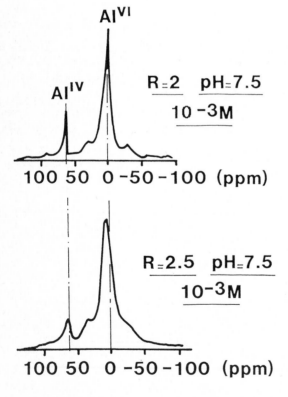

Fig. 6. ^{27}al MAS NMR spectra of solids precipitated after diluting to 10^{-3} M, R = 2 and R = 2.5 at pH 7.5.

Table 5. Al_{13} content in the solids precipitated after diluting, at pH = 6.5 - 7.5 and 8.5,, r = 2.0 and r = 2.5 as calculated by ^{27}Al MAS NMR experiments.

Flocculant	pH								
	6.5			7.5			8.5		
	10'	1 h	24 h	10'	1 h	24 h	10'	1 h	24 h
r = 2.0	80 | 100 %	~ 80	N/A	80 | 100 %	80	80	60	45	35
r = 2.5	80 | 1 00 %	~ 75	N/A	80	60	60	60	35	25

that Al_{13} content is between 80 to 100 % of aluminum for r = 2.0, pH 6.5 and T ≤ 1 h and also for r = 2.5, pH = 7.5 and T ≤ 1 h. Al_{13} is a stable specie. It is not sensitive, during a certain time, to the hydrolysis process when pH and dilution abruptly increase.

The structure of such colloids only constituted by Al_{13} was determined from SAXS measurements. First as shown above we have measured the size of isolated Al_{13} in r = 2. Also this solution provides $I_o(s)$ of the equation 7. Eqn. 4 gives :

$$\ln I(s) = \ln I(0) - (4 \pi^2/3) R_g^2/s^2$$

R_g = 9.8 A. Since this polymer is of spherical shape, the radius is then

$$R = (5/3)^{1/2} R_g = 12.6 \text{ A}$$

The ionic radius of this polymer in the crystalline phase is equal to 5.4 A. The results must be interpreted in terms of "hydrated" ions. The Cl^- ions are close to the polymer Al_{13} and therefore play a substantial role in the scattering (ref. 9).

The surface area σ/V of Al_{13} is determined from the experimental value of the radius R = 12.6 A is σ/V = 2380 m^2/cm^3.

Investigation of solution r = 2.5, $T_v \leq 1$ h)

The ^{27}Al NMR in liquid state and solid state showed that only Al_{13} are present (Fig. 2 and 5). Either in isolated or aggregated state.

From eqn. 4 the experimental radius of gyration is Rg = 93 A. Nevertheless its angle validity domain is too short to assure a physical meaning for this value. The structure of these colloids only can be evaluated from comparison between experimental data and modelling of aggregates by computer. As shown Fig. 7a, the $\log_{10} - \log_{10}$ plot of the I(s) scattering function shows a linear portion for $s \geq 1/2\pi x$ (departure from the linear law at large S values). Experimentally x = 10 A. This value is

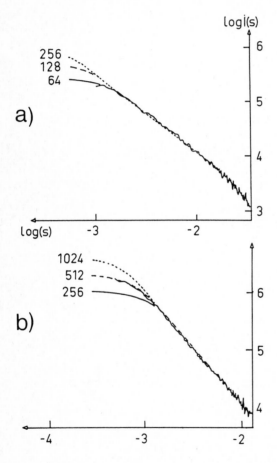

Fig. 7. a) $\log_{10} I(s)$ - $\log_{10}(s)$ plot for r = 2.5 solution. Three theoretical curves are shown using aggregate models (D = 1.43) with 64 Al_{13} (-), 128 Al_{13} (--), 256 Al_{13} (----). The best fit is obtained with 64 Al_{13}. b). $\log_{10} I(s)$ - $\log_{10}(s)$ plot for r = 2.5 solution. Three theoretical curves are shown using aggregate models (D = 1.8) with 256 Al_{13} (-), 512 Al_{13} (---) and 1024 Al_{13} (----). The best fit is obtained with 512 Al_{13}.

in good agreement with the value of the radius of Al_{13} measured in r = 2 solution. This behavior shows that the aggregation of the Al_{13} subunits yields to fractal aggregates. The fractal dimension calculated from the slope of the linear portion (eqn. 7) is D = 1.45.

Investigation of r = 2.6 -Tv $\leq$ 24 h

The MAS NMR (Table 4) data showed that the aggregates are formed of Al_{13} subunits whatever the aging time. The $\log_{10}$ I(s) - $\log_{10}$ (s) plot (Fig. 8b) for r = 2.5 shows a linear portion limited for smallest s values at S = $1/2\pi ro$ with ro = 10 A. This behavior, as above, is characteristic of the fractal aggregation of Al_{13}. The fractal dimension is D = 1.85.

The modelling aimed to i) prove the fractal structure as deduced from S.A.X.S. measurement, ii) calculate the number of Al_{13} subunits in the different aggregates. A cluster-cluster model (ref. 15-18) was used to simulate the scattering curves. From generated clusters of 64 to 1024 particles, the P(r) function was calculated by simply establishing the histogram of interparticle distances within the aggregate. Then the

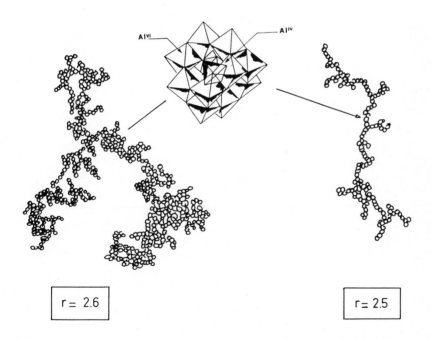

Fig. 8. Schematic representation of the aggregates present in R = 2.6 solution (D = 1.8, $N_{Al_{13}}$ = 512) and R = 2.5 solution (D = 1.43, $N_{Al_{13}}$ = 64).

theoretical I(s) scattering functions are refined to fit the experimental data. For r = 2.5 and r = 2.6 fractal aggregates of respectively 64 and 512 Al_{13} were obtained (Fig. 7a and 7b). A picture of the bloth clusters with D = 1.43 and D = 1.86 is shown Fig 8. One see that for D = 1.43 the aggregate is much more linear (loose structure) than for D = 1.86. This implies that the kinetics of aggregation are different. The origin of this kinetic of aggregation is the surface charge and its distribution.

The addition and pH increase shocked such colloids away from thermodynamic equilibrium and yielded large flocs of Al_{13} with a fractal dimension D ~ 1.8 from r = 2.5 solution and D ~ 2 from r = 2.6 solution.

The surface charge is the driving force of the aggregation of Al_{13}. Titration curves between pH = 6 and pH = 9 (shown in Fig. 9) exhibit a ZPC for pH = 8.4.

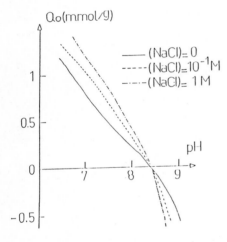

Fig. 9. Experimental amount of H^+ consumed versus pH, by the surface of aluminum colloids in [NaCl] = 0 (-), [NaCl] = 10^{-1} mol.l^{-1} (---), [NaCl] = 1 mol.l^{-1} (-.-.-.).

The electrokinetic potential of the colloids (Fig. 10) shows the same value of I.E.P. NaCl is an indifferent electrolyte because ZPC = I.E.P. The both curves are fitted from the calculated superficial parameters in the table 6. The calculated N_s value is 1.1 site . nm^{-2}.

From the pK_j it is possible to plot the distribution of the active superficial sites (Fig. 11). Thus, two significant remarks may be made :

1. The relative amount of charged sites responsible for the ζ-potential does not exceed 4 %.

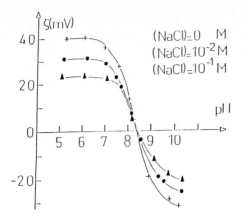

Fig. 10. Electrochemical potential versus pH of aluminum colloids in [NaCl] = 0 (+), [NaCl] = 10⁻² mol.l⁻¹ (o), [NaCl] = 10⁻¹.l⁻¹ (▲). The full lines -- are the calculated curves.

2. The neutral -Al-OH distribution curve interacts -Al-OH$_2^+$ and -Al-O$^-$ curves respectively at pH values different of pK_1 and pK_2. This is due to an electrical field at the solid-liquid interface expressed by the EY factor in the different superficial equations.

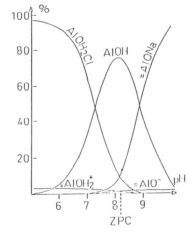

Fig. 11. Distribution of the active surface sites versus pH in [NaCl] = 0.

The high stability of Al_{13} structure versus pH allows us to consider Al_{13} and the aggregation products as solids. When pH increases only the surface charge varies. In the more acidic pH, i.e. pH = 4.0, the charge of Al_{13} is 7^+ as indirectly determined from exchange with Na^+ or Ca^{++} cation on a montmorillonite (ref. 19). When pH increases the colloids exhibit a

Table 6. Optimal calculated pK at various ionic strengths and for $\gamma_{Na_2} = \gamma_{Cl}$. The calculated active site number is ns = 1.1 site/nm^2. These values allow to fit the zeta - pH and Q_0 - pH curves.

NaCl	Activity coefficients $\gamma_{Na} = \gamma_{Cl}$	pK_1	pK_2	pK_3	pK_4	pK_5	pK_6
0	0.064	6.65	10.15	3.18	3.38	0.004	−1.18
0.01	0.004	6.65	10.15	3.16	3.35	0.035	−0.4
0.1	0.079	6.77	10.02	3.19	3.38	0.15	−0.34
Mean values		6.69	10.11	3.18	3.37	0.048	−0.63

a decrease in surface charge. The surface accessible for H^+ ions is roughly the surface area of desolvated (r = 6 Å), i.e. the geometric surface : $S = 4\pi r^2 = 4.5$ nm^2 or for d = 2 g/cm^3, S = 2000 m^2/g. The surface charge σ by nm^2, expressed by $[H^+]/nm^2$ decreases from pH = 4 to ZPC (Fig. 12). Between pH 4 and pH 6 σ varies from ∼ 1.55 to 0.45 $[H^+]/nm^2$. In the same time the isolated Al_{13} particles form aggregates more and more close. For pH > 6 the decreasing of σ is linear and the structure of the aggregates does not change very much, except the particle size.

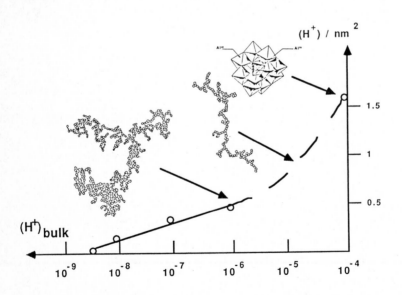

Fig. 12. $[H^+]$ amount consumed by the aluminum colloids surface versus $[H^+]$ bulk concentration and the related Aluminum aggregates.

CONCLUSION

Aluminum particles formed from the hydrolysis of $AlCl_3 \cdot 6H_2O$ with NaOH, where $r = (OH)/(Al) \geq 2$, contain primarily Al_{13}. The structure of Al_{13} is stable up to an aging time of 1 hr. With increasing pH these aluminum particles form aggregates whose size and structure depend on the surface charge. When the charge is high the aggregates of Al_{13} are very open. The region on the Al_{13} surface where another Al_{13} can link is presumably limited. When pH increases, σ decreases and the number of anchoring sites on each Al_{13} increases. The resulting structure is much denser.

The removal efficiency of mineral particles depends strongly on the structure and surface charge of these flocculants. The flocculant obtained in $r = 2.5$ with a fractal dimension of 1.43 yields dense, rapidly settling flocs.

REFERENCES

1   J.Y. Bottero, J.M. Cases, F. Fiessinger and J.E. Poirier, J. Phys. Chem., 84 (1980) 2933-2937.
2   A. Guinier and G. Fournet, Small-angle scattering of X-rays, John Wiley and Son, 1955.
3   M.A.V. Axelos, D. Tchoubar, J.Y. Bottero and F. Fiessinger, Journal de Physique, 46 (1985) 1957-1961.
4   M.A.V. Axelos et al., J. Phys. (1986).
5   B.B. Mandelbrot, Form, chance and dimension, Freeman, San Francisco, 1977.
6   D. Plee, F. Borg, L. Gatineau and J.J. Fripiat, J. Amer. Chem. Soc., 107 (1985) 2362-2366.
7   E. Oldfield and R.J. Kirkpatrick, High resolution NMR of inorganic (to be published).
8   E. Rakotonarivo, J.Y. Bottero, J.M. Cases and F. Thomas, (Submitted for publication).
9   J.Y. Bottero, D. Tchoubar, J.M. Cases and F. Fiessinger, J. Phys. Chem., 86 (1982) 3667-3670.
10 G. Johansson, Acta Chem. Scand., 14 (1960) 771-774.
11 J.W. Akitt, W. Greenwood, B.L. Khandelwal and G.R. Lester, J. Chem. Soc. Dalton Trans. (1972) 604-612.
12 S. Ghose and T. Tsang, Amer. Mineral, 58 (1973) 748-754.
13 R.J. Kirkpatrick, K.A. Smith, S. Schramm, G. Turner and Wang-Hong Yang, Ann. Rev. Earth Planet Sci., 13 (1985) 220-227.
14 J.Y. Bottero, M.A.V. Axelos, D. Tchoubar, J.M. Cases, J.J. Fripiat and F. Fiessinger, J. Colloids and Interface Science, 117 (1987) 47-62.
15 R. Jullien and R. Botet, Aggregation and fractal aggregates, World Scientific, 1987.
16 D. Weitz and M. Oliveira, Phys. Rev. Lett. 52, 116 (1984) 1433-1435.
17 R. Ball and R. Jullien, Journal de Physique, 45.L (1984) 1031-1035.
18 P. Meakin, J. Colloid and Interface Science, 102(2) (1984) 505-512.
19 J.Y. Bottero, M. Bruant and J.M. Cases, Clay Minerals, 1987 (in press).
20 D. Tchoubar and J.Y. Bottero (In preparation).

Interfacial Phenomena in Biotechnology and Materials Processing,
edited by Y.A. Attia, B.M. Moudgil and S. Chander
Elsevier Science Publishers B.V., Amsterdam, 1988 — Printed in The Netherlands

HYDROSOLS FROM LOW-RANK COALS: LOW TEMPERATURE OXIDATION

EDWIN S. OLSON, JOHN W. DIEHL, and MICHAEL L. FROEHLICH

University of North Dakota Energy Research Center, Box 8213 University Station, Grand Forks, ND 58202

SUMMARY

Low-rank coals have been converted to a hydrosol in high yields by blending at high speed in sodium hydroxide at temperatures less than 45°C. The stable dispersions of sodium humate are highly reactive to alkylating agents and oxygen at low temperatures. Air injection during the blending does not change the particle size distribution of the isolated humic acids significantly. Oxidation results in the formation of carboxylic acid groups but does not proceed to the the benzenepolycarboxylic acid level. Air oxidation of the humate hydrosol at 100°C did not further change the nature of the material.

INTRODUCTION

Coal has been recognized as a colloid since the early part of the century. This 1911 description of coal as a colloid was contributed by Potomie (ref. 1). The extracts which can be derived from coals by extraction with base (humic acids) or solvents have also been regarded as colloids by some chemists; however, there has been controversy involved in the measurement of the sizes and molecular weights of the extract particles (ref. 2). Dynamic light scattering techniques can provide particle size data which is more appropriate for the low-rank coal derived-materials than the vapor phase osmometry techniques used previously (ref. 3). Measurements of the sizes of the particles of sodium humate from North Dakota lignite using dynamic light scattering show that this material is a dispersion of particles in the one-quarter to about one micron range.

An objective of our studies was to investigate the use of colloidal dispersions of coals in various processes which may take advantage of the small (submicron) particle size. It was hypothesized that diffusion-controlled processes would benefit from the improved access of solvent and reagents to the coal material. The liquefaction of colloidal dispersions of coal is in progress and will be reported elsewhere. In this paper, the air oxidation of the colloidal coal is described, since the diffusion of air into the coal structure may be a limiting process. The effects of air oxidation on the nature and yields of humic acids was investigated under conditions of high mass flow and high shear.

EXPERIMENTAL

Preparation of hydrosols

Beulah lignite was blended in 10 g batches with 250 ml of 5% sodium hydroxide solution in a 1.25 l kitchen-style Osterizer at the high speed setting (approximately 24,000 rpm) (ref. 4). The temperature rose to 45°C during the two-hour blending time. The following atmospheric conditions were used in various studies: 1) operation in a nitrogen dry bag, 2) operation with the rubber lid in place, and 3) operation with an injection port in the base through which air was blown. The rubber lid was not air tight and allowed some minimal amount of air to reach the coal dispersion during the blending period. The blended lignite was centrifuged at 3000 rpm in an IEC Model K centrifuge. The supernatant dispersion was carefully decanted from the nondispersed residue.

Conversion yields for the formation of the hydrosol form were measured by coagulating the dispersion with concentrated hydrochloric acid (to pH = 1), centrifuging the coagulated humic acids, washing 12 times with water, adding ethanol, evaporating on a rotary evaporator, and drying in a vacuum oven at 110°C. Yields of humic acids from 10 g of Beulah lignite under the three conditions were 1.9 g (29% on a moisture-free basis) for the nitrogen conditions, 5.7 g (86% mf basis) for the minimal air conditions, and 5.2 g (79% mf basis) for the injected air conditions. A second test with air injection gave lower conversion to the hydrosol--4.1 g (62%). The corresponding amounts of nondispersed residue were 4.0 g (61% mf basis), 1.3 g (20% mf basis) and 1.2 g (18%, mf basis), respectively.

Ultimate or elemental analyses (on a moisture- and ash-free or maf basis) of the humic acid products obtained using the three different conditions are shown in Table 1, along with that of the original coal.

TABLE 1

Ultimate analytical data for Beulah lignite and humic acids

Sample	Moisture/Ash Free %				
	C	H	N	S	O(by dif.)
Beulah lignite	72.5	3.9	0.95	1.33	21.3
Air inj. humic acid	65.2	4.1	0.84	0.75	29.1
Min. air humic acid	63.1	4.0	0.77	0.75	31.5
Nit. humic acid	67.5	4.2	0.86		26.7

The supernatants obtained after centrifugation of the acidified humic acids described above were not colored and were not believed to contain much coal material. Although a soluble colorless fulvic acid product could have been present in small amounts, no way could be found to obtain such material from the supernatant brine. The supernatant from the air injection experiment was further studied by concentrating on a rotary evaporator, extracting with methanol and treating the methanol extract with diazomethane in ether. The methylated extract was analyzed by gas chromatography/mass spectrometry, however no peaks were observed, thus no benzenepolycarboxylate methyl esters were present.

Oxidation at 100°C

The reaction of the hydrosol with air at 100°C was carried out by refluxing the hydrosol from Beulah lignite in a two neck flask with a set of condensers attached to one neck. Air was bubbled rapidly through a tube below the aqueous dispersion. After cooling, acid was added to coagulate the hydrosol giving a precipitate of the humic acid. No change in the amount of humic acid occurred as a result of the air oxidation at 100°C. No benzenepolycarboxylic acids were detected by the extraction, methylation and GC analysis procedure described above.

Instrumental Solid-state CP/MAS spectra were obtained at 25 MHz on a Chemagnetics M-100S spectrometer. A contact time of 1 ms and a pulse delay of 1 s were used on all samples. These parameters are similar to those used by Hatcher for CP/MAS studies of humic acids from peat, soil and water (hh). Dried samples were spun at 4.0 kHz using ice cooled nitrogen gas in 9.5 mm Kel-F rotors. Chemical shifts were externally referenced to the aromatic peak of hexamethylbenzene at 132.20 ppm. Each spectrum was the result of 6000-8000 repetitions; 20-30 Hz line broadening was applied to each FID before Fourier transformation. Hartmann/Hahn match was set using HMB and involved RF field strengths of about 12 kG and 48kG for ^{1}H and ^{13}C respectively. Peaks were integrated over the following segments or ranges of the spectra: aliphatics, -20 to 90 ppm; aromatics, 90 to 165 ppm; carboxylic acids and other carbonyls, 165 to 250ppm. Aromatic to aliphatic peak intensity ratios were calculated simply from these values.

Photoacoustic-Fourier transform infrared (PAS-FTIR) spectra (8 cm^{-1} resolution) of the coals, humic acids and residues were obtained with a Nicolet 20SXB spectrometer with a photoacoustic accessory cell obtained from M-TEC which was purged with helium. Samples were vacuum-dried and undiluted.

Dynamic light scattering studies were performed on the sodium humate dispersions with a photon correlation spectroscopy system (Model 4700) at Malvern Instruments, Inc. A scattering angle of 45° was used to maximize signal from the distribution. Electrophorectic mobilities of the humate dispersions were determined with a Malvern Zetasizer II which measures the frequency spectrum of scattered He-Ne laser light using a photon counting detector and digital correlator.

RESULTS AND DISCUSSION

In order to carry out investigations of the reactivity of low-rank coals in a colloidal form, a good method for producing the colloidal coal was needed. The yields of extracts from low-rank coals have generally been low using conventional methods, except for the naturally highly oxidized lignites (leonardites). The low-rank coals generally have structures containing many carboxylic acid and phenolic hydroxyl groups, consequently they give higher conversions to the humic acids when treated with aqueous 5% sodium hydroxide than do the higher rank coals. Nevertheless only 4% yields of humic acids can be extracted from the lignites by stirring for 12 hours. The conversions may be higher at higher temperatures or if the coal is highly oxidized prior to the extraction. Leonardites can give humic acids in 90% yield (ref 5).

The conversion of low-rank coals to the colloidal (hydrosol) form in aqueous base has been vastly improved by the use of a kitchen-style blender (ref. 4, 6). North Dakota lignites can be converted to the humate hydrosol at temperatures less than $45^{\circ}C$ in yields as high as 90% using high speed blending. These humate dispersions are stable to centrifugation and appear dark but clear to eye examination; however, they exhibit Tyndall effects similar to those observed for the humates obtained in low yields using magnetic (low shear) stirring. Somewhat larger yields of the hydrosol were obtained from the low-rank coals when the blender was not placed in a nitrogen dry bag. Thus a trace of air was able to penetrate the lid of the blender and oxidize the coal material to a small extent, increasing the suspendability of the coal particles. The amount of conversion of the coal to the hydrosol was measured by adding acid to the centrifuged hydrosol to precipitate or coagulate the humic acid, which was then measured by centrifugation, washing to remove salt, drying and weighing.

The size of the particles in the humate dispersion obtained from blending Beulah ND lignite under nitrogen was determined using a Malvern photon correlation spectroscopy system. The z-average size of the dispersed particles was 410 nm, and a bimodal distribution over the range 250 to 1300 nm was exhibited (ref. 6). Electrophoretic mobility of the alkaline humate dispersion was measured using a Malvern Zetasizer II instrument in which the frequency spectrum of scattered laser light was analyzed using a photon counting detector and digital correlator. A broad symmetrical distribution of mobilities were measured for the humate particles produced under inert atmosphere conditions. A mean value of -2.33 microns/sec/volt/cm was obtained. This corresponds to a mean zeta potential value of -32.6 mvolts. The distribution had a range of +13 to -78 mvolts with a %width of 53.5.

The reactivity of the humate dispersion was investigated by studying the reactions with air oxygen under blending conditions. The oxidation of coal is believed to be a diffusion controlled reaction and the acceleration of the oxidation should be observed as a result of reduced particle size (submicron) and high mass flow conditions of the blender. A port was installed in the base of the blender for injection of air into the alkaline coal dispersion under high shear blending conditions. The humic acids obtained from the oxidation were brown, as compared with the humic acids obtained under nitrogen, which were black. The oxidized hydrosol was examined with the dynamic light scattering technique. Particle size analysis (photon correlation) of the oxidized hydrosol obtained with air injection blending of Beulah lignite showed that no change in the Z-average particle size had occurred (415 nm). The distribution of particle sizes was bimodal, as in the original hydrosol.

The conversion was determined by isolation of the humic acid as before. Yields of humic acids (62 to 79%) from the air injection blending were slightly less than those obtained while operating the blender under a minimal air condition with the lid of the blender in place (86-95%), and higher than the yields obtained using a nitrogen atmosphere (29-45%). Mass balances indicated that the air injection experiment resulted in a small loss of coal material (8%). This loss may be due to loss of carbon dioxide or other gases or to loss of soluble and colorless macromolecular material, such as fulvic acids, formed during the oxidation.

Oxidation of coals with oxygen in alkaline media at high temperatures (greater than $180^{\circ}C$) produces valuable benzenepolycarboxylic acids as major products. The possibility of formation of water soluble benzenepolycarboxylic acids during the air injection oxidation of the humate hydrosol in the blender was investigated by stirring the supernatant, obtained after acidification and centrifugation of the humic acids, with diazomethane in ether, followed by gas chromatography/mass spectrometry analysis. No benzenepolycarboxylic acids were detected as the methyl esters, indicating that the activation energy for oxidation of aromatic rings to the benzenepolycarboxylic acids was not achieved in these low temperature conditions (less than $45^{\circ}C$). In another experiment, the blending time with air injection was extended to six hours; however no water soluble benzenepolycarboxylic acids were obtained.

Finally, the temperature of the reaction of humate hydrosols with air was increased to $100^{\circ}C$ by injecting air into a refluxing dispersion. In this experiment, the dispersion was prepared by blending with air at $45^{\circ}C$, and then reacted with air in another vessel at $100^{\circ}C$ (without blending). Again no

benzenepolycarboxylic acids were observed. Since no change in particle size
was observed during oxidation and no small molecular weight fragments could be
found, any oxidation which occurred during air injection blending must be
associated with the high molecular weight humic acids, and no significant loss
of carbon structure occurred during the oxidation. At some higher temperature
degradation of the colloidal coal will occur as it does with large coal
particles, and the rates of reaction may be different, owing to diffusion
control, but we have not yet examined the higher temperature reactions.

The elemental analysis of the isolated humic acid from the air injection
method showed that less oxygen (obtained by difference) was incorporated into
the organic structure of this material than into the product obtained using
minimal air blending, but obviously more than in the humic acid obtained using
nitrogen atmosphere. The explanation for this apparent reversal in oxygen
content may be the loss of carbon dioxide or highly soluble, high oxygen
content fulvic acid during the air injection blending. The elemental analysis
of a humic acid isolated from a naturally occurring leonardite was found to be
C, 63.5; H, 3.5; N, 1.3; S, 0.7; O, 31.1 (ref. 5). This composition closely
resembles that of the minimal air blended humic acid, rather than that of the
air injection humic acid. However, since the conditions for formation of the
leonardite were different (dry, no strong base), it is risky to compare these
products too closely.

The nature of the oxidation of the humic acids which does occur during the
air injection blending below $45^{\circ}C$ was investigated by photoacoustic-Fourier
transform infrared spectroscopy (PAS-FTIR) and ^{13}C CP/MAS NMR spectroscopy.
The peak corresponding to the carbons of the carboxylic acids at 175 PPM can
easily be seen to increase in the series of spectra of humic acids obtained
under increasing oxygen conditions (top to bottom in Figure 1). Subject to
the limitation that the population of observed carbons may not be exactly the
same in each sample, the integrated intensities of the peaks in the carbon NMR
spectra can give a roughly quantitative comparison of the number of carboxylic
acid, aromatic and aliphatic carbons. The ratio of the integrated intensities
corresponding to carboxylic acid carbons to aromatic carbons to aliphatic
carbons in the original Beulah lignite (Figure 1A) is 8:54:37 and in the humic
acids obtained initially by blending under nitrogen (Figure 1B) is 11:56:33.
The ratio changed to 19:47:35 in the oxidized ($45^{\circ}C$) humic acids (Figure
1D). The initial drop in the aliphatic carbon content in comparing the coal
with the nitrogen humic acid is due to the partitioning which occurs in
forming and separating the hydrosol. The highly aliphatic liptinite macerals
tend to concentrate more in the non-dispersable residue which is removed by
centrifugation, thus the humate has a lower concentration of aliphatic
liptinites, and a higher concentration of carboxylic acids which are more

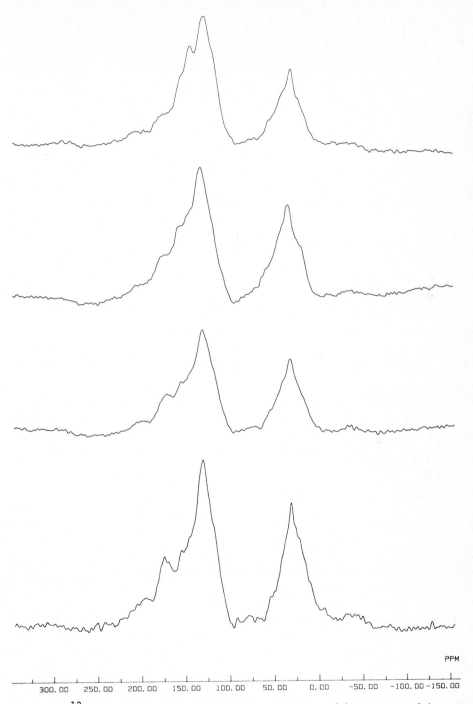

PPM

Fig. 1. ^{13}C CP/MAS NMR spectra of Beulah products. (A) Lignite. (B) Humic acids from blending under nitrogen. (C) Humic acids from blending with minimal air. (D) Humic acids from blending with air injection.

concentrated on the particles which form the electric double layer stabilized dispersion. When the hydrosol was oxidized with air, the carboxylic acid content increased substantially, mostly at the expense, it appears, of the aromatic and/or quinone carbons, rather than the aliphatic carbons The alkoxy ether peaks in the NMR spectrum at about 75 PPM were difficult to integrate and no quantitative results are available, however they appear to have been eliminated in the spectrum of the humic acid from the air injection experiment.

Information on the change in carboxyl groups is also available from the PAS-FTIR spectra of the humic acids obtained under different conditions (Figure 2). The carbonyl stretching band in the FTIR spectrum of the humic acid from air injection blending (Figure 2A) exhibits a two-fold increase over the peak in the humic acid obtained under blending in nitrogen (Figure 2C), owing to the formation of carboxylic acid groups during the oxidation/blending. The ratio of the area of the aromatic (1600 cm^{-1}) ring stretching peak to the aliphatic C-H bending peak (1450 cm^{-1}) also decreased in the humic acids obtained during air injection blending. Deconvolution of the aromatic band from the overlapping acid and quinone carbonyl absorption bands is required for quantitative results, but this has not been possible to carry out accurately. The changes which occur during the oxidation of the hydrosol during air injection blending are more rapid and extensive than those which occur with large coal particles.

Although useful benzenepolycarboxylic acids were not obtained in these experiments, the humates themselves have several uses, such as drilling muds and binders for briquettes. A knowledge of the factors which determine their structure and amounts produced will be of value. These observations will also pertain to the study of the dispersion of leonardites and chemically oxidized coals which is aided by fungal cultures. Thus the digestion of oxidized coals may result from stabilization of the dispersion by fungal metabolites, rather than enzymatic action.

Furthermore, it should be pointed out that the humic acids resulting from the coagulation of the hydrosol with acid have a very low mineral content (ref 6). In the case of Beulah lignite, the ash content drops from 8.4% (moisture-free basis) in the original coal to 0.83% in the recovered humic acids obtained with minimal air blending. The ash content of the humic acids obtained with air injection blending was higher, 3.0 % (mf basis). The residue which was separated from the hydrosol by centrifugation showed an increased mineral content (19% ash, mf basis). The sulfur content of the recovered products from blending Beulah lignite has also been reduced from 1.3% (maf basis) in the original coal to 0.75% in the humic acid and 0.73% in the residue. Further work in developing a new coal cleaning method by

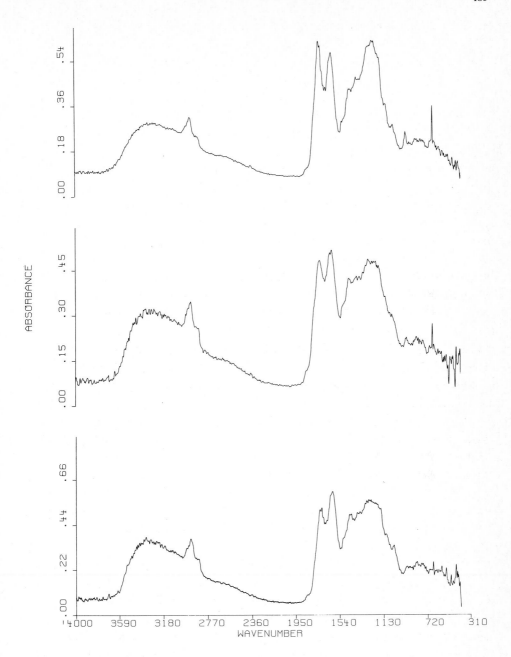

Fig. 2. Photoacoustic-Fourier transform infrared spectra of humic acids from Beulah lignite. (A) Humic acids from blending with air injection. (B) Humic acids from blending with minimal air. (C) Humic acids from blending under nitrogen.

converting the low-rank coals to hydrosols in a blender is in progress. The higher ash content results obtained for blending with air injection suggests that minimal exposure to air should give cleaner products.

In conclusion, a method has been developed which is effective in producing low-rank coal hydrosols in quantities large enough for testing in various processes. Air oxidation of the hydrosol proceeds rapidly and results in the formation of additional carboxylic acid groups, but does not proceed to the extent where benzenepolycarboxylic acids or other small molecular weight species are formed.

ACKNOWLEDGMENTS

The dynamic light scattering measurements of the particle size distribution and the electrophoretic mobility measurements were contributed by Arthur McNeill of Malvern Instruments, Inc. Ash analyses were provided by Dana Maas and Ray DeWall. The research was supported by the US Department of Energy under Cooperative Agreement Number DOE-FC21-86MC10637.

REFERENCES

1   H. Potonie, Abh d. K. Press. Landesantalt, 55 (1911) 2.
2   D.W. van Krevelen, Coal, Elsevier Publ. Co., Amsterdam, 1961.
3   P.H. Given, in: M.L. Gorbaty, J. W. Larsen and I. Wender (Eds.), Coal Science, Vol. 3, Academic Press, Orlando, FL, 1984, pp. 63-252.
4   E.S. Olson, J.W. Diehl, and M.L. Froehlich, ACS Div. of Fuel Chem. Preprints, 32(1) (1987) 94-97.
5   W.W Fowkes and C.M. Frost, Bur. of Mines Report of Invest., 5611 (1960), 1-12.
6   E.S. Olson, J.W. Diehl, and M.L. Froehlich, Fuel (Submitted for publication).

Interfacial Phenomena in Biotechnology and Materials Processing,
edited by Y.A. Attia, B.M. Moudgil and S. Chander
Elsevier Science Publishers B.V., Amsterdam, 1988 — Printed in The Netherlands

ENTRAPMENT AND ENTRAINMENT IN THE SELECTIVE FLOCCULATION PROCESS,
PART 1: MECHANISMS AND PROCESS PARAMETERS

Y.A. ATTIA and SHANING YU

The Ohio State University, Mining Engineering Division, Department of
Metallurgical Engineering, 148 Fontana Laboratories, 116 W. 19th Avenue,
Columbus, Ohio 43210 USA

SUMMARY
    The selective flocculation process is a promising technique for separating
fine and ultrafine minerals. However, this process is always more or less
constrained by the entrapment and entrainment of unwanted minerals in the
flocs, which lowers floc quality. In this series of papers, entrapment and
entrainment, in selective flocculation, are studied. In the first part of
this study, the mechanisms of entrapment and entrainment and the important
process parameters which influence entrainment and entrapment in the selective
flocculation process are examined. Important parameters include: the
concentration of dispersed-phase minerals, slurry viscosity, particle size and
shape of the minerals to be separated, mechanical agitation rate, conditioning
time, floc size and structure, and so on. In the companion paper, possible
methods for minimization of entrapment and entrainment are reported.

INTRODUCTION

    The selective flocculation process is a promising technique for the separa-
tion of fine and ultrafine minerals. It consists of four sub-processes: 1.
dispersion of mineral slurry, 2. selective flocculant adsorption and floc
formation, 3. floc conditioning, and 4. floc separation. (Fig. 1 illustrates
the sub-processes involved in selective flocculation.) In the selective
flocculation process, the most important consideration is the difference in
physico-chemical properties of various particles in the mineral slurry. The
technique is essentially based on the preferential adsorption of an organic
flocculant onto the particles to be flocculated (ref. 1). Selective polymer
adsorption is the key step to achieving selective flocculation and the design
and preparation of selective polymeric flocculants proved feasible. Many
examples are available in literature (refs. 2-8). However, in addition

492

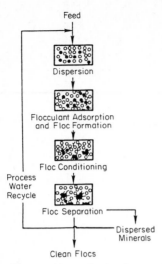

Fig. 1. Illustration of the four sub-processes of selective flocculation.

to flocculant selectivity, the separation efficiency of the selective
flocculation process is almost always constrained by other aspects, such as
the degree of libration of minerals, floc-separator efficiency, and entrapment
and entrainment of unwanted particles in the flocs. This article is concerned
only with entrapment and entrainment.

Entrapment and entrainment are the phenomena that some of the mineral
species in the system mis-report to the fraction where they should not be,
(e.g. dispersed-particles in the floc fraction) due to mechanical and/or
physical reasons. Both entrapment and entrainment could considerably lower
the quality of the clean floc product, and could even result in failure of
separtion. To achieve successful selective flocculation separation,
minimization of entrapment and entrainment should be the objective.
Therefore, the mechanisms of entrapment and entrainment need to be well
understood. Knowledge of the mechanisms and possible ways of minimizing
entrapment and entrainment will be useful for the design, evaluation, and
simulation of the selective flocculation process. It should be mentioned that
only a meagre number of literature reports (refs. 1, 9-11) could be found on
the entrapment and entrainment problems of selective flocculation.

The purpose of this study is to define the mechanisms of entrapment and
entrainment, and to discuss the important process parameters which influence
the entrapment and entrainment of the dispersed-phase impurities into the
flocs. The discussion of these factors is based on theoretical analyses and
experimental results. In the companion paper (ref. 12), some possible methods
for the minimization of entrapment and entrainment are reported.

ENTRAPMENT AND ENTRAINMENT

Definitions of Entrapment and Entrainment

As illustrated in Fig. 2, the mineral slurry system undergoing selective flocculation can be simply considered as consisting of two types of mineral components. The mineral component to be selectively flocculated is called floc-phase species, all others are termed dispersed-phase species. After adding the selective flocculant, the slurry is separated into two distinguishable fractions (after allowing it to settle): the supernatant (dispersion layer) containing the dispersed-phase species and the floc layer which is rich in floc-phase species. In either the floc layer or the supernatant, there always exists a certain amount of species not reporting to their respective fractions. Those mis-reporting components in the floc fraction, due to mechanical and physical reasons, are defined as entrapped and entrained particles. This type of entrapment and entrainment is usually a more serious problem than the loss of floc-phase particles in the dispersion fraction. Hence, the focus of this article will be on this type of entrapment and entrainment.

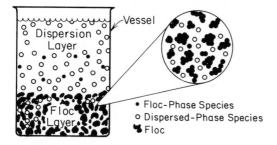

- Impurities are Trapped Between the Flocs in the Floc Layer, i.e., Inter Floc Association.

- Entrainment Takes Place Mainly During Floc Separation.

Fig. 2.   Illustration of a selective flocculation system and entrainment.

FLOC FORMATION   FLOC GROWTH

Fig. 3.   Illustraion of the process of entrapment.

Mechanisms of Entrapment

Entrapment usually takes place during the floc formation stage and during the floc growth and conditioning stage. Fig. 3 illustrates the entrapment phenomenon. Entrapment can take place according to the following mechanisms:

1. Fluid entrapment: the dispersed-phase particles are trapped in the flocs with the entrapped fluid.

2. Mechanical entrapment: where the dispersed-phase impurity particles are trapped in the flocs during their formation and growth. In this mode of entrapment, it is assumed that no adsorption of polymeric flocculant on the entrapped particles takes place.

Mechanisms of Entrainment

Entrainment takes place mainly during floc separation. The mechanisms of entrainment of the dispersed-phase species into the floc layer are currently thought to be through a combination of the following:

1. Fluid entrainment: the dispersed-phase species particles enter the floc layer with entrained fluid. The entrained particles are situated between the flocs, rather than inside the individual floc structure, as is the case in entrapment.

2. Gravitational entrainment: this is caused by the co-settling of coarse or heavy dispersed-phase particles with the large size flocs having the same terminal settling velocity.

3. Mechanical entrainment: it is an enmeshing of impurities during the formation and the movement of the floc layer in the slurry.

4. Attachment entrainment: similar to slimes' coating phenomenon, in which the impurity particles physically attach on the surface of flocs and move with them.

Of all these mechanisms, fluid and mechanical entrainment are usually the major contributors to entrainment. Gravitational entrainment is only serious when the slurry contains coarse and heavy particles to be dispersed, and the differential settling is used as the floc separation technique. This type of entrainment takes place mainly during floc separation.

FACTORS INFLUENCING ENTRAPMENT AND ENTRAINMENT IN SELECTIVE FLOCCULATION

Entrapment and entrainment in selective flocculation processes are influenced by many factors. The important parameters include the concentration of dispersed-phase species, floc size and structure, solids content, specific gravity of the dispersed-phase minerals, slurry viscosity, particle size and shape of the minerals to be separated, mechanical agitation rate, conditioning time, and floc separation mechanisms. These parameters are discussed in some detail below.

Effect of the Concentration of Dispersed-phase Species

It is reasonable to assume that the concentration of the dispersed-phase species in the entrained and entrapped liquid is similar to that of the supernatant (i.e. the dispersion layer). Therefore, it is anticipated as a general rule that the higher the concentration of the dispersed-phase (or impurity) species in the system, the higher the chance for entrapment and entrainment.

Figs. 4 and 5 show the effect of feed grade on the floc quality (indicating the amount of entrapment and entrainment) for selective flocculation of copper minerals and coal slurries, respectively. The results indicate that the lower the feed grade (i.e. the higher the concentration of the dispersed-phase particles), the lower the quality (or the higher the entrapment and entrainment as indicated by the higher ash content in the coal system). The data in Fig. 4 was obtained from Refs. 11 and 13. Although these were obtained under different conditions of flocculant and dispersant concentrations, the overall trend is still clearly consistant with the general rule stated above. Fig. 5 shows the de-ashing ratio of the coal flocs as a function of the ash (dispersed-phase impurities) content of the coal feed slurry. The de-ashing ratios were obtained from the widely different conditions, which explains the wide scatter of data in Refs. 1, 11 and 14. This data was obtained under data points in the figure. However, the overall trend observed in the case of the copper system (Fig. 4) is still valid here and consistent with the general rule stated above.

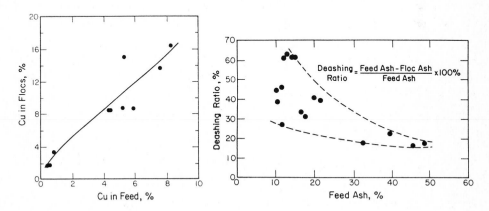

Fig. 4. Effect of feed grade on floc quality in selective flocculation of copper minerals.

Fig. 5. Effect of feed ash on the de-ashing ratio of selective flocculation of coal minerals slurry.

Effect of Specific Gravity of Dispersed-phase Particles

As stated before, for the selective flocculation system having coarse and heavy minerals, gravitational entrainment could cause a serious problem when settling is used for floc separation. Heavy particles may co-settle with the lighter flocs, thus causing poor separation. The effect of specific gravity on entrainment can be readily demonstrated by calculating the terminal settling velocity of the flocs and that of the heavy dispersed-phase particles using Stoke's law. If we assume a laminar flow condition, the terminal settling velocity, v_t, of an equivalent spherical particle with size d and specific gravity, ρs, can be expessed as:

$$v_t = \frac{d^2(\rho_s - \rho_1)g}{18\eta} \tag{1}$$

where $\rho 1$ is specific gravity of the liquid medium of the system, η is viscosity, and g is the gravitational acceleration constant. This formula indicates that the particles with higher specific gravity have faster settling velocities. From Equation 1, the ratio of floc diameter to dispersed-phase particle diameter, at an equivalent terminal settling velocity, can be derived.

$$\frac{d \ (flocs)}{d \ (dispersed)} = \left[\frac{\rho_s(dispersed) - 1}{\rho_s(floc) - 1} \right]^{1/2} \tag{2}$$

For example, for a coal-pyrite slurry system, in which the specific gravity of coal flocs is estimated at 1.13 and that of pyrite is 5.0, the equi-settling diameter ratio of coal flocs to pyrite equals 5.55. That is, a 20 micron pyrite particle will settle at the same rate as an 111 micron coal floc. The terminal settling velocity distribution curves of pyrite particles and coal flocs versus size are calculated and shown in Fig. 6.

Effects of Particle Size and Shape

Similar to the effect of specific gravity, large sizes of the dispersed-phase particles also cause gravitational entrainment. If we suppose a laminar system, as expressed in Equation 1, the terminal settling velocity is proportional to the square of the particle size. Therefore, the larger particles will settle at a faster rate than a floc with the same size.

The particle size has an important role in particle dispersion. In addition to gravitational settling, the role of surface forces becomes more important below a critical particle size, and the dispersion becomes easier to achieve below that critical particle size. Fig. 7 shows the effect of pyrite particle size on selective dispersion of pyrite in the presence of a

flocculant using polyacrylic acid-xanthate (PAAX) as the dispersant (ref. 15).
The results indicated that the dispersion was much more efficient for the
smaller-sized (up to 10 microns) pyrite particles than for relatively coarser
ones (up to 37 microns).

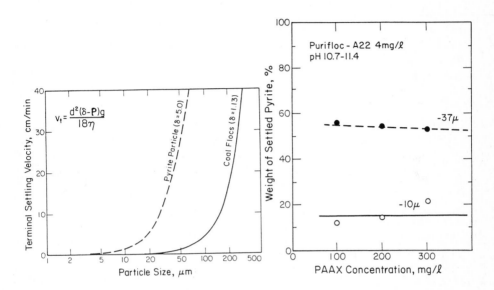

Fig. 6. The terminal settling
velocities of coal flocs and
pyrite in aqueous solution,
versus size, assuming a
5% solids content, 20% ash
(sp.gr = 2.7) and 2% pyrite
(sp.gr. = 5.0).

Fig. 7. Flocculation-dispersion
behavior of pyrite (-37 and -10
microns) in the presence of polyacrylic
acid/xanthate (PAAX, dispersant to
pyrite) and 4 mg/l of Purifloc-A22
(flocculant) at pH 10.7-11.4 (ref. 15).

Since the settling velocity of particles is relative to size, shape and
roughness, the effect of these parameters on gravitational entrainment is
obvious. For example, the spherical-shaped particles settle slower than the
needle-shaped ones, but faster than the plate-shaped ones, and, the
rough-surface particles settle slower than those of smooth ones.

Effects of Solids Content

It is known that in many separation systems, high solids content causes an
increase in slurry viscosity (especially for fine particle systems) and slurry
density which decreases the terminal settling velocity due to hindered settling of particles. Particularly for a selective flocculation system, the high

solids content may lead to the difficulty in achieving: 1. liberative dispersion (that means all of the particles in the suspension are individually or separately suspended in the slurry) and 2. selective formation of flocs and the floc layer. These factors lead to high entrapment and entrainment of impurities in the floc layer. Therefore, high solids content generally leads to poor separation performance.

Many experimental results have provided evidence that the separation

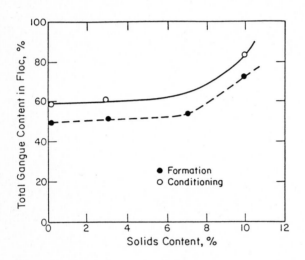

Fig. 8. Effect of solids concentration on the formation and conditioning of malachite flocs (ref. 10).

efficiency of the selective flocculation process usually decreases, while increasing the solids content of the slurry. Fig. 8 shows the effect of solids concentration on total gangue content of malachite flocs during their formation and conditioning (ref. 10). It can be seen from the plot that the total gangue (dispersed-phase particles) content of the flocs during formation remained constant, at about 52% at the solids concentration of 3 to 7%, but increased to 71% at a solids concentration of 10%. The reason was attributed to denser gangue environment around the flocs during their formation. During the conditioning, the total gangue associated with the flocs remained constant, at about 52% below a solids concentration of 7%. Above 7% solids, the gangue association increased appreciably.

Another example was the effect of solids content on the selective flocculation of coal, as shown in Fig. 9 (ref. 16). In these tests, a totally hydrophobic polymer (FR-7) was employed as the selective flocculant for coal and sodium metaphosphate (SMP) as dispersant for the ash-forming minerals.

The dosage of the reagents was kept constant. The results indicated that the floc grade decreased with increasing solids content. At low solids content, e.g. about 1.5 to 2%, more than 60% of the total ash could be rejected. The coal recoveries achieved were between 80-90%, and the floc products were very clean. When the suspension contained high solids content (e.g. 10%), the separation was poor, and only a small amount (< 20%) of ash minerals could be removed. The main reason for the poor separation at higher solids content was considered to be the high entrapment and entrainment of ash minerals into the coal floc portion (ref. 16).

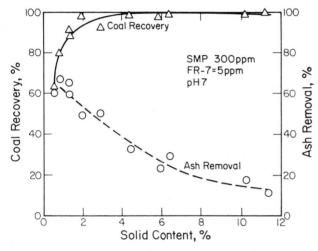

Fig. 9. Effect of solids content on selective flocculation of coal slurry (ref. 16).

Effects of Mechanical Agitation/Mixing

The reasons mechanical agitation is needed in the selective flocculation process are: 1) homogeneous mixing of the slurry particles; 2) dispersion of the polymeric flocculant and the speeding up the rate of flocculation by increasing particle collisions; and 3) the breaking up formed flocs to release entrapped species.

Two examples illustrated in Fig. 10 show the effect of mechanical agitation on total shale content in coal flocs and quartz (gangue) content in malachite flocs, respectively (refs. 10 and 11). It can be seen that, as agitation increased, the quantity of associated gangue decreased appreciably above 750 rpm. Both entrapped and entrained gangue decreased proportionally. It was found that at 1000 rpm, the floc sizes were the smallest. The higher agitation level must have reduced gangue entrapment and entrainment in flocs (ref. 10).

500

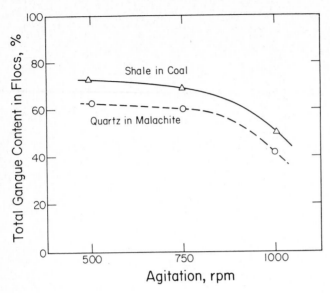

Fig. 10. Effect of mechanical agitation speed on entrapped and entrained gauges content in coal/shale (ref. 11) and malachite/quartz systems, respectively (ref. 10).

Effects of Floc Size and Structure

Floc formation takes place as soon as the flocculant is adsorbed on the surfaces of the minerals. When flocs form, some dispersed-phase particles will probably entrap in the flocs by any of the mechanisms stated earlier. Generally, the effects of floc size and structure on entrainment and entrapment can be summarized as follows:

1. The larger the floc size, the more likely that the dispersed-phase species will be entrapped within the flocs.

2. The more compact the floc structure, the less likely that impurities will be entrapped in the flocs. Conversely, fluffy flocs with loose and open structure are more likely to entrap more impurities.

3. The more compact the floc layer, the less likely that impurities will be entrained in the floc layer.

4. The stronger the floc structure the more difficult it is for already entrapped impurities to diffuse out of the flocs into the bulk slurry.

Floc size and structure depend on many parameters. For example, the dosage and type of flocculant, water quality (pH, ionic type and strength), floc formation rate, mechanical agitation and mixing, conditioning time, separation technique and so on. However, it can be generally stated that relatively compact and small floc sizes would be desirable for minimization of entrainment and entrapment.

CONCLUSIONS

1. Entrapment and entrainment are important phenomena which adversely affect the performance of the selective flocculation process. Entrapment usually takes place as a result of two mechanisms: a) fluid entrapment and b) mechanical entrapment. The main mechanisms for entrainment of dispersed-phase species into the floc layer are: a) fluid entrainment, b) gravitational entrainment, c) mechanical entrainment during the motion of the floc layer and d) weak attachment between the dispersed-phase particles and flocs. Generally, entrapment takes place during floc formation and floc growth, while entrainment occurs mainly during floc separation.

2. Factors contributing to entrapment and entrainment include: high solids content, poor mixing, and the presence of a coarse and heavy unwanted concentration of dispersed-phase particles.

3. Effective methods are needed to minimize the problems of entrapment and entrainment.

ACKNOWLEDGEMENT

The fellowship of Shaning Yu, provided by the Ohio Mining and Mineral Resource Research Institute, is gratefully acknowledged.

REFERENCES

1 Y.A. Attia and D.W. Fuerstenau, Principles of Separation of Ore Minerals by Selective Flocculation, in: (Ed.) N. Li, Recent Development in Separation Science, Vol. IV, Ch. 5, CRC Press, Florida, 1978, pp. 51-69.
2 Y.A. Attia and J.A. Kitchener, Development of Complexing Polymers for the Selective Flocculation of Copper Minerals, in Proceedings of 11th Int., Mineral Proc. Congr., Cagilari, Italy, 1975.
3 Y.A. Attia, Synthesis of PAMG Chelating Polymers for the Selective Flocculation of Copper Minerals, Int. J. Miner. Process. 4, 1977, pp. 191-208.
4 G. Baudet, et al., Synthesis and Characterization of Selective Flocculants Based on Xanthate Derivatives of Cellulose or Amyloze, Ind. Miner. Mineralurigie, No. 1, 1978, pp. 19-35.
5 Y.A. Attia, Fine Particle Separation by Selective Flocculation, Separation Science and Tech., 17(3), 1982.
6 T. Shizuka and H. Hotta, Coal-deashing Process, US Patent No. 4,437,861, March 20, 1984.
7 Y.A. Attia, R.G. Sinclair, R.A. Markle, M. Cousin and R.O. Keys, Preparation of Polymethyl Vinyl Oxime and Other Polyoximes for the Selective Flocculation of Cassiterite from Troumaline and Quartz, in: Y.A. Attia (Ed.) Flocculation in Biotechnology and Separation Systems, Elsevier, 1987.
8 Y.A. Attia, S. Yu and S. Vecci, Selective Flocculation Cleaning Upper Freeport Coal with Totally Hydrophobic Polymeric Flocculant, in: Y.A. Attia (Ed.), Flocculation in Biotechnology and Separation Systems, Elsevier, 1987.
9 Y.A. Attia, H.N. Conkle and S.V. Krishnan, Selective Flocculation Coal Cleaning for Coal Slurry Preparation, in: Proceedings of 6th Int. Symp. on Coal Slurry Combustion, Orlando, June, 1984.
10 S.V. Krishnan and Y.A. Attia, Floc Characteristics in Selective Flocculation of Fine Particles, in: B.M. Moudgil and P. Somasundaran

(Eds.) Flocculation, Sedimentation and Solidation, Sea Island, Georgia, Jan. 27-Feb. 1, 1985, pp. 229-248.

11 Y.A. Attia, S.V. Krisnan and D.M. Deason, Final Report on Selective Flocculation Technology, Group Program, Battelle Columbus Laboratories, Columbus, OH, Aug. 1983.

12 S. Yu and Y.A. Attia, Entrapment and Entrainment in Selective Flocculation Process, Part 2: Possible Methods for the Release of Entrapment and Entrainment, in: Y.A. Attia, B.M. Moudgil and S. Chander (Eds.), Interfacial Phenomena in Biotechnology and Materials Processing, Elsevier, 1988.

13 Y.A. Attia, Selective Flocculation of Copper Minerals, Ph.D. Thesis, University of London, 1974.

14 K.H. Driscoll, Polymer Mixing in Selective Flocculation and the Feasibility of Dissolved Air Flotation as a Floc Separation Technique, Master's Thesis, The Ohio State University, 1987.

15 Y.A. Attia and D.W. Fuerstenau, Feasibility of Cleaning High Sulfur Coal Fines by Selective Flocculation, 14 Int. Mineral Proc. Congress, Toronto, Canada, Oct., 1982.

16 Y.A. Attia and S. Yu, Production of Super-clean Coals from a High Sulfur Coal by Selective Flocculation, in: P. Chugh and R. Caudle (Eds.), Processing and Utilization of High Sulfur Coals II, Elsevier, 1987.

Interfacial Phenomena in Biotechnology and Materials Processing,
edited by Y.A. Attia, B.M. Moudgil and S. Chander
Elsevier Science Publishers B.V., Amsterdam, 1988 — Printed in The Netherlands

ENTRAPMENT AND ENTRAINMENT IN THE SELECTIVE FLOCCULATION PROCESS,
PART 2: POSSIBLE METHODS FOR THE MINIMIZATION OF ENTRAPMENT AND ENTRAINMENT

SHANING YU and YOSRY A. ATTIA

The Ohio State University, Mining Engineering Division, Department of
Metallurgical Engineering, 148 Fontana Laboratories, 116 W. 19th Avenue
Columbus, Ohio, 43210 U.S.A.

SUMMARY
In.order to achieve high separation efficiency in selective flocculation,
entrapment and entrainment of dispersed-phase impurities into the flocculated
fraction need to be minimized. In the first part of this study, the
mechanisms of entrapment and entrainment and the important process parameters,
which influence these phenomena, are discussed. Generally, minimization of
entrapment and entrainment may be achieved by either of the two following
routes: 1. optimization of process operating conditions to prevent entrapment
and entrainment and 2. employment of suitable techniques for releasing
entrapped and entrained particles. Possible methods for the first route
include: a) optimization of process parameters, b) floc conditioning using a
rotary inclined conditioner, and c) employment of suitable floc separation
techniques. The second route involves re-dispersion of the flocs to release
the entrapped and entrained impurities, followed then by a re-flocculation
process. In this paper, these possible methods for the minimization of
entrapment and entrainment are discussed.

INTRODUCTION

As mentioned in the companion paper in this study (ref. 1), the separation

efficiency of the selective flocculation process is constrained by entrapment

and entrainment, i.e. the dispersed-phase impurities mis-report to the floc

products. Usually, the entrapment occurs by two mechanisms. First, the

dispersed-phase impurity particles are trapped with fluid (fluid entrapment),

and secondly the dispersed-phase particles are trapped in the small flocs

(mechanical entrapment). On the other hand, the entrainment of dispersed-

phase impurities into the floc layer has four mechanisms: 1. fluid entrain-

ment, 2. gravitational entrainment of coarse and heavy dispersed-phase par-

ticles, 3. mechanical entrainment, and 4. attachment entrainment. Entrap-

ment and entrainment can affect the quality of the floc product adversely and

may, in an extreme case, cause failure of separation. To achieve highly

efficient separation, the task is to minimize the entrapment and entrainment

of impurities in the flocculated fraction. Only a few reports are available which cover the topic of minimization and release of entrapped and entrained particles in the selective flocculation process (refs. 2-6). In this paper, the purpose is to discuss and summarize some possible methods for minimization and release of entrapped and entrained particles.

Generally, minimization of entrapment and entrainment is handled in two ways. The first is to prevent and/or minimize entrapment and entrainment during the selective flocculation and separation process. The other is to employ suitable techniques to release the entrapped and entrained impurities from the flocs. These two techniques are discussed and examples are given below.

PREVENTION AND MINIMIZATION OF ENTRAPMENT AND ENTRAINMENT

The principal ways of reducing entrapment and entrainment during selective flocculation separation of mineral slurries are considered to be: 1. optimization of process parameters which influence these phenomena before floc formation and growth; 2. conditioning of the flocs during their formation and growth; and 3. use of suitable separation techniques to separate the flocs from the slurry.

Optimization of Process Parameters

In the companion paper (ref. 1), the influences of process parameters on entrapment and entrainment were studied. As stated, the factors contributing to entrapment and entrainment include high solids content, poor mixing, the presence of coarse and heavy unwanted particles, large floc size and loose structure, and a relatively high concentration of dispersed-phase particles.

A study on optimizing polymer mixing conditions in selective flocculation of ultrafine Upper Freeport coal was made recently by Driscoll and Attia (ref. 7). In that study, the process parameters included polymer dosage, time of polymer addition, slurry solids content, pH, polymer stock solution concentration, polymer dispersion time, shear rate during mixing, and floc conditioning time. Under less non-optimum conditions, the amount of ash content in the flocs was typically 1.5 to 2 times as high as that obtained under optimized conditions. Under optimized conditions, in two steps of selective flocculation, the ash content was greatly reduced from about 12% in the feed down to about 2.7%, and the coal recovery was as high as 90%. Entrapment and entrainment of ash particles in the coal flocs were reduced considerably when optimum conditions were used.

Use of Inclined Rotating Conditioner

Floc conditioning is an important sub-process of selective flocculation. Typically, flocs are formed in the first few seconds after the polymer

solution is added to the suspension. This is followed by a deliberate
conditioning period with controlled conditions in order to "grow" the flocs
with desirable characteristics, such as minimum entrapment and entrainment of
dispersed-phase species and the ability to withstand subsequent separation
from the suspension without breakup (ref. 8).

Fig. 1 illustrates two types of conditioners: the rotating inclined
conditioner and the stationary stirring tank conditioner. When compared with
the stationary conditioner, the inclined rotating conditioner has the
advantages of: 1. a relatively larger area for floc settling; 2. a short
settling distance for flocs which may reduce the mechanical entrainment; 3.
release of some entrapped impurities by shearing forces produced by the
rotation; and 4. allowing floc growth while minimizing entrainment and
and entrapment through control of floc size and structure. Therefore, the
inclined rotating conditioner should be more suitable for reducing entrapped
and entrained impurities in the floc layer. Table 1 compares the results of
total gangue contents obtained using a stationary mixing conditioner and a
rotating inclined conditioner at two conditioning times and two flocculant
dosages for the selective flocculation of malachite (ref. 4). At a polymer
dosage of 3 ppm and conditioning time of 5 minutes, the rotation method
produced flocs of somewhat lower gangue content. However, when the
conditioning time was extended to 10 minutes, both methods resulted in a
decrease of gangue content in flocs to about 43%. At the higher flocculant
dosages of 20 ppm, lower gangue contents were obtained by both methods
compared to those at a lower flocculant dosage of 3 ppm. At this high dosage,
the flocs formed were small, compact and cohesive. The higher polymer dosage
may be contributing to a multitude of polymer-mineral bridges which yields
tightly-held and compact flocs. These small and compact flocs may be
visualized as entrapping and entraining fewer gangue particles than large
open-structured flocs. A significantly lower gangue content of 16.7% was
obtained with rotation for 5 minutes. However, this increased to 25% when
conditioned for 10 minutes, still lower than that obtained by stationary mixer
conditioning. The rotating inclined conditioner was seen to have a distinct
advantage over the stationary mixing conditioner.

Use of Suitable Separation Techniques

According to Attia et al. (refs. 4 and 5) entrainment is usually
responsible for the major portion of total impurities in the floc product.
Therefore, it is important to adopt suitable approaches for floc separation
because entrainment usually takes place during floc separation.

Floc separation is usually performed in sedimentation thickeners. This
type of equipment is the simplest and the most common means employed for floc

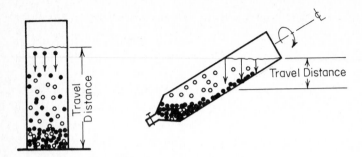

Fig. 1. Comparison of (a) stationary stirring tank conditioner and (b) inclined rotating conditioner.

separation. Because the settling area of thickener is large, fluid entrainment is usually high. Also, sedimentation thickening is always subject to gravitational entrainment when heavy particles are presented.

In addition to sedimentation thickening, floc separation could be conducted using different techniques (refs. 5 and 8). Possible methods are: 1. separation in a centrifugal field, 2. sieving, 3. elutriation, 4. fluidized bed elutriation, 5. conventional froth flotation, 6. column flotation, 7. dissolved air flotation, and 8. combined separation techniques. It was found (ref. 5) that the best four potential methods were: modified thickening, column flotation, modified elutriation and dissolved air flotation. Some of these separation techniques are discussed briefly.

TABLE 1
Comparison of total gangue contents in malachite flocs obtained by mixer conditioning and by rotation conditioning (ref. 4).

Polymer Dosage mg/l	Conditioning Time, minutes	Total Gangue Content in Flocs, %	
		Mixer Conditioning	Conditioning Rotation
3	5	60.0	52.0
3	10	42.2	43.4
20	5	51.4	16.7
20	10	32.3	25.0

(i) <u>Separation in a centrifugal field</u>. High field intensity is the main advantage of a centrifugal classifier, which provides the highest specific

throughput for relatively sharp separation of fine particles, with different
size or density. Typical devices for this type of separation are centrifuges
or hydrocyclones. In the centrifugal field, the floc fraction may be
pressured to form a more compact layer. The entrained impurities in the floc
layer can be considerably fewer than in a gravitational field. However, the
entrainment of coarse and heavy dispersed-phase minerals in a centrifugal
field, analogous to gravitational entrainment, would still take place. This
type of device can only be used when the amount of coarse and heavy
dispersed-phase minerals in the slurry is not high.

   (ii) Floc flotation. To overcome gravitational entrainment of coarse and
heavy dispersed-phase minerals into floc fraction, the flocs can be separated
by various flotation techniques.

      (a) Froth flotation. Froth flotation, an effective method for
separating fine minerals, is also considered a method for floc separation. In
the Tilden process, froth flotation was used to separate gravitationally
entrained "coarse" silica from selectively flocculated non-magnetic taconite
(ref. 9). Selective flocculation is also used to separate selectively
flocculated clay flocs from fine potash ores (ref. 10). Thus, the commercial
feasibility of floc separation by froth flotation has been proven. However,
froth flotation is not always the best means for floc separation. For
example, the high shear-rate usually employed in the froth flotation cell can
cause an extensive breakup of flocs leading to significant losses. In
addition, the presence of ultrafine impurity particles which are entrained
with the flocs, may interfere with separation efficiency. Since flotation is
not effective for ultrafine particles, removal of dispersed-phase impurities
should be optimized in a prior selective flocculation step. In essence, froth
flotation plays the role of cleaning the flocs.

      (b) Column flotation. To overcome the deficiency of conventional
froth flotation, it is better to use column flotation (refs. 11-14). There are
many types of flotation columns, but it is believed that the counter-current
column flotator with two zones (the collection zone and the cleaning zone,
ref. 13) is more suitable for floc separation. As illustrated in Fig. 2, the
collection zone is located at the lower portion of the column and behaves like
a conventional flotation cell. Another region located at the upper portion of
the column is characterized by a deep packed bubble bed called the bubbling
zone. Since the air bubbles are thermodynamically unstable, they will
spontaneously collapse and/or coalesce. This results in the rejection of the
entrained hydrophilic particles back to the collection zone. The bubble bed,
plus the liquid drainage, essentially suppress entrainment of impurity fines
into the clean product. Therefore, the bubble bed is also called the cleaning
zone (ref. 13). In addition, the gentle hydrodynamic conditions in the column

508

flotation cell make column flotation more suitable for separation of flocs
from suspensions than do conditions in conventional froth flotation cells.

The feasibility of using column flotation for separating coal flocs has
recently been studied by the authors (ref. 15). Their preliminary
experimental results indicated that column flotation is a promising approach
for cleaning and separating the selectively flocculated coal. However,
because dispersed-phase shale particles are ultrafine, their entrainment with
air bubbles and fluid in the column cannot be suppressed totally. Column
flotation is probably more suitable for the rejection of pyrite minerals which
are usually gravitationally entrained into coal flocs. This concept is being
investigated for coal disulfurization by the authors.

(c) Dissolved air flotation. Another method for reducing entrainment
is using the technique of dissolved air flotation. It has been successfully applied on a large scale for water effluent treatment (refs. 16 and 17)
but only recently first employed by Attia et al. (refs. 5, 7, and 18) for the
separation of coal flocs from a flocculated suspension. A conceptual diagram
of a dissolved air flotation apparatus is illustrated in Fig. 3. It consists
of a sealed vessel connected to a long cylindrical column filled with packing.
The water and air flow counter-currently through the column, and saturation
takes place as the air and water meet under pressure in the packed zone of the
column. In such a process, the flocs might be conditioned with a flotation
reagent to increase floc-bubble attachment and bubble stability. More often,
micro-bubbles are entrapped in the floc structure, thus rendering them
bouyant. Consequently, the air which is dissolved to saturation under
pressure in water, is released into a flotation cell. As the pressure is
suddenly reduced to atmospheric, the dissolved air precipitates as very fine

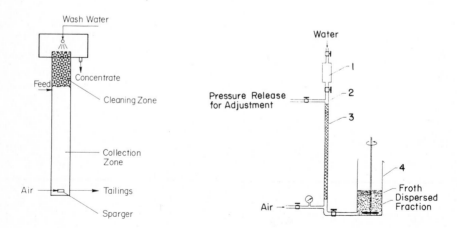

Fig. 2. Diagram of a flotation column   Fig. 3. Diagram of a dissolved
        characterized as two zones.   air flotation apparatus.

bubbles. These micro-bubbles attach preferentially to the flocs, while leaving the remainder of the particles in suspension. Heavy and relatively coarse dispersed-phase particles can settle, without interferring with the floc separation. Since no intensive agitation is applied in the flotation cell, flotation entrainment of the dispersed-phase particles to the overflow, due to water entrainment, could be reduced. In the Driscoll and Attia study (ref. 7) on coal floc separation, the ash content was considerably reduced from about 12% to less than 3% by applying two steps of dissolved air flotation. The coal recovery achieved was over 95%. Furthermore, dissolved air flotation is thought to be an ideal approach for rejection of gravitationally entrained particles from flocs. This technique has been demonstrated by the senior author's group (ref. 7) to enhance the removal of pyrite from coal flocs.

RELEASE OF ENTRAPPED AND ENTRAINED IMPURITY PARTICLES

The release of entrapped and entrained particles is usually achieved by using re-dispersion of flocs, followed by re-flocculation, i.e. repeating the selective flocculation process several times until the desired cleaning or concentration is achieved. In the case of a slimes-floc coating type of entrainment, liberative dispersion techniques might be applied. These techniques include the application of ultrasonic irradiation and the use of specific dispersants.

Principles of Releasing Entrapment and Entrainment

Generally, the strategy for the release of the entrained impurity species (not including gravitationally entrained ones) is to diffuse entrained dispersed-phase particles from the floc layer into the dispersion layer. This diffusion might follow Flick's first diffusion law:

$$J = D\frac{dC}{dx} \qquad\qquad (1)$$

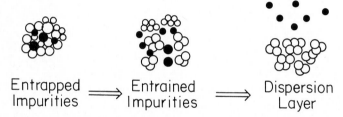

Entrapped Impurities $\Longrightarrow$ Entrained Impurities $\Longrightarrow$ Dispersion Layer

Fig. 4. Release of entrained and entrapped particles.

where D is the diffusion coefficient of the ultrafine particles, J is the flux of the entrained fine particles from the flocs to the dispersed-phase, C is the concentration of the ultrafine "impurity" particles in the flocculated phase, and dC/dX is the gradient of the dispersed-phase particle concentration across the floc layer. This formula shows that if the concentration of dispersed-phase species entrained in the flocs layer is higher than in the dispersion layer, the dispersed-phase species could diffuse from the floc layer to the dispersion layer (see Fig. 4).

Release is more difficult for entrapped impurity particles than for entrained ones because particles are locked in the flocs. Therefore, it is necesary to breakup the flocs. The general strategy for the release of entrapped particles includes two steps. The first step in the transformation of entrapped species to entrained ones is, the breaking up of flocs. The second step is to diffuse the entrained species further, into the dispersion layer, as illustrated in Fig. 4.

Re-dispersion and Re-flocculation

Since a certain amount of impurities are always entrained and/or entrapped in the floc layer, a cleaning process is usually required for obtaining a high quality floc product. In the cleaning step, the flocs are broken up by the use of agitation. All of the species are re-dispersed in the slurry. This is followed by re-flocculation through the addition of selective flocculant, leaving the released impurities in the dispersion. In some cases, multiple stages of the re-dispersion and re-flocculation processes are needed to achieve the desired cleaning level.

Fig. 5 (a) and (b) shows the multiple stages of selective flocculation of copper ores (ref. 19) and coal (ref. 20), respectively. The breakup of flocs (or floc re-dispersion) was obtained by mechanical agitation. The results indicated that removal of entrapped and entrained gangue particles was enhanced as the number of re-dispersion and re-flocculation steps were increased. The degree of removal of entrapped and entrained particles is indicated by increasing the grade of copper in the flocs and then decreasing the ash content in the coal flocs.

Ultrasonic Vibration

In the re-dispersion and re-flocculation process, some of the impurities released by re-dispersion achieved by mechanical agitation can be re-trapped. Also, some ultrafine particles could be attached weakly to the flocs, a phenomenon akin to slimes coating in mineral processing. To overcome these problems, the use of ultrasonic vibrations could be helpful (ref. 6).

The operation principles of the ultrasonic apparatus can be briefly

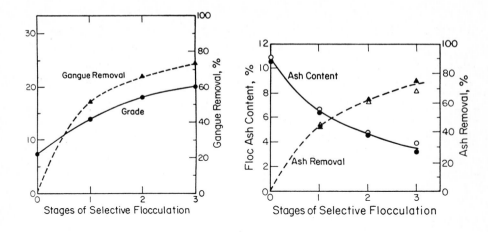

a) Copper ores (ref. 19)   b) Coal (ref. 20)

Fig. 5. Multiple stages of a selective flocculation process for the release of entrained and entrapped impurities.

described (ref. 21). The ultrasonic apparatus operates at very high frequencies (e.g. around 55,000 cyles per second). The sound waves are generated by the transducer, which changes high frequency electrical energy into mechanical energy or vibrations. The vibrations cause alternating high and low pressure waves in the liquid. This action forms millions of microscopic bubbles which expand during the low pressure wave and form small cavities. When used in the selective flocculation process, the high pressure waves produced by the ultrasonic apparatus collapes these cavities or implodes. This occurrence may result in: 1. a mechanical "scrubbing" action that loosens the attachment-entrained impurities on the floc surfaces; 2. the breakup of big sized flocs into small ones thereby transferring some entrapped impurites into entrained species; 3. a supply of energy to enable fluid-entrained species to diffuse into the dispersion layer; and 4. the formation of smaller and more compact flocs.

Attia and Yu (ref. 8) applied ultrasonic dispersion in the selective flocculation of coal. For comparison, three parallel series of tests were conducted using: 1. ultrasonic vibration before the floc formation, 2. ultrasonic vibration after the formation of coal flocs, and 3. no ultrasonic vibration. It was found that there was no difference between the results of tests 1 and 3. That is, ultrasonic dispersion before the formation of flocs did not affect the release of the entrapped and entrained impurities. This

was probably due to the occurrence of re-entrainment and re-entrapment.
However, if ultrasonic vibration was applied after the formation of coal
flocs, i.e. during re-dispersion of flocs, as in series 2, more ash minerals
could be rejected to the tailings with only a minor loss in coal recovery. In
these tests, nearly 85% of the ash minerals and more than 70% of the pyrite
were efficiently rejected to the tailings. The super-clean coal level was
attained. After four or five stages of selective flocculation separation, the
ash content was reduced from 12 or 13.4% to 2.7 or 3.0%. The total sulfur was
reduced from 1.6% down to 0.75 - 0.94%, pyritic sulfur from 1.10% to 0.43%.
The coal recovery obtained was over 80%. Results indicated that employing
ultrasonic vibration was effective for releasing some trapped ash-forming and
pyrite ultrafine particles from flocs after coal floc formation.

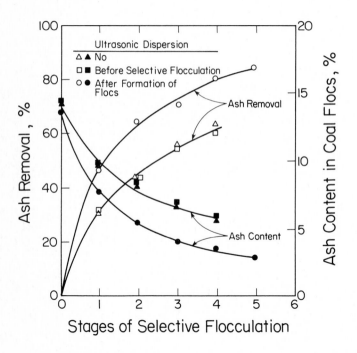

Fig. 6. Comparison of selective flocculation of coal slurries: (a) with and
without ultrasonic dispersion, (b) before and (c) after, the formation of
flocs. Conditions: 300 ppm smp, pH 7, FR-7 dosage = 5.3.1.1.1 ppm as stages,
solids content = 2.9% with no ultrasonic dispersion, and, with ultrasonic
dispersion, before the formation of coal flocs, and = 2.8% after the formation
of coal flocs (ref. 6).

CONCLUSIONS
  The following conclusions can be drawn from the results mentioned above:

1. To improve selective flocculation efficiency, it is necessary to minimize and release entrained and entrapped impurity mineral particles.

2. Minimization of entrapment and entrainment may be achieved by two routes. The first is optimization of the process operating conditions to prevent entrapment and entrainment during floc formation, floc conditioning and floc separation. Possible methods for achieving this include: a) optimization of process parameters, b) floc conditioning using a rotating inclined conditioner, and c) use of suitable floc separation techniques. The second route involves employment of suitable techniques for releasing entrapped and entrained particles in the flocs. The general approach is re-dispersion of the flocs followed by re-flocculation. Re-dispersion is usually performed by mechanical agitation.

3. Ultrasonic vibration is more effective in releasing entrapped and entrained particles from the flocs rather than in preventing entrapment and entrainment during the formation and growth of the flocs.

4. Further research is needed to develop more efficient methods of minimization of entrapment and entrainment and to elucidate their exact mechanisms.

ACKNOWLEDGEMENT

The fellowship of Shanning Yu, provided by the Ohio Mining and Mineral Resource Research Institute, is acknowledged.

REFERENCES

1   Y.A. Attia and S. Yu, Entrapment and Entrainment in the Selective Flocculation Process, Part 1: Mechanisms and Process Parameters, in: Y. Attia, B.M. Moudgil and S. Chander (Eds.) Interfacial Phenomena in Biotechnology and Materials, Elsevier, 1988.
2   Y.A. Attia and D.W. Fuerstenau, Principles of Separation of Ore Minerals by Selective Flocculation, in: N. Li, (Ed.), Recent Development in Separation Science, Vol. IV, Ch. 5, CRC Press, Florida, 1978, pp. 51-69.
3   Y.A. Attia, H.N. Conkle and S.V. Krishnan, Selective Flocculation Coal Cleaning for Coal Slurry Preparation, in: Proceedings of 6th Int. Symp. on Coal Slurry Combustion, Orlando, June, 1984.
4   S.V. Krishnan and Y.A. Attia, Floc Characteristics in Selective Flocculation of Fine Particles, in: B.M. Moudgil and P. Somasundaran (Eds.), Flocculation, Sedimentation and Consolidation, 1985, pp. 229-248.
5   Y.A. Attia, S.M. Krishnan, and D.M. Deason, Final Report on Selective Flocculation Technology, Group Program, Battelle Columbus Laboratories, Columbus, OH, Aug. 1983.
6   Y.A. Attia, S.V. Krishnan and D.M. Deason, Final Report on Selective Flocculation Technology, Group Program, Battelle Columbus Laboratories, Columbus, OH, Aug., 1983.
7   K.H. Driscoll and Y.A. Attia, Effects of Process Parameters on the Selective Flocculation Cleaning of Upper Freeport Coal, in: Y.Attia, B.M. Moudgil and S. Chander (Eds.), Interfacial Phenomena in Biotechnology and Materials Processing, Elsevier, 1988.
8   S. Yu and Y.A. Attia, Review of Selective Flocculation in Mineral Processing, in: Y.A. Attia (Ed.) Proceedings of Flocculation in Biotechnology and Separation Systems, Elsevier, 1987.

9   D.W. Frommer, Preparation of Nonmagnetic Taconites for Flotation by
    Selective Flocculation, 8th Int. Mineral Process Congr., Paper D-9,
    Leningrad, USSR, 1968.
10 A.F. Banks, Selective Flocculation-Flotation of Slimes from a Sylvinite
    Ore, in: Beneficiation of Mineral Fines, NSW Workshop Report, 1979.
11 J. Groppo, Column Flotation Shows Higher Recovery with Less Ash, Coal
    Mining, Aug., 1986. pp. 36-38.
12 V.I. Tyurnikova and M.E. Naumov, Improving the Effectiveness of Flotation,
    English Edition, Technicopy Limited, England, 1981.
13 G. Dobby and J. Finch, Flotation Column Scale-up and Simulation, 17th
    Canadian Mineral Processors, Operators Conference, Ottawa, Canada, Jan.,
    1985.
14 S. Yu, Particle Collection in a Flotation Column, Master's Thesis, McGill
    University, Montreal, PQ, Canada, 1985.
15 Y.A. Attia and S. Yu, Feasibility of Separation of Coal Flocs by Column
    Flotation in: K.V.S. Sastry (Ed.), Column Flotation, AMIE/SME, Jan., 1988.
16 P. Mihaltz and L. Czako, Study of the Purification of Fat- and
    Protein-containing Waste Water by Means of Chemical Coagulation and
    Flotation, Dev. Food Sci., No. 9, 1984.
17 I. Rousev and V. Grigirov, New Industrial Technology and Equipment for
    Mechanical and Physical Chemical Purification of Waste Water from Meat
    Production Plants, Dev. Food Sci., No. 9, 1984.
18 Y.A. Attia, V. Kogan and K. Driscoll, Separation of Flocs from Suspensions
    by Dissolved Air Flotation, in: Y.A. Attia (Ed.), Flocculation in
    Biotechnology and Separation Systems, Elsevier, 1987.
19 Y.A. Attia, Selective Flocculation of Copper Minerals, Ph.D. Thesis, the
    University of London, 1974.
20 Y.A. Attia, S. Yu and S. Vecci, Selective Flocculation Cleaning of Upper
    Freeport Coal with a Totally Hydrophobic Polymeric Flocculant, in:
    Y.A. Attia (Ed.), Flocculation in Biotechnology and Separation Systems,
    Elsevier, 1987.
21 Anon., Ultrasonic Cleaning and Vapor Degreasing in Industry, Branson
    Cleaning Equipment Co., 1980.

Interfacial Phenomena in Biotechnology and Materials Processing,
edited by Y.A. Attia, B.M. Moudgil and S. Chander
Elsevier Science Publishers B.V., Amsterdam, 1988 — Printed in The Netherlands

USE OF FLOCCULANT FOR FILTRATION AND DEWATERING OF FINE COALS

S.R. FANG, Y.S. CHENG and S.H. CHIANG

Department of Chemical/Petroleum Engineering, University of Pittsburgh, Pittsburgh, Pennsylvania 15261

SUMMARY
The filtration and dewatering characteristics of three Chinese coals were investigated. The rate of filtration and the extent of dewatering were experimentally determined by using a laboratory vacuum filter. The filtration and dewatering were enhanced by adding different types of flocculants and surfactants. The effects of flocculant ionic properties, concentrations as well as mixing conditions were studied.

Among the flocculants studied, Accoal-Floc 204, an anionic flocculant, was found to be most effective. The addition of 4 ppm of this anionic flocculant under the best operating conditions resulted in more than an order of magnitude increase in the cake permeability (from 20 mD to 220 mD) and a significant decrease in the final moisture content (from 25.1% to 15.5% by weight) for the Xinglong coal. On the other hand, the use of surfactants in these Chinese coals did not lead to a significant decrease in the final moisture content of the filter cakes.

INTRODUCTION

Current coal mining and cleaning processes produce a large quantity of fine coal in a slurry form. In order to minimize transportation cost and maximize heating value, dewatering of the fine coal slurry is necessary. Among various operations, vacuum filtration and dewatering is widely used in the coal industry. However, the use of vacuum filter for the dewatering of fine coal is often found to be inadequate.[1] The final moisture content of filter cakes made of fine coal particles (less than 28 mesh) is usually greater than 25 percent by weight. A further reduction of the moisture by thermal drying consumes a great deal of energy and also may cause serious air pollution problems. Therefore, it is most desirable to explore other means for enhanced dewatering of fine coals.

A number of techniques have been proposed in the past to improve vacuum filtration and dewatering of fine coals.[2] These techniques can be classified into two categories: slurry pretreatment and cake post-treatment. The former involves adding filter aids or chemical reagents into the slurry, while the latter deals with washing filter cakes by surface active agents. Gala reported significant improvement in the dewatering of a Pittsburgh seam coal when selected surfactants were used. He showed that these experimental results can be interpreted in terms of the adsorption

characteristics of the surfactants.[3]   As a followup to Gala's study,
Venkatadri investigated surfactant washing on dewatering of the same
coal.[4]   In addition, a flocculant was used to pretreat the coal slurry and
further improvement in the dewatering characteristics was observed.[5]

As in the United States, coal fine dewatering has been a serious problem
in China.   For a number of Chinese coals, direct vacuum filtration was found
to be inadequate in reducing the final moisture content of filter cakes to
less than 30 percent.[6]   Three Chinese coals were chosen for the present
study.   The effects of flocculant and surfactant addition on the dewatering
of these coals were investigated.   The results were compared to Pittsburgh
seam coal.

BACKGROUND AND THEORY

Filtration and Dewatering

Filtration, by definition, is an operation separating solid particles
from a fluid by passing the slurry through a filter medium.   In cake
filtration, solid material accumulates on the surface of a medium, so that
filtration is through the bed of the deposited solids after a short initial
period.   At the end of the cake filtration, a 100% saturated cake is formed,
and the dewatering process begins.   During the dewatering period, air
penetrates into the cake and displaces the water within the pore space of
the filter cake when the desaturation forces are high enough to overcome
capillary forces.   The most important factor in the cake filtration is the
single phase permeability (or conductivity) of the filter cake, while the
measure of dewatering is considered to be the final moisture content of the
filter cake.   A number of ways to calculate single phase permeability are
available.[7]   When a cake is incompressible, the single phase permeability
can be directly calculated using the experimentally determined filtration
rate during the cake formation period and the Ruth equation.

Enhanced Filtration and Dewatering

The enhancement of filtration and dewatering of fine coal can be
achieved by adding two types of chemical reagents, flocculant and
surfactant.   The basic function of flocculant is to aggregate fine
particulate matter to form flocs so that the apparent particle sizes are
enlarged thus increasing the filtration rate of the slurry.   The function of
surfactant, however, is to reduce liquid/air interfacial tension and to
change the wettability of coal surfaces, leading to altered capillary
retention forces and hence the cake drainage (dewatering).

EXPERIMENTAL

Equipment

A specially designed apparatus was used in this study to experimentally determine the rate of the filtration and dewatering process. The equipment, as shown in Figure 1, consists of a plexiglas cylinder, a load cell and a transducer system. Filter cakes are formed in the cylinder, and filtrate is collected in a container placed on top of the load cell. The amount of filtrate collected in the container is continuously monitored by the load cell and associated recording system. Details of this equipment are described elsewhere.[3]

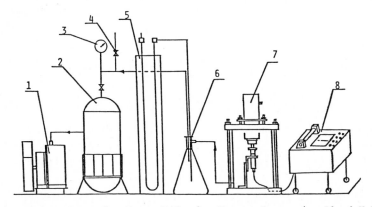

1 - Vacuum Pump, 2 - Surge Tank, 3 - Vacuum Gauge, 4 - Bleed Valve,
5 - Hg Monometer, 6 - Vacuum Trap, 7 - Filtration Unit, 8 - Recorder

FIGURE 1. Schematic Diagram of Experimental Apparatus

Materials

Three typical Chinese coals were obtained directly from Taiyuan, Pingding and Xinglong coal preparation plants. These coal samples represent clean coal products, which have approximately 10% ash with a nominal particle size less than 500 μm. A Pittsburgh coal (Bruceton mine), supplied by the Pittsburgh Energy Technology Center, Department of Energy, was a raw coal channel sample containing 6% ash with a particle size comparable to the Chinese coals.

The chemical additives used in this study include three flocculants and two surfactants. The flocculants, Accoal-Floc 204, Accoal-Floc 16 and Magnifloc 494C, were supplied by the American Cyanamid Company. The surfactants, Aerosal-OT and Trinton-X114 were obtained from the Rohm and Haas Company. The key properties of these additives are listed in Table 1.

TABLE 1
Chemical Additives Tested

Name	Type	M.W.
Accoal-Floc 204	Anionic	$4-6 \times 10^6$
Accoal-Floc 16	Nonionic	$4-6 \times 10^6$
Magnifloc 494C	Cationic	$2-4 \times 10^6$
Aerosol-OT	Anionic	444.5
Triton-X114	Nonionic	536

Procedure

To obtain a representative sample for each experimental run, the coal fines were thoroughly mixed and subdivided into small portions by coning and quartering. The moisture contents of original samples were determined by drying a known amount of coal fines. A 100 gram coal slurry was then prepared by mixing a predetermined amount of coal and distilled water. The slurry concentration was maintained at 25% solid by weight for all experimental runs. After stirring for 30 minutes, the slurry was poured into the plexiglas cylinder and a vacuum of 67 kPa was immediately applied. The filtrate passed through a filter paper (Watman #1) and was collected in a container resting on the top of the load cell assembly. The filtrate weight versus time was recorded by an HP-1062A dual channel recorder. After 10 minutes, the vacuum line was disconnected, and the filter cake was weighed and dried at $50^{\circ}C$. The final moisture content of the filter cake was then determined by material balance, and the single phase permeability of the cake was calculated using the Ruth equation.[7]

RESULTS AND DISCUSSION

Characterization of Coal Samples

The three Chinese coals, and the Pittsburgh coal (Bruceton Mine coal) were analyzed experimentally. The ash contents, sulfur contents and densities of these four coal samples are listed in Table 2. Since the Chinese coals have very wide particle size distributions, analysis of particle size for each coal sample was conducted using three objective lenses of different magnifications with the aid of a computerized Leitz TAS image analyzer. The results obtained at the three magnifications were then statistically combined to give the overall particle size distribution. Figure 2 shows the particle size distributions of these four coal samples.

TABLE 2
Characteristics of Coal Samples

Coal	Ash Content (%)	Sulfur Content (%)	Density
Xinglong	11.1	0.25	1.43
Pingding	10.4	0.56	1.43
Taiyuan	9.0	0.42	1.40
Pittsburgh	6.0	1.13	1.30

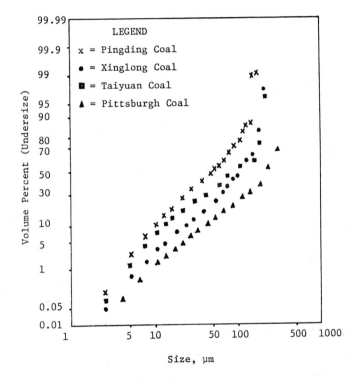

FIGURE 2: Particle Size Distribution

Flocculant Aided Filtration and Dewatering

As shown in Table 3, direct vacuum filtration and dewatering of the coals, especially some of the Chinese coals, does not yield a sufficiently low cake moisture content. The difficulties in dewatering are primarily due to the presence of ultrafine particles (< 10 μm) and clay materials. In order to increase the filtration rate and to reduce the final moisture content of filter cakes, three flocculants, Accoal-Floc 204, Accoal-Floc 16 and Magnifloc 494C, were tested. The effects of these flocculants are discussed below.

TABLE 3
Results of Direct Vacuum Filtration and Dewatering

Coal	Permeability (mD)	Final Moisture (Wt.%)
Xinglong	21.0	25.1
Pingding	22.8	21.5
Taiyuan	53.2	17.5
Pittsburgh (-32 mesh)	200.4	17.4

(1) Effect of Ionic Property of Flocculant. Although there are many commercial synthetic polyelectrolyte flocculants, they can be grouped into a few different types. For example, according to their ionic properties, flocculants can be classified into three types: anionic, nonionic and cationic. The anionic flocculants are usually sodium acrylate or partially hydrolyzed polyacrylamide. The nonionic flocculants are almost exclusively polyacrylamide or its derivatives, and the cationic flocculants generally contain quaternary ammonium groups.[8] In selecting a flocculant, ionic property and molecular weight are two important factors to be considered. Previous studies show that flocculant with a molecular weight of 4-6 million is most effective for a system containing a large percentage of mineral materials and ultra fine particles.[9] With a similar molecular weight, three flocculants, Accoal-Floc 204 (anionic), Accoal-Floc 16 (nonionic) and Magnifloc 494C (cationic), were selected for this study.

These flocculants were applied to all three Chinese coals, and their effects on single phase permeability and final moisture of filter cakes are shown in Figures 3-5 and 6-8, respectively. It is noted that a similar trend has been found in all three Chinese coals. The anionic flocculant is most effective in both increasing the single phase permeability and reducing the final moisture content of filter cakes. The nonionic flocculant becomes more effective at higher dosages, while the cationic flocculant is least effective in nearly all cases.

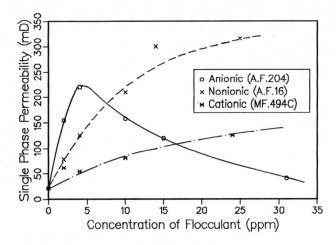

FIGURE 3. Effect of Ionic Property of Flocculant on Single
Phase Permeability of Xinglong Coal Cakes

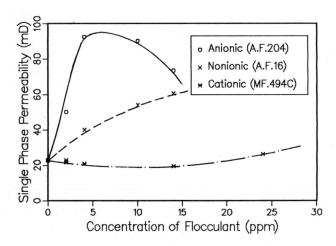

FIGURE 4. Effect of Ionic Property of Flocculant on Single
Phase Permeability of Pingding Coal Cakes

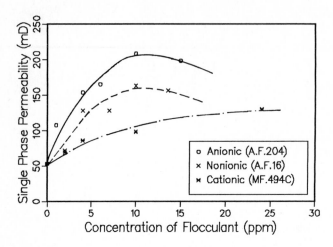

FIGURE 5. Effect of Ionic Property of Flocculant on Single
Phase Permeability of Taiyuan Coal Cakes

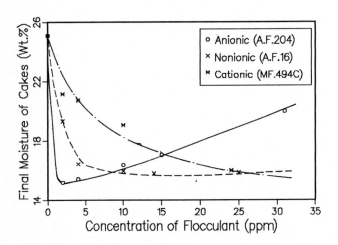

FIGURE 6. Effect of Ionic Property of Flocculant on
Final Moisture of Xinglong Coal Cakes

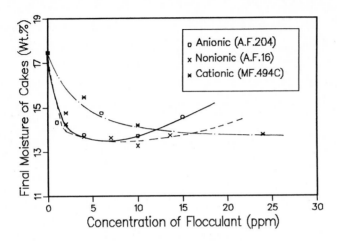

FIGURE 7. Effect of Ionic Property of Flocculant on
Final Moisture of Taiyuan Coal Cakes

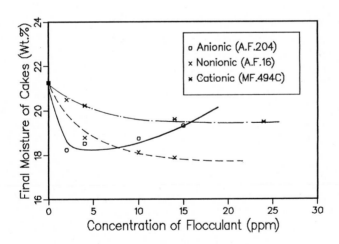

FIGURE 8. Effect of Ionic Property of Flocculant
on Final Moisture of Pingding Coal Cakes

The effectiveness of the flocculant is directly related to the nature of the flocculant and coal particles. Coal particles are generally negatively charged at the surface.[5] However, recycled water used in coal preparation plants generally contains a significant amount of calcium and magnesium ions.[9] These poly-valent cations bind with ionized anionic flocculant more tightly than sodium ions, and thus effectively neutralize the negative charge of the ionized anionic flocculant. It is believed that this tight binding provides a poly-valent cation bridge for the flocculation of particles with negative zeta potential.[10] Practically, Reuter and Hartan recently reported a general guideline for selecting a proper flocculant in mineral processing.[11] They claimed that a medium anionic flocculant was most effective for a slurry with a neutral pH. The experimental findings from this work obey their general rule and also agree very well with experimental observations from other investigators.[12]

(2) Effect of Concentration of Flocculant. The dosage of flocculant is a main factor in slurry pretreatment. An optimum concentration exists for a given flocculant and a given coal sample. Beyond the optimum dosage, the filtration performance deteriorates. This is because that as the concentration of the flocculant increases, the system reaches a point where particles are protected by the presence of an adsorbed polymer layer. Therefore, overdosing may lead to restabilization of the slurry and thus the deterioration of the flocculation.

It is very encouraging to find that only 4 ppm of the most effective flocculant, Accoal-Floc 204, results in an order of magnitude increase in single phase permeability and about 38% decrease in fine moisture content of filter cakes for the most difficult dewatering coal, the Xinglong coal (see Figures 3 and 6). The optimum dosages of Accoal-Floc 204 for the Pingding and Taiyuan coals are 4 ppm and 10 ppm, respectively. Figure 9 shows a comparison of the effectiveness of the anionic flocculant between the Xinglong coal and the Pittsburgh coal. The flocculant seems much more effective with the Xinglong coal. This may be attributed to the fact that the Xinglong coal contains a larger portion of small particles and more ash materials than the Pittsburgh coal. When filtration rate and final moisture content of filter cakes are controlled by these two factors, the addition of an effective flocculant can result in a large change in cake structure, therefore resulting in a significant improvement in filtration and dewatering.

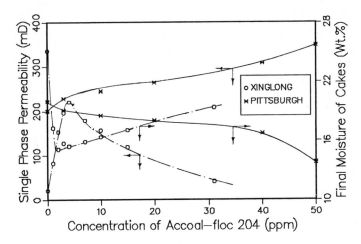

FIGURE 9. Comparison of Chinese & Pittsburgh Coals

(3) <u>Effect of Mixing Conditions</u>. The effectiveness of flocculant is believed to depend on the mixing conditions. In general, the flocculation rate is related to flocculant dose, probability of particle collision, specific surface of the particle, fluid motion velocity gradient and flocculation time.[13] Due to the limited availability of coal samples, the effect of mixing conditions on filtration and dewatering was studied only by changing the flocculation time while keeping other variables constant.

A series of experimental runs were conducted by using Accoal-Floc 204 and the Taiyuan coal. In each test, the concentration of the flocculant was kept at 10 ppm. After the flocculant was added, the slurry was agitated at 380 rpm for a predetermined period of time. The effects of the mixing time on the permeability and the final moisture content of the cakes are shown in Figure 10. It is clear that a 30 second mixing time appears to be optimum under which both a high single phase permeability and a low final moisture content were obtained. It is generally postulated that a shorter mixing time does not allow the flocculation process to come to completion, while a longer mixing time may result in a breaking-up of the flocs. Therefore, a lower single phase permeability and a higher final moisture of filter cakes were resulted from either shorter or longer mixing time.

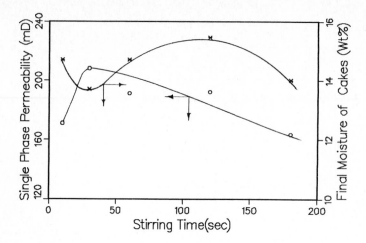

Figure 10. Effect of Stirring Time on Filtration
& Dewatering of Taiyuan Coal Cakes

<u>Surfactant Aided Filtration and Dewatering</u>

In order to further reduce the moisture content of the filter cakes, two commerical surfactants, Triton-X114 (nonionic) and Aerosol-OT (anionic), were used in this study. Two different operating procedures, premixing and washing, were applied, and the results are presented below.

(1) <u>Effect of Premixed Surfactant</u>. The premixing experiments were performed by the same procedure as before except that a predetermined amount of surfactant was added and was mixed with the coal slurry for 30 minutes. Tables 4 and 5 show the effects of premixed Aerosol-OT and Triton-X114 on the Taiyuan coal. As a comparison, previous work[5] on a Pittsburgh coal is also included in these tables. As opposed to the previous work, only premixed Aerosol-OT has a marginal effect on decreasing the final moisture content of the Taiyuan coal filter cake.

TABLE 4
Effect of Aerosol-OT (Premixed)

Taiyuan Coal (This Work)		Pittsburgh Coal (Previous Work)	
Conc. (ppm)	Final Moisture (Wt.%)	Conc. (ppm)	Final Moisture (Wt.%)
0	17.5	0	17.4
50	14.8	62	15.3
150	15.2	190	12.7
300	18.3	350	12.8
		500	13.7

TABLE 5
Effect of Triton-X114 (Premixed)

Taiyuan Coal (This Work)		Pittsburgh Coal (Previous Work)	
Conc. (ppm)	Final Moisture (Wt.%)	Conc. (ppm)	Final Moisture (Wt.%)
0	17.5	0	17.4
50	18.7	62	15.4
150	17.6	160	13.8
300	16.3	250	12.1
		500	12.7

(2) Effect of Surfactant Washing. In contrast to the premixing experiment, the surfactant washing was carried out by introducing a given amount of surfactant wash liquor onto the top of the filter cake at the end of the filtration period. In this study, the concentration of the wash liquor was varied from 150 ppm to 500 ppm of the selected surfactant, while a wash ratio (defined as volume ratio of wash liquor to the void space of the filter cake) of 2 was utilized for all tests. Similar to the premixed surfactants, the surfactant washing only resulted in a minor decrease in final moisture content of the Pingding coal cake (see Table 6).

TABLE 6
Effect of Surfactant Washing

COAL SAMPLE:	Pingding
SURFACTANT:	Aerosol-OT
WASH RADIO:	2

SURFACTANT CONC. (ppm)	MOISTURE CONTENT (Wt.%)
0	21.26
150	20.13
250	20.13
500	20.51

(3) Effect of Combination of Flocculant and Surfactant. It has been reported that surfactant washing of flocculated fiter cake reduced the final moisture content of filter cakes by as much as 60%[5]. The surfactant washing was also applied in this work to investigate the combined effect. In each experimental run, 4 ppm Accoal-Floc 204 was added to the Taiyuan coal slurry. The slurry was then mixed for 30 seconds at a stirring speed of 380 rpm. Immediately following the filtration period, a given amount of wash liquor (wash ratio=2) with a desired surfactant concentration was used to wash the filter cake. The total processing time (filtration plus

washing) was kept at 10 minutes. As shown in Table 7, the best result is a reduction of about 1.5 percentage points (from 13.8% to 12.3%) in final moisture content obtained by using 100 ppm Triton-X114 solution as the wash liquor.

TABLE 7
Effect of Combination of Flocculant and Surfactant

Surfactant	Concentration (ppm)	Moisture Content (Wt.%)
–	0	13.8
Aerosol-OT	50	13.9
Aerosol-OT	250	13.0
Aerosol-OT	500	13.5
Triton-X114	50	13.5
Triton-X114	100	12.3
Triton-X114	250	13.5

In summary, although the use of surfactant is very effective to enhance the dewatering of the Pittsburgh coal, it is only marginally effective for the Chinese coals tested. On the other hand, the results show that the flocculant plays a dominant role in the dewatering of Chinese coals. Even in the case of the combined effect, the major portion of reduction in the final moisture content of the fiter cake can be attributed to the addition of Accoal-Floc 204. The difference in dewatering behavior between the Chinese coals and the Pittsburgh coal is likely caused by different particle size distributions and mineral contents of the ash material. The Chinese coals contain much more ultra fine particles (up to 10% of the total weight smaller than 10 microns) than the Pittsburgh coal (2% of total smaller than 10 microns). Furthermore, the Chinese coals have higher ash contents. The addition of flocculant increases effective particle size leading to a change in structure of the filter cake and thus can greatly improve filterability of ultra fine materials. The use of surfactant, however, only changes the surface properties of the particles. When a filter cake contains a large portion of small particles and thus small pores, the surface property modification alone is not sufficient to cause a significant reduction in the capillary forces. Therefore, the addition of surfactant is less effective when ultra fine particle and mineral materials control the filtration and dewatering process. A similar observation was also found in filtration and dewatering of fine coal refuse[9].

CONCLUSIONS

1. The addition of flocculant can greatly improve filtration and dewatering of fine coal.

2. Among various flocculants tested, an anionic flocculant is found to be most effective in enhancing filtration and dewatering of fine coal.

3. Concentration of flocculant is a major factor in flocculation, and an optimum dosage exists for a given flocculant and coal sample.

4. Effectiveness of flocculant strongly depends on mixing conditions used for the pretreatment of coal slurry.

5. Effect of surfactant varies with coal samples.

ACKNOWLEDGEMENT

The financial support provided by the U.S. Department of Energy under contract No. DE-AC2285PC81582 is gratefully acknowledged.

REFERENCES

1. R.J. Wakeman, Residual Saturation and Dewatering of Fine Coals and Filter Cakes, Powder Technology, Vol. 40, 1984, pp. 53-63.
2. R.J. Wakeman, V.P. Mehrotra and K.V.S. Sastry, Mechanical Dewatering of Fine Coal Refuse Slurries, Bulk Solids Handling, Vol. 1, 1981, pp. 281-293.
3. H.B. Gala, Use of Surfactants in Fine Coal Dewatering, Ph.D. dissertation, School of Engineering, University of Pittsburgh, 1982.
4. R.A. Venkatadri, G.E. Klinzing, and S.-H. Chiang, Filter Cake Washing with Chemical Reagents, Filtration and Separation, Vol. May/June, 1985, pp. 172-177.
5. R.A. Venkatadri, Effect of Surface Active Agents on Filtration and Post Filtration Characteristics, Ph.D. dissertation, School of Engineering, University of Pittsburgh, 1984.
6. Private Contact with Officers, The Ministry of Coal Industry of China.
7. H.B. Gala, and S.-H. Chiang, Filtration and Dewatering - Review of Literature, U.S. Department of Energy Report NO. OOEIT/14291-1, Vol. September, 1980.
8. C. Orr, Filtration Principles and Practices, Marcel Dekker, Inc., 270 Madison Avenue, New York, New York, 10016, 1977.
9. Y.S. Cheng, S.R. Fang, J.W. Tierney and S.-H. Chiang, Application of Enhanced Vacuum Filtration to Dewatering of Fine Coal Refuse, paper presented at the Fifth Symposium on Separation Science and Technology for Energy Application, Tennessee, October 26-29, 1987.
10. M.E. Lewellyn and S.S. Wang, Organic Flocculants in Dewatering Fine Coal and Coal Refuse: Structure vs. Performance, Macromolecular Solution: Solvent Property Relationship in Polymers, Pergamon, New York, 1982, pp. 134-150.
11. J.M. Reuter and H.G. Hartan, Structure and Reaction Kinetics of Polyelectrolytes and Their Use in Solid-liquid-Processing, paper presented at the World Congress Particle Technology, Part IV, pp. 269-286, Nurnberg, Germany, April 16-18, 1986.
12. M.J. Pearse and A.P. Allen, The Use of Flocculants and Surfactants in the Filtration of Mineral Slurries, Filtration & Separation, Vol. Jan./Feb., 1983, pp. 22-27.
13. J. Mackiewicz, The Development of Flocculation Effects in Filter Theory, Chemical Engineering Communication, Vol. 23, 1983, pp. 305-314.

Interfacial Phenomena in Biotechnology and Materials Processing,
edited by Y.A. Attia, B.M. Moudgil and S. Chander
Elsevier Science Publishers B.V., Amsterdam, 1988 — Printed in The Netherlands

DETECTION OF LOW CONCENTRATION OF PROTEINS BY SOLUBILIZATION AND

DESOLUBILIZATION WITH THE ANIONIC SURFACTANT DODECYL SULFATE

A. Lopez-Valdivieso[1], E. J. Platt[2], K. Karlsen[2] and G. L. Firestone[2]

[1]Department of Materials Science and Mineral Engineering University of
California, Berkeley, CA 94720

[2]Department of Physiology-Anatomy & Cancer Research Laboratory, University of
California, Berkeley, CA 94720

SUMMARY
     A highly sensitive procedure that virtually eliminates nonspecific
adsorption of radiolabeled proteins during immunoprecipitation has been devised
using Staphylococcus aureus containing protein A (Staph A). Immunoprecipitates
(antigen-antibody complexes) were solubilized from Staph A pellets into dodecyl
sulfate micelles at 23 $^{\circ}$C. Antigen-antibodies complexes were then desadsorbed
from the micelles and rebound to new Staph A by dilution of solubilized
material in buffer solutions with 1% Triton X-100 and 0.5% sodium deoxycholate.
Using this procedure, the immunoprecipitation of several different classes of
radiolabeled proteins from total extracts were tested, including membrane bound
viral glycoproteins and cytoplasmic proteins. In each case, the nonspecific
background was lowered from approximately 2250 to less than 25 parts-per-
million with a final recovery of 30-50% that depended on the specific antigen
and antibody preparation.

INTRODUCTION

     The expression of specific protein products from radiolabeled tissues has

been commonly examined by a procedure that makes use of the high adsorption

capacity of protein A molecules on staphylococcus aureus (Staph A) for specific

IgG and IgM isotypes (1-3). Basically, the procedure involves the interaction

of a small amount of radiolabeled antigens with an excess of antibodies,

followed by incubation with enough Staph A so as to bind all appropiate

antibodies regardless of whether they contain bound antigens. The advantage of

this procedure is that an immunoprecipitate per se need not be formed to

separate immunocomplexes from cellular polypeptides not recognized by the

antibodies. Thus, small absolute amount of radiolabeled antigens can be

rapidly and selectively adsorbed to Staph A immunopellets and quantitatively fractioned away from the bulk polypeptides by simple low-speed centrifugation. Subsequent electrophoretic analysis or liquid scintilation can then be readily accomplished. However, when the antigen of interest is expressed at low levels nonspecific background binding of bulk radiolabeled polypeptides may pose a significant problem for explaining electrophoretic results. In this case, the desired immunoadsorbed antigen may be displayed only as a minor band, indistinguishable from the bands representing bulk radiolabeled polypeptides (3). Hence, this work aimed at developing an immunoadsorption technique capable of analyzing low concentration of polypeptides at low background. Research has been emphasized on the desorption of antigen-antibody complexes from Staph A by dodecyl sulfate (SDS) micelles and the recovery of the complexes on fresh Staph A using sodium deoxycholate and in solution conditions below the critical micelle concentration of SDS (4,5). The technique has been tested using mouse mammary tumor virus (MMTV) proteins expressed in infected rat hepatoma cells since in induced cells the level of viral polypeptides is 0.05% of total synthesized protein, while in glucorticoid tucated cells MMTV proteins represent less than 1% of total cellular proteins in hormone-treated cells (6,7).

MATERIALS AND METHODS

Materials

Dubelcco's modified Eagle's medium and horse serum were obtained from the cell culture facility at the University of California, San Francisco. L-[^{35}S]Methionine (1000 Ci/mmol) from American Corporation, dexamethasone and sodium dodecyl sulfate from Sigma Chemical Company, Pansorbin from Calbiochem, En^{3}Hance from New England Nuclear, and Kodak X-ray film from Merry X-Ray Chemical Corporation were used in this study. The MMTV antiserum was generouly provided by L. J. T. Young and R. DM Cardiff (Department of Pathology, University of California at Davis).

Cell, Method of Culture, and Radiolabeling Procedure

M1.54 is a cloned line of MMTV-infected rat hepatoma cells and contains 10 stably integrated MY.TV proviruses (7-9). Cultures were propagated as monolayers in Dulbecco's modified Eagle's medium suplemented with 10% horse serum at $37^{\circ}C$ in a control atmosphere of air/CO_2 (95%:5%). When involved, dexamethasone was added to the culture medium to a final concentration of 1 uM. Dexamethasone-induced and uninduced cells were radiolabeled with 100-200 uCi [^{35}S]methionine/4 ml medium for 16 h in methionine-deficient medium suplemented with 0.5% dialyzed fetal calf serum. At the end of the radiolabeling period, monolayers were washed several times in phosphate-buffered saline, gentle dislodged from the culture plates in phosphate-buffered saline containing Tris-EDTA (10 mM:1 nM, pH 7.5), and centrifuged at 600xg for 3 min; cell pellets were solubilized as described below.

Triton X-100/Deoxycholate Solubilization of Cells

Radiolabeled cells were solubilized by homogenizing cell pellets (10^7) in 1 ml buffer of 1% Triton X-100, 0.5% deoxycholate, 5nM EDTA, 250 mM NaCL, 25 mM Tris-HCl, pH 7.5, at $4^{\circ}C$. The cell homogenates were centrifuged at 20,000xg for 10 min at $4^{\circ}C$. The supernatant fractions were utilized as sources of solubilized polypeptides. Such fractions were analyzed for total radiolabeled proteins by precipitating a small aliquot with 10% trichloroacetic acid. An identical procedure was used to solubilize nonradiolabeled cells.

Immunoadsorption on Staph A

Radiolabeled and nonradiolabeled proteins were adsorbed on Pansorbing, a formaldehyde-fixed S. aureus (Cowen I strain) containing cell surface-associated protein A (Staph A). Cell extracts were first preasorbed in the absence of antibodies using 300 ul of solubilized material and 10 ul of a 10% solution (w/v) of fixed Staph A in solubilization buffer (preadsorbed radiolabeled extracts). Simultaneosly, the Staph A employed for the immunoadsorption was preadsorbed with nonradiolabeled cells extracts (preadsorbed Staph A). After 15 min, the preadsorbed radiolabeled extracts and the preadsorbed Staph A were each harvested by centrifugation for 3 min in an Eppendorf centrifuge at 23°C. Subsequently, the radiolabeled extracts were added to 100 ul solubilization buffer containing 50 mg/ml bovine serum albumin (BSA) and 10 ul of a 1/10 dilution (in phosphate buffer saline) of either anti-MMTV or preimmune serum. After incubation for 10 min at 23°C, 10 ul of 10% Staph A (that had been preincubated with unlabeled cell extracts in solubilization buffer) was added and the mixture incubated for an additional 5 min. The entire reaction mixture was layered over 600 ul of 1 M sucrose in solubilization buffer and centrifuged for 3 min, and the nonadsorbed material was removed by aspirating down to the sucrose interface. The tubes were then washed by overlaying the sucrose cushion with 2 M urea in solubilization buffer containing 500 mM NaCl, and aspirating down to the Staph A pellets. Finally, the pellets were washed with solubilization buffer containing 250 mM NaCl and then with a solution containing 5 mM EDTA, 10 mM Tris-HCl and at pH 7.5. The washed Staph A pellets were either stored at -20°C or immediately reimmunoprecipitated as described below.

Dodecyl Sulfate Solubilization and Readsorption on Staph A

Antigen-antibody complexes were desorbed from the Staph A with 30 ul of 1% SDS in phosphate-buffered saline by incubating for 10 min at 23°C, followed by a 3 min centrifugation. The supernatant fractions were added to 500 ul solubilization buffer containing 10 mg/ml BSA and 0.8 ul of the appropiate immune or preimmune serum. After 15 min, 10 ul of 10% preadsorbed Pansorbin was added and the reaction allowed to proceed for 5 min at 23°C. The Staph A immunocomplexes of Pansorbin were collected by centrifugation and subsequently washed first with solubilization buffer and then with a solution containing 10 mM Tris-HCl, 5mM EDTA and at pH 7.5. The Staph A pellets obtained were either stored at -20°C or prepared immediately for SDS-polyacrylamide gel electrophoresis.

Dodecyl Sulfate Electrophoresis and Autoradiography

Staph A pellets were mixed with 25 ul SDS-gel sample buffer (10) and 5 ul of 0.5 M dithiotrietal (DTT) at 100°C for 2 min. The mixture was centrifuged for 2 min in order to collect a supernatant for electrophoresis experiments, which were carried out using SDS-polyacrylamide gels containing 9.5% acrylamide. Proteins were subsequently fixed and stained by incubation overnight with 0.4% Coomasie brilliant blue in 10% acetic acid and 25% isopropyl alcohol. Gels were then destained in 10% acetic acid for 4 h, impregnated with En^{3}Hance for 1 h, incubated in water for 1 h, dried at 60°C, and analyzed by fluorography on Kodak X-Omat AR-5 film at -70°C

Specific Conductance Measurements

The critical micelle concentration of the surfactant dodecyl sulfate was determined by measuring the specific conductance of 100 ml solutions prepared with phosphate buffered saline at various SDS concentrations. The determinations were carried out at 23°C using a Beckman RC-9 conductivity bridge and a cell with a constant of 1 cm^{-1}. The conductance of the solutions was calculated in millimhos per centimeter.

RESULTS AND DISCUSSION

Immunoadsorption of Mouse Mammary Tumor Virus Polypeptides (MMTV) on Staph A

When the objective is to detect low concentration of proteins after long X-ray film exposures using the well established immunoadsorption procedure on Staph A devised by Kessler (1,2), a major problem that can arise is a significant background because of the nonspecific binding of proteins to Staph A. To improve this procedure, this study involved MMTV-infected rat hepatoma cells that express an absolute level of MMTV proteins of about 1% of total synthesized protein after treatment with dexamethasone, a synthetic glucocorticoid. Also, in uninduced cells, in which MMTV polypeptides represents 0.05% of total cell protein. The MMTV-infected rat hepatoma cell line, M1.54, (6,11) was treated with and without 1 uM dexamethasone for 24 h in the presence of [^{35}S]methionine and MMTV polypeptides immunoadsorbed with polyclonal anti-MTV or preimmune serum from Triton X-100/deoxycholate solubilized extracts. The results obtained are shown in Figure 1 by lanes A through D, which display a fluorograph after 1 month exposure of immunoadsorbed MTV proteins from 1 million cpm trichloroacetic acid-precipitable radioactivity. The viral proteins are denoted by arrows; however, as can be seen a significant amount of the nonspecific background are detected. In fact, most of the predominant bands seen in the lanes represent nonspecifically bound proteins since the preimmune lanes given by P display the same pattern of protein species.

Detection of Low-Concentration Protein with Dodecyl Sulfate (SDS) Micelles

Next, research work was emphasized on examining the effect of the anionic surfactant dodecyl sulfate on reducing the nonspecific background in electro-phoretic results so that the low concentration of MMTV proteins could be unambiguosly detected. Immunoadsorbed proteins on Staph A-anti-MTV pellets were incubated with 30 ul phosphate buffered saline containing a known amount of SDS. Then, the extract was diluted with 600 ul of buffer containing 1% Triton X-100 and 0.5% deoxycholate, fresh antibodies and Staph A, and the immunoadsorption was continued as outlined above. The results given in Figure 2 indicate the radioactivity obtained in the final Staph A versus the SDS concentration (in weight percent) used to initially elute immunocomplexes. It can be observed that a maximum of recovered radioactivity is obtained at the narrow SDE concentration about 1%. The fluorographic results obtained from the final Staph A pellets are shown in Figure 1 by lanes E through H, where it can be noted that the low concentration MMTV proteins expressed in uninduced M1.54 are clearly delineated in the absence of nonspecific background (compare lanes E and F). Remarkably, with the addition of dodecyl sulfate in the immunoadsorption procedure, exposures of X-ray film of up to 2 months did not display nonspecific background in the electrophoresis gels.

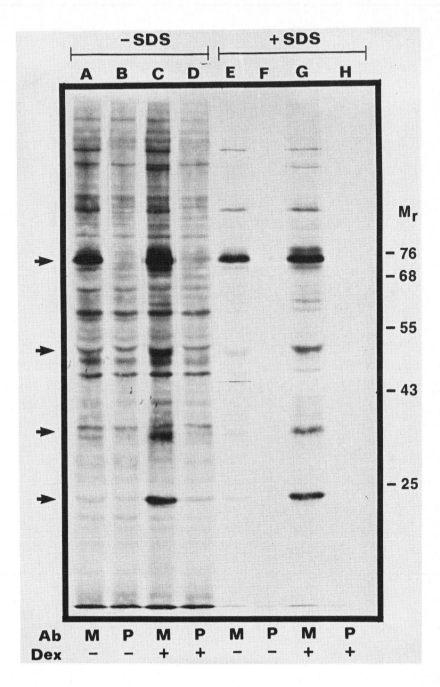

Fig. 1. The effect of dodecyl sulfate (SDS) solubilization on the nonspecific background of Staph A immunoadsorption. Lanes A through D are results in the absence of SDS. Lanes E through H are results in the presence of SDS. M and P represents anti-MMTV antibodies and preimmune serum, respectively.

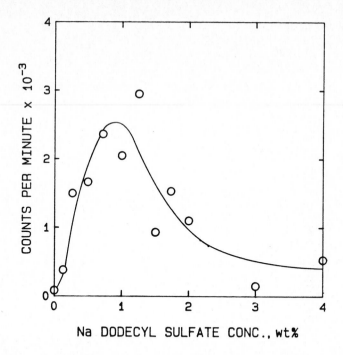

Fig. 2. The effect of sodium dodecyl sulfate concentration in the elution and rebinding of MMTV proteins to Staph A.

The specific conductance of phosphate-buffered saline solutions was measured as a function of dodecyl sulfate concentration. The results are given in Figure 3, where the change in slope of the specific conductance line indicates that monomer dodecyl sulfate molecules aggregate forming micelles (12). This transition of the surfactant from a monomer state to that of micelles occurs at the so called the critical micelle concentration (cmc). According to the results shown in Figure 3, the cmc in our system takes place at 0.1% dodecyl sulfate. By comparing Figures 2 and 3, it can be noted that the release of antigens from Staph A-anti-MMTV pellets occurs under conditions in which dodecyl sulfate micelles are present in solution. This suggest that the desorption from the pellets happens because of the penetration of a portion of radioactive protein into the micelle. Similarly, the readsorption of protein on fresh Staph A occurs because the conditions are such that micelles are no longer present in the system. The readsorption on fresh Staph A was carried out with solutions in which the concentration of dodecyl sulfate was lowered from 1% to less than 0.05%, that is below the cmc.

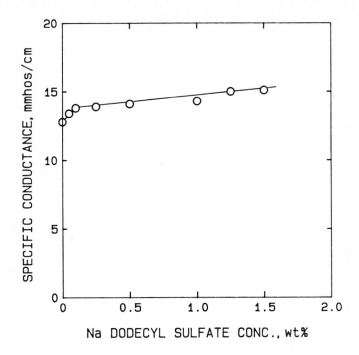

Fig. 3. The specific conductance of phosphate buffered saline as a function of sodium dodecyl sulfate concentration at $23^{\circ}C$.

The nonspecific background and recovery of MMTV proteins from Staph A immunoadsorptions carried out with and without the SDS-elution and readsorption step were quantitated. Dexamethasone-induced cells were radiolabeled with [^{35}S]methionine, and Triton X-100/deoxycholate solubilized extract representing 2 million counts per minute (cpm) trichloroacetic acid precipitable radioactivity were immunoadsorbed with either anti-MMTV or preimune sera. Pelletable radioactivity with the preimmune serum represents the nonspecific background, and the recovery of MMTV proteins was determined by substracting immunoadsorbed cpm with preimmune serum from the radioactivity pelleted in the presence of anti-MMTV serum. The results shown in Table 1 indicate that the nonspecific background binding to Staph A is reduced from 2250 ppm to about 25 ppm before and after the SDS-elution step, respectively.

TABLE 1. EFFECT OF SODIUM DODECYL SULFATE (SDS) ELUTION ON THE RECOVERY OF
IMMUNOADSORBED MTV PROTEINS AND NONSPECIFIC BACKGROUND.

SDS Elution	Immunoadsorbed Radioactivity[a] (counts per minute)		MTV-Specific Radioactivity[b]	Recovery[c] (percent)
	Anti-MTV Serum	Preimmune Serum		
Without	15,100 ±500	4,590 ±400	10,500	100
With	4,220 ±300	48 ±20	4,170	30

a. Immunoadsorptions were carried out from 2 x 10^6 trichloroacetic acid
precipitable radioactivity (^{35}S-methionine) from extracts of hormone induced
cells.
b. Calculated as the difference in immunoadsorbed radioactivity in immunocomplexes formed with anti-MTV serum and immunocomplexes formed with preimmune
serum.
c. The recovery of MTV-specific radioactivity from samples analyzed prior to
the SDS-elution step is considered 100% and the recovery after the SDS-elution
step is calculated from the ratio of MTV-specific radioactivity detected with
SDS elution to the amount detected without the SDS-elution procedure.

The overall recovery of MTV proteins after the SDS-elution and
readsorption steps varied between 35 and 50%. In addition, the nonspecific
background is approximately 0.2% in the two Staph A steps. Because of the low
amount of total radioactivity in the SDS-solubilized fraction, the background
after the second Staph A immunoadsorption falls below the sensitivity of a 2-
month exposure on X-ray film. Thus the trade-off in our procedure is at least
a 50% loss in specific recoverable polypeptides for a significant reduction
(over 100-fold) in overall nonspecific background. The net effect is a significantly larger signal-to-noise ratio in the immunoadsrption reaction,
resulting in a more sensitive analysis of immunoadsorbed proteins.

CONCLUSION

A highly sensitive, low background immunoadsorption technique has been
developed for analyzing the expression of low concentrations of radiolabeled
proteins. The technique is based on the ability of dodecyl sulfate molecules
to desorb immunocomplexes from Staph A pellets under conditions above the cmc
of dodecyl sulfate and to allow rebinding on fresh Staph A under conditions
below the cmc.

REFERENCES

1 S. W. Kessler, Rapid isolation of antigens from cells with a staphylococcal protein A-antibody adsorbent: parameters of the interaction of antibody antigen complexes with protein A, J. Immunol. 115 (1975) 1617-1624.

2 S. W. Kessler, Cell membrane antigen isolation with the staphylococcal protein A-antibody adsorbent, J. Immunology 117 (1976) 1482-1489.

3 R. D. Ivarie, and P. P. Jones, A rapid sensitive assay for specific protein synthesis in cells and in cell-free translations: Use of staphylococcus aureus as an adsorbent for immune complexes, Analytical Biochem. 97 (1979) 24-35.

4 E. J. Platt, K. Karlsen, A. Lopez-Valdivieso, P. W. Cook, and G. L. Firestone, Highly sensitive immunoadsorption procedure for detection of low abundance proteins, Analytical Biochem. 156 (1986) 126-135.

5 N. J. John, and G. L. Firestone, A sensitive two-step immunoadsorption procedure to analyze low abundance polypeptides by lowering nonspecific background, BioTechniques 4 (1986) 404-406.

6 G. L. Firestone, F. Payvar, and K. R. Yamamoto, Glucocorticoid regulation of protein processing and compartmentalization, Nature 300 (1982) 221-225.

7 G. L. Firestone, and K. R. Yamamoto, Two classes of mutant mammary tumor virus-infected HTC cell with defects in glucocorticoid-regulated gene expression, Molecular and Cellular Biology 3 (1983) 149-160.

8 G. M. Ringold, P. R. Shank, H. E. Varmus, J. Ring, and K. R. Yamamoto, Integration and transcription of mouse mammary tumor virus DNA in rat hepatoma cells, Proc. Natl. Acad. Sci. USA 76 (1979) 665-669.

9 D. S. Ucker, G. L. Firestone, and K. R. Yamamoto, Glucocorticoids and chromosomal position modulate murine mammary tumor virus transcription by affecting efficiency of promoter utilization, Molecular and Cellular Biology 3 (1983) 551-561.

10 U. K. Laemmli, Cleavage of structural proteins during the assembly of the head of bacteria T4, Nature 227 (1970) 680-685.

11 G. L. Firestone, The Role of protein glycosylation in the Compartmentalization and processing of mouse mammary tumor virus glycoproteins in mouse mammary tumor virus-infected rat hepatoma cells, J. Biological Chemistry 258 (1983) 6155-6161.

12 D. G. Hall and G. J. T. Tiddy, Surfactant solutions: Dilute and concentrated, in: E. H. Luncanssen-Reynders (Ed.), Anionic Surfactants: Physical Chemistry of Surfactant Action, Marcel Dekker, New York, 1981, pp. 55-108.

Interfacial Phenomena in Biotechnology and Materials Processing,
edited by Y.A. Attia, B.M. Moudgil and S. Chander
Elsevier Science Publishers B.V., Amsterdam, 1988 — Printed in The Netherlands

POLYMERS IN FLOCCULATION AND AGGLOMERATE BONDING

R. HOGG and D. T. RAY

Mineral Processing Department, 108 Steidle Building, University Park, PA 16802

SUMMARY

The role of polymer adsorption in suspension destabilization, floc growth, and floc breakage is discussed. Both adsorption and flocculation are found to depend on the conditions of polymer addition. Conditions which favor rapid floc growth are generally associated with relatively low adsorption densities, presumably due to the reduced available surface area following floc growth. The growth of flocs appears to depend on the polymer concentration in solution while their strength depends on the polymer content of the floc. As a consequence, agitation conditions and rates of polymer addition can be used to effect some degree of control over floc size and strength.

The use of commercial, high molecular weight flocculants as binders for compacted, nonplastic powders has also been investigated. It is shown that by controlling the polymer application so as to favor agglomerate strength, i.e. by maximizing polymer adsorption, green strengths comparable to those obtained using conventional binders can be obtained at substantially reduced organic matter content.

INTRODUCTION

Despite their extensive use as flocculants for fine-particle suspensions, the exact role of soluble polymers in flocculation processes has yet to be completely elucidated. Flocculation is a complex process involving several, more-or-less simultaneous, sub-processes including:

- particle destabilization
- floc growth
- floc degradation (breakage)

Floc growth is, of course, the primary objective of a flocculation process, destabilization is generally necessary to permit flocs to grow, while floc degradation is an inevitable consequence of agitation of a suspension.

Polymeric flocculants play a role in each of these sub-processes through adsorption of the polymer on particle surfaces and incorporation in the growing flocs. While previous analyses of polymer-induced flocculation have led to proposed mechanisms for particle destabilization due to adsorption (1), the observation that polymers promote the rapid formation of very large flocs suggests an additional important role in floc binding, minimizing floc breakage (2). In this paper we will attempt to evaluate the function of polymers in flocculation, particularly their influence on floc strength. In addition we will demonstrate the use of the same polymers as binders in agglomerated powders.

FLOCCULATION PROCESSES

Floc growth

When a solution of a polymeric flocculant is added to an agitated suspension of fine particles, rapid floc growth begins immediately as shown in Figure 1. Floc growth continues as long as polymer is added to the

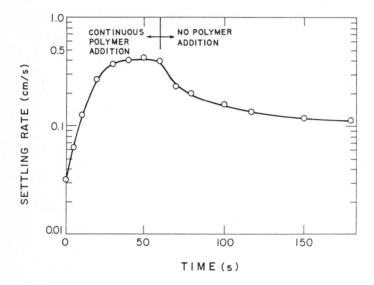

Fig. 1. Flocculation of an agitated clay suspension. Nonionic polymer added continuously at 5.3 mg/L.min for one minute. Agitation, at 1000 rpm, continued over the entire test period.

suspension, but the rate of growth decreases and the floc size approaches a limiting value. If polymer addition is discontinued, further agitation of the suspension leads to a progressive reduction in floc size, presumably due to irreversible floc breakage. In a previous publication (3) we have demonstrated that floc size can be closely correlated with sludge settling rate. Since the latter can be measured much more easily than floc size, we will generally use settling rate as an indicator of floc size.

The effect of the rate of addition of polymer to the suspension is illustrated in Figure 2. It has been demonstrated (3) from these data that:

> the initial rate of floc growth is directly proportional to the rate of polymer addition, except at very high addition rates where it appears that mixing of the polymer with the suspension may be inadequate.

545

· the limiting floc size (settling rate) increases with rate of
polymer addition.

The effects of agitation on floc growth are shown in Figure 3. The
results indicate that:

· the initial growth rate is essentially independent of agitation
speed. This suggests that growth rates are determined by the amount
of polymer available, i.e., by collision efficiencies (1,4) rather
than by collision frequencies which would be expected to increase
with increasing shear rate (5,6).

· the limiting floc size decreases with increasing shear, suggesting
that floc growth may be limited by floc breakage.

Based on the results shown in Figures 1-3, it is concluded that optimal
conditions for flocculation are found when agitation of the suspension is just
sufficient to ensure adequate mixing of the polymer and agitation is continued

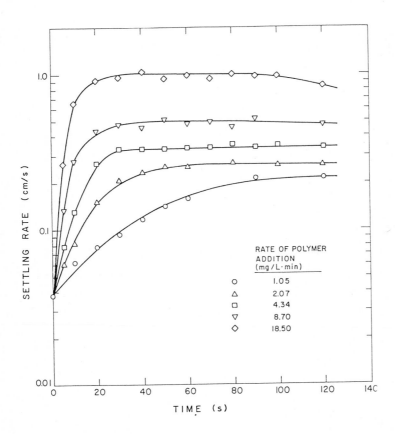

Fig. 2. The effect of rate of polymer addition on the flocculation of an
agitated clay suspension.

only so long as polymer is being added. Further optimization, within the above constraints, can be achieved by adjusting the rate and extent of polymer addition to give the desired floc size.

Particle Destabilization

Examination of Figures 1-3 reveals that, for these particular conditions, the apparent initial floc sizes and settling rates are significantly higher than would be expected for well-dispersed systems of about 1 μm particles. In fact these suspensions were not well dispersed but exhibited some degree of "natural" flocculation – presumably due to low surface charge for Al_2O_3 at pH 9 and face-edge interactions (7) for kaolin at pH 5. Similar experiments, conducted at pH values where the suspensions were initially highly stable, showed that flocculation was quite different. Polymer dosages up to 25 times higher than those shown in Figure 1-3 were necessary to provide "good" floccu-lation. The results of such experiments are summarized in Table 1.

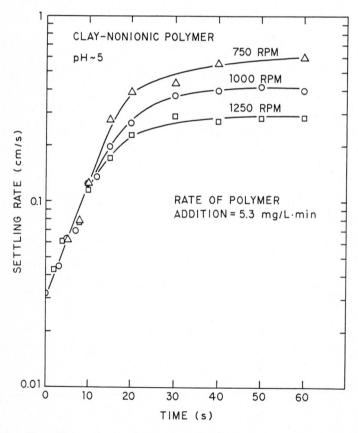

Fig. 3. The effect of agitation on the flocculation of a clay suspension during continuous polymer addition.

TABLE 1
Effects of surface charge and initial suspension stability on the performance
of polymeric flocculants

Material	pH	Initial State of Dispersion	Surface Charge	Minimum Dosage to Reduce Supernatant Turbidity to <100 NTU (mg/L)		
				Cationic	Nonionic	Anionic
Clay	3-4	Poor	-(+edge)	5	3	2
	7-8	Moderate	-	20	20	nf
	11	Good	-	30	45	nf
Al_2O_3	4-5	Good	+	nf	50	120
	9	Poor	0	*	5	5
	11	Good	-	25	**	nf

nf   No flocculation
 *   Turbidity increases with polymer addition.
**   Some flocculation, but turbidity remains high.

It is often assumed that the effectiveness of polymeric flocculants
depends, to a considerable extent, on the charge on the polymer relative to
that on the particles; cationic polymers should be effective for negatively
charged particles and so on. The results given in Table 1 indicate that this
is only partially correct. The cationic polymer is, indeed, effective for
negatively charged particles and incapable of flocculating positively charged
Al_2O_3 at pH 4-5. The anionic polymer, on the other hand, is very effective
when the particle charge is low (Al_2O_3 at pH 9) or mixed positive and negative
(clay at pH 3-4), but will only flocculate positively charged Al_2O_3 particles
(pH 4-5) at extraordinarily high dosage.

A considerably clearer correlation exists between the effectiveness of
the polymers and the initial state of dispersion of the particles. When the
initial suspension is unstable (clay at low pH, Al_2O_3 at pH 9) each of the
polymers causes good flocculation at low dosage, with the exception of the
cationic polymer on Al_2O_3 which appears to be acting as a dispersant rather
than a flocculant. It is quite possible that, for the latter case, there may
be a narrow range of (low) dosage where flocculation is enhanced.

Based on these results, it is concluded that, with the possible exception
of low molecular weight polyelectrolytes, polymers are ineffective reagents
for destabilizing suspensions. Their great value lies in their ability to
produce large flocs from already unstable suspensions.

Floc Breakage

Degradation of flocs during agitation without polymer addition was
illustrated in Figure 1. More detailed investigations (2,8) have revealed

548

that degradation is similar to other size reduction processes such as the
grinding of solid particles and that the rate of agglomerate breakage
increases with increasing floc size and decreases with increasing polymer
content of the flocs. It appears that these fine-particle suspensions undergo
essentially irreversible floc breakage when subjected to agitation in the
absence of fresh polymer addition.

Polymer Adsorption

It is generally agreed that flocculation of fine particles by polymers
involves adsorption of the polymer at particle surfaces. In an attempt to
clarify the effects of physical and chemical variables described above, we
have conducted a series of adsorption measurements under flocculation conditions.

Typical "adsorption isotherms" corresponding to two different rates of
polymer addition to a suspension of Al_2O_3 particles at pH 9.5 are shown in
Figure 4. These results show that the extent of polymer adsorption is strongly dependent on rate of addition - increasing the rate actually reduces the
amount adsorbed. Noting that the higher addition rate corresponds to more
rapid growth of larger flocs (see Figure 2), it appears that polymer adsorption occurs on the external surfaces of the growing flocs so that rapid growth
reduces polymer uptake.

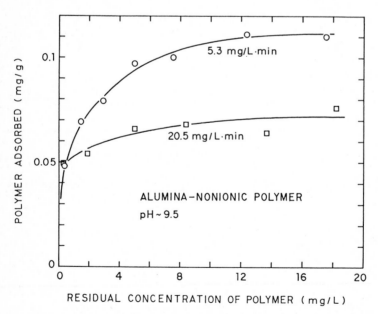

Fig. 4. The effect of rate of continuous polymer addition on the adsorption
of polymer on Al_2O_3 particles under flocculation conditions (no additional
mixing following polymer addition).

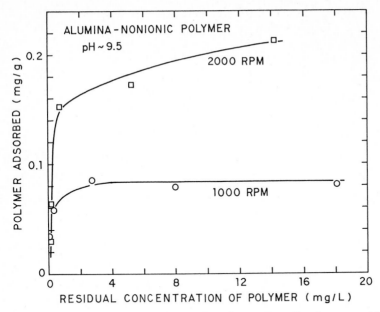

Fig. 5. The effect of agitation on the adsorption of polymer on Al_2O_3
particles under flocculation conditions.

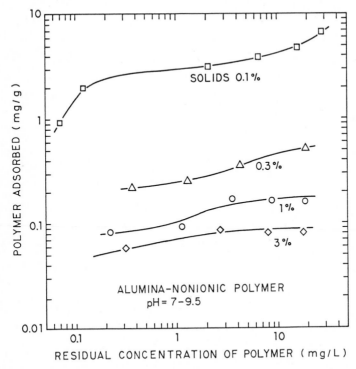

Fig. 6. The effect of solids concentration on the adsorption of polymer on
Al_2O_3 particles under flocculation conditions.

The effects of agitation and solids concentration on adsorption are illustrated in Figures 5 and 6. Again, it can be seen that the smaller flocs formed with more intense agitation (see Figure 3) are associated with increased polymer adsorption. Similarly, the more rapid floc growth which occurs at high solids concentration causes a reduction in the amount of adsorption.

Floc degradation, such as that shown in Figure 1, obviously leads to an increase in external surface area and would be expected to promote polymer adsorption. This is confirmed by the results shown in Figure 7. The upper curve in the figure is for a total polymer dosage of 5.3 mg/1 added over a period of one minute. Adsorption continues to increase up to 5 minutes at which time the amount adsorbed (about 0.16 mg/g) is almost equal to the total amount added (5.3 mg/1 per 30 g/1 of solid = 0.18 mg/g). The lower curve corresponds to continuous polymer addition at the same rate but over the entire 4 minute period. Despite the much higher dosage, the amount adsorbed is actually less (about 0.11 mg/g) than when degradation of the flocs is permitted.

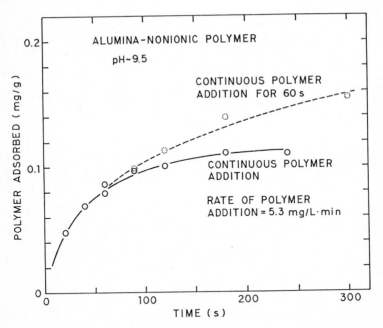

Fig. 7. The effect of continued agitation on the adsorption of polymer on Al_2O_3 particles at 3% solids by weight.

    Dashed curve: Polymer addition for first 60 sec. only – total dosage
                    5.3 mg/L.
    Solid curve: Continuous polymer addition for entire test period – final
                    dosage 21.2 mg/L after 240 sec.

It was shown above (see Table 1) that initially well-dispersed suspensions are difficult to flocculate using high molecular weight polymers. An example of adsorption under these conditions is given in Figure 8 for Al_2O_3 particles at pH 4.5. Clearly, the results are quite different from those obtained at pH 9 (Figure 6). Adsorption densities are very much higher and the results for 0.1 and 3% solids appear to fall on a single curve. Again, it appears that the extent of adsorption is controlled by the available surface area, which is very high for the well-dispersed particles at pH 4.5.

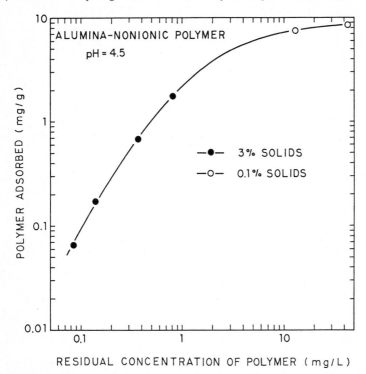

Fig. 8. Adsorption of polymer on well-dispersed Al_2O_3 particles at pH 4.5. Note that values for low and high solids concentrations fall on a single curve.

From these adsorption studies, it is concluded that, while polymer adsorption may be prerequisite to effective flocculation, the actual extent of adsorption, under flocculation conditions, is controlled by the extent of flocculation, not the converse as is often supposed. Lyklema (9) has pointed out that polymer adsorption typically approaches saturation at about 1 to 2 mg/m^2. The results shown above range from less than 0.01 mg/m^2 for a highly flocculated suspension to about 0.9 mg/m^2 for the dispersed, dilute suspension at pH 4.5. These values are again consistent with the limitation of adsorption to the external surfaces of flocs.

Agglomerate Bonding

Our investigations of floc breakage have indicated that increased polymer content tends to inhibit breakage (2) and that a primary function of polymers in flocculation may be to act as binding agents allowing flocs to grow to larger sizes. This observation raises the question of whether flocculant polymers could be used specifically as binders for compacted powders.

A series of experiments has been conducted in which suspensions of Al_2O_3 particles were flocculated with a nonionic polymer, filtered and the resulting cakes air-dried to less than 1% moisture. The dried filter cakes were granulated by forcing the material through a 70 US mesh screen and the

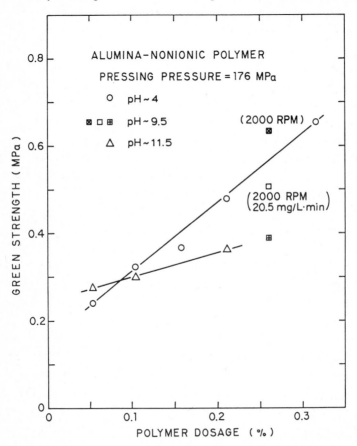

Fig. 9. Green strength of compacted Al_2O_3 powder using non-ionic polycrylamide as a binder. Polymer was added at 1.05 mg/L.min to a 20 wt% suspension of solids agitated at 1000 r.p.m. unless otherwise stated. Polymer dosage is expressed as wt% relative to the solids.

granules were compacted in a ½ inch cylindrical die. The pellets were oven dried at 100°C for two hours and stored in a desiccator. Green strength measurements were obtained by diametral compression testing in an Instron Universal Testing Machine.

The results of the green strength measurements are shown in Figure 9. Analysis of these results indicates that pellet strength depends on extent and uniformity of polymer adsorption on the particle surfaces. It was shown above that polymer adsorption depends on the degree of flocculation; in these particular experiments, a further complication is added in that the filtration step can substantially increase the amount of adsorption by exposing internal floc surfaces to the polymer solution. The distribution of polymer adsorbed during filtration might, however, be expected to be considerably less uniform than that obtained during agitation of the suspension.

Polymer adsorption is generally highest on well-dispersed suspensions, e.g. Al_2O_3 at pH 4 (Figure 8) and, indeed, these conditions lead to the highest green strengths. The Al_2O_3 suspension is also well dispersed at pH 11 (see Table 1), but adsorption is found to be lower than at pH 4, perhaps because the slightly anionic character of the polymer causes some repulsion from the negatively charged particle surfaces. The green strengths corresponding to polymer addition at pH 11 reflect this reduced adsorption level. At pH 9, the suspension is unstable and flocculates rapidly on polymer addition leading to very low adsorption (Figure 6). Green strengths are higher than would be expected on the basis of adsorption, probably due to additional polymer extraction during filtration. The higher green strengths observed for low rate of polymer addition or highest agitation rate are in accord with previous observations of higher adsorption (Figure 4). Although the polymer contents of the pH 9 pellets were about the same, it is postulated that higher adsorption obtained during agitation leads to a more uniform distribution of the polymer and, therefore, to greater strength.

The green strengths obtained in this study are comparable to values reported using more conventional binder systems. Nies and Messing (10), for example, have reported green strengths of about 0.9 MPa for the same Al_2O_3 powder using a polyvinyl alcohol binder at about 3% by weight. This can be compared to the maximum value obtained in this study of about 0.7 MPa for polyacrylamide at only 0.3%. Based on these results, it appears that flocculant type polymers, added to a stable aqueous suspension under highly turbulent conditions, could have considerable potential for use as binders for compacted powders.

CONCLUSIONS

The following general conclusions can be drawn from the work described in this paper:

- High molecular weight polymers can be very effective for promoting floc growth but not for destabilizing dispersions.

- Flocculation by polymers is significantly affected by physical conditions.

- For flocculating suspensions, polymer adsorption is determined by the external (floc) surface area.

- Floc growth depends on adsorption on floc surfaces.

- Floc breakage depends on the polymer content of the flocs.

- Under appropriate conditions, polymer flocculants can be effective binders for compacted powders.

ACKNOWLEDGEMENTS

The research described in this paper was supported by the National Science Foundation under Grant No. CPE-8121731.

REFERENCES

1   R. H. Smellie, Jr., and V. K. La Mer, Flocculation, subsidence and filtration of phosphate slimes, VI, a quantitative theory of filtration of flocculated suspensions. J. Coll. Sci. 23 (1958) 589-599.
2   D. T. Ray and R. Hogg, Agglomerate breakage in polymer-flocculated suspensions, J. Coll. Interf. Sci. 116(1) (1987) 256-268.
3   R. Hogg, R. C. Klimpel, and D. T. Ray, Agglomerate structure in flocculated suspensions and its effect on sedimentation and dewatering, Minerals and Metallurgical Processing, 4 (1987), 108-113.
4   R. Hogg, Collision efficiency factors for polymer flocculation, J. Coll. Interf. Sci., 102(1) (1984) 232-236.
5   H. R. Kruyt, Editor, Colloid Science I, Elsevier, Amsterdam, 1952.
6   R. Hogg, R. C. Klimpel and D. T. Ray, Growth and structure of agglomerates in flocculation processes, Agglomeration '85, Proceedings of the 4th International Symposium on Agglomeration, C. E. Capes, Editor, CIM/ISS, Toronto, Canada, 1985.
7   H. van Olphen, Introduction to Clay Colloid Chemistry, 2nd Edition, Wiley, NY, 1977.
8   L. A. Glasgow and R. A. Luecke, Mechanisms of deaggregation for clay-polymer flocs in turbulent systems, I&EC Fundamentals, 19(1980) 148-156.
9   J. Lyklema, The colloidal background of flocculation and dewatering, Plenary Lecture, Engineering Foundation Conference on Flocculation and Dewatering, Palm Coast, Florida, January 1988.
10   C. W. Nies and G. L. Messing, Effect of glass-transition temperature of polyethylene glycol-pasticized polyvinyl alcohol on granule compaction, J. American Ceramic Society, 67(4)(1984) 301-304.